Space Exploration

Other titles in
Chambers Compact Reference

- 50 Years of Rock Music
- Great Inventions Through History
- Great Modern Inventions
- Great Scientific Discoveries
- Masters of Jazz
- Movie Classics
- Musical Masterpieces
- Mythology
- The Occult
- Religious Leaders
- Sacred Writings of World Religions

To be published in Autumn 1992

- Catastrophes and Disasters
- Crimes and Criminals
- Saints

To be published in 1993

- Great Cities of the World
- Movie Stars
- Operas
- World Folklore

Space Exploration

J K Davies

Chambers

EDINBURGH NEW YORK

Published 1992 by W & R Chambers Ltd
43–45 Annandale Street, Edinburgh EH7 4AZ
95 Madison Avenue, New York N.Y. 10016

British Library Cataloguing in Publication Data

A catalogue record for this book is available from the British Library.

Library of Congress Cataloging-in-Publication Data applied for

ISBN 0 550 17013 8

Chambers Compact Reference Series Editor   Min Lee

Cover design   Blue Peach Design Consultants Ltd
Typeset by Alphaset Graphics Limited, Edinburgh
Printed in England by Clays Ltd, St Ives, plc

Acknowledgements

In writing this book I have drawn on other, more comprehensive, texts and on my files, amassed over two decades of collecting newspapers, press releases, public relations handouts and photographs. I also made extensive use of *Spaceflight*, the magazine of the British Interplanetary Society and *Space*, published by the Shepard Press, which both provided valuable background information.

In addition I should thank those who have responded so generously to specific requests for material required for this book. They include: British Aerospace, Centre National d'Etudes Spatiale (French space agency), DLR (German space agency), Dornier, ESA, EOSAT, Fairchild Space Company, Grumman Corporation, Hughes Aircraft Company, ILC, INTELSAT, LTV Corporation, Martin-Marietta Astronautics, MBB, NASA Ames Research Center, NASA Dryden Flight Research Center, NASA Goddard Spaceflight Center, NASA Jet Propulsion Laboratory, NASA Cape Kennedy Space Center, NASA Marshall Space Center, Rockwell International, Spar Aerospace, Telesat Canada, TRW.

I should also give a special thanks for the generosity of the Audio Visual Division of NASA Headquarters in Washington, Mike Gentry of the Johnson Space Center and Jim Wilson of the NASA JPL for their help with photographs. My wife gave generously of her time to check the text and to advise on its suitability for the non-specialist reader. Min Lee of Chambers guided me through the complex process of making my manuscript into part of her series.

Illustration credits

Page	
xii	courtesy NASA
5	courtesy NASA
9	courtesy NASA
12	courtesy NASA
15	courtesy S Vermeer – ESA
18	courtesy Rolls-Royce
21	© Novosti
24	courtesy SPAR Aerospace
26	courtesy NASA
29	courtesy China Great Wall Industry Corp via *Space* magazine
34	courtesy ESA
38	courtesy Amsat Corp
38	courtesy British Aerospace
45	courtesy ESA/D Ducros
53	courtesy CNES/Glavcosmos
53	courtesy NASA/CNES
57	courtesy NASA
60	courtesy NASA/JPL
64	courtesy NASA
67	courtesy NASA
70	courtesy DFVLR
72	courtesy ESA
75	courtesy NASA/Smithsonian Institution/Lockheed Corporation
80	courtesy NASA
83	courtesy Hughes Aircraft Company
89	courtesy NASDA
96	© Novosti
99	courtesy NASA/Grumman Corporation
101	courtesy NASA
104	courtesy NASA
107	courtesy NASA/Martin-Marietta
115	courtesy NASA/JPL
116	courtesy NASA/JPL
120	© Novosti
125	courtesy NASA

Page	
128	courtesy British Aerospace
133	© Novosti
136	© TASS
142	courtesy NASA
145	courtesy NASA
148	courtesy NASA
152	courtesy NASA
155	courtesy NASA
160	courtesy NASA/JPL
160	courtesy NASA
167	© Novosti
170	courtesy NASA
173	courtesy NASA
181	© Novosti
189	© Novosti
194	courtesy NASA
197	courtesy NASA
197	courtesy NASA
202	courtesy ESA
208	courtesy British Aerospace
213	courtesy Rockwell International
218	courtesy NASA
223	© Novosti
223	courtesy NASA
226	courtesy NASA/Hughes Aircraft Company
229	courtesy NASA
234	courtesy NASA
237	courtesy NASA
240	courtesy Martin-Marietta
243	courtesy NASA
247	© TASS
250	courtesy NASA
252	© Novosti
254	© Novosti
257	courtesy Martin-Marietta
260	© Novosti
262	courtesy Rockwell International

Contents

Preface

When I was asked to write this contribution to the Compact Reference series my first reaction was that there are already many books on space travel and that it might be difficult to find anything new to say. Then I realized that, despite all that has been published, there is still a place for a small, inexpensive summary of the last 35 years of space exploration. This book attempts to provide that by presenting a brief essay on about 100 related topics. Naturally the boundaries are rather arbitrary, and it is impossible to describe the entirety of projects such as Apollo or Salyut in just a few pages, so I have tried to present the essential facts, and have left it for the reader to follow up the minutiae in more detail if he or she pleases. As such I hope this little work will provide a useful reference for those who already have extensive spaceflight libraries, and will serve as an introduction for anyone who is just developing an interest in the subject.

Author's Note

In recent years there has been pressure to modify various well-used words or expressions in order to remove what some consider a masculine bias from our language. Despite this, spacecraft have traditionally been referred to as manned or unmanned and attempts to formulate alternative, genderless, terms (such as uncrewed and unstaffed) have not been very successful. Accordingly, I have used traditional terminology throughout this book. However, as Valentina Tereshkova showed many years ago, space travel is not restricted to men and the use of such language should not discourage anyone who wants to play a part in the exploration of space from exploiting their talents to the full. Remember, the sky is no longer the limit.

John Davies
Summer 1992

To my teachers, from all walks of life.

Apollo 11 astronaut Edwin Aldrin on the lunar surface, July 1969

Introduction

The first recorded story of a journey to the Moon may have been the 'True History' (so called because it contained not a word of truth) in which Lucian of Samosata (c.117–180) described the adventures of a ship's crew swept to the lunar surface by a waterspout. There were other early examples of what we might today call science fiction, such as Johann Kepler's story 'Somnium' and the description by Bishop Goodwin, in 1683, of a flight to the Moon in a vessel pulled by geese. These flights of fancy were followed by the more sober stories of Jules Verne, whose books *From the Earth to the Moon* (1865) and *Around the Moon* (1870) recount the adventures of three men launched into space by the Columboid, a giant cannon. In 1901 H G Wells published his novel *The First Men in the Moon*, using a gravity-defying material called Cavorite to propel two astronauts to the lunar surface and return one of them to the Earth.

By 1920 the works of Konstantin Tsiolkovski, Robert Goddard and Hermann Oberth had begun to put space travel on a sound theoretical footing and led to the establishment of space travel societies around the world. These included the American Interplanetary Society (which was renamed the American Rocket Society in 1934), the British Interplanetary Society, founded in 1933, the Verein für Raumschiffahrt (VFR–German society for space travel) founded in 1927, and groups called GIRDs (societies for the study of rocket propulsion) in Moscow and Leningrad. Some of these groups, notably the VfR and the Soviet GIRDs, actually developed rockets; others, such as the British Interplanetary Society, concentrated on theoretical studies of space travel. Unfortunately, the shadow of war which hung over Europe in the late 1930s caused much of the world's rocket development to be swallowed into military programmes and these culminated in the German V2 ballistic missile used in the later stages of World War II.

Although the V2 was not a very successful weapon (it was expensive to produce, had a small warhead and was not very accurate), it did show the potential of long-range missiles and led to a post-war race to develop larger and larger rockets which, combined with the newly proven atomic bomb, promised to revolutionize strategic military thinking. Missile technology continued to develop throughout the early 1950s and led to the first really practical proposals for the launching of artificial satellites.

The USA announced that it would develop and launch a satellite in International Geophysical Year (1957–8), an 18-month period of international programmes to study the Earth and its environment, but to the surprise of most of the world, the US Vanguard satellite was beaten into space by the Soviet Sputnik. Although US President Eisenhower claimed to be unworried by the launch of Sputnik, the fact that the USSR, long perceived in the West as technically backward, could achieve this feat before the USA sent political and military shock waves around the world. The USA quickly mobilized its scientific and technical resources to launch a satellite called Explorer 1, but by then the two superpowers were locked in a highly visible space race with national prestige as the prize.

The USSR, which had been forced to develop large ballistic missiles because it lacked the technology to produce lightweight nuclear bombs, held the upper hand in this race for several years. Its superior rocket power allowed it to achieve a string of space records including the first animal in orbit, the first Moon probe, the first impact on the Moon, the first photographs of the far side of the Moon, the first probes to the planets and the first man in space. In response, in May 1961, US

President John F Kennedy declared a race to the Moon, promising that an American would land on the Moon and return within a decade, a commitment fulfilled by the crew of Apollo 11 in July 1969.

In fact, US space technology had passed that of the USSR in about 1965 during the two-man Gemini programme which developed the techniques NASA needed for its Moon missions, but this would not become apparent until the failure of the Soviet Moon programme and the triumph of Apollo. However, having beaten the USSR to the lunar surface, the US manned space programme went into a decade of decline during which the USSR established the world's first space stations with its Salyut (and later Mir) programmes. Only in the area of unmanned planetary exploration did the USA maintain its convincing superiority over the USSR, with NASA's Mariner and Viking spacecraft exploring the inner solar system while Pioneer 10 and 11, then Voyager 1 and 2, returned stunning pictures of the frozen worlds of the outer planets.

The 1970s also marked the emergence of new space powers. After rather mixed fortunes during the 1960s, Europe, under the direction of the European Space Agency (ESA), became a force to be reckoned with in the arena of unmanned spaceflight, while several other nations, notably Japan, China and India, developed their own national space programmes. Another important development was the increasing use of space technology for international communications and the routine use of satellites for services such as weather forecasting and navigation.

By the early 1980s NASA's hopes were pinned on its much delayed, reusable Space Shuttle, and indeed the first two dozen flights of the Space Shuttle seemed to justify the enormous cost of its development. Space Shuttle astronauts launched satellites, performed dramatic spacewalks and repaired or rescued ailing satellites in orbit. Although it could never sustain the promised launch rate of up to one flight per week, the Space Shuttle looked set to dominate the space programmes of the 1980s. However, all was not well with the Space Shuttle programme and in January 1986 the Space Shuttle Challenger exploded, killing its crew of seven and setting the US space programme back by several years.

The Soviet space programme, which had by now achieved its aim of setting up a permanently manned space station, ran into problems during the political and economic turmoil at the end of the 1980s and in the early 1990s. Although cosmonauts had learned to stay in space for up to a year, and the Salyut and Mir stations had hosted guest cosmonauts from more than a dozen countries, there was increased criticism of the economic value of space stations and the usefulness of the Soviet Buran space shuttle.

As we enter the early 1990s the outlook for space exploration is none too favourable. The international space station Freedom is behind schedule and subject to constant criticism, whilst the Soviet space programme, especially in the area of manned spaceflight, may not survive the break-up of the USSR. European space efforts are under pressure as the cost of developing an independent manned space programme begins to bite in the exchequers of a Europe in which political divisions are changing almost daily. Only in the East, in Japan and China, are there signs of continuing enthusiasm for space technology.

However, the present problems may turn out to be only a temporary setback. In 1989 US President George Bush set the USA on course for a return to the Moon and promised that it would then go onwards to Mars. With superpower rivalry reduced by the collapse of the USSR it is possible that a more integrated international space programme may emerge before the year 2000. If so, we may yet see a multinational astronaut crew en route to the planet Mars before the first footsteps on the Moon are a half-century old.

Apollo

First flight February 1966, last flight July 1975

America's moonship

A key element in the Moon programme, Apollo was a versatile spacecraft that was also used for other projects.

The decision to develop a three-man spacecraft to follow on from the one-man Mercury capsule was announced by NASA on 29 July 1960. The new spacecraft was intended to be able to operate close to the Earth or around the Moon and was already named Apollo when President John F Kennedy committed the USA to a manned lunar mission. In November 1961, following a design competition, North American Aviation (now Rockwell International) was selected to build the new spacecraft which could operate independently or in conjunction with a special lunar landing module (see p97). The North American design featured two separate parts with quite different functions. The command module provided a pressurized compartment in which astronauts could live and work. The unpressurized service module provided power, propulsion and life-support systems for the command module throughout mostof the flight and was cast off just before re-entry.

To speed up the development programme, early manned Apollo craft were built to a so-called Block 1 standard, able to operate in Earth orbit, but lacking many of the design refinements for a complete lunar mission. These extra systems were installed in the later, more sophisticated, Block 2 spacecraft. The Block 1 and Block 2 spacecraft also differed externally; for example in the positioning of radiators and the umbilical connection to the command module. However, following the fire which claimed the lives of three astronauts in January 1967 (see p192), the Apollo design was critically re-examined and no Block 1 versions ever carried a crew.

After the fatal fire and subsequent redesign, Apollo emerged as a very capable spacecraft and 15 manned flights were made. These were the 11 missions in the Apollo programme, three missions in which a modified Apollo was used to ferry crews to the Skylab space station and the joint Apollo–Soyuz mission in1975.

Command Module (CM)

The conical command module, 3m high and 3.8m in diameter at its base, was the central element of the Apollo spacecraft. It provided a home and control centre for the astronauts throughout their mission and was the only part of the entire Apollo/Saturn system that could return safely to Earth. It had to be rugged enough to survive the forces of launch and re-entry, to protect the crew against extremes of temperature, to resist collisions with meteorites and the sudden impact of splashdown, yet it had to be as light as possible.

The forward section contained a pair of small thrusters for attitude control during re-entry, the parachutes required for landing and a tunnel which could be used to enter the lunar module. At the end of the tunnel was an airtight hatch and a removable docking probe used to link the command and lunar modules. The main cabin crammed couches for three astronauts, instrument panels, storage bays, navigation systems, spacesuits, food supplies and personal hygiene equipment into a volume of just under 6cu m. The couches, which were made of steel frames covered with fireproof, fibreglass cloth, rested on crushable struts which absorbed the impact of landing. Control devices, such as the pilot's hand controller, were attached to the arm rests. The centre couch could be folded away to give the astronauts more room to move about. There were five windows: two faced forwards to allow the astronauts to see during docking with the lunar module, and three were used for general observations. A single hatch, opposite the centre couch, was used to enter and leave the CM on the ground. The bottom section contained another 10 re-entry control jets, their fuel tanks and the heatshield. Manufactured from phenolic epoxy resin, the heatshield was ablative: during re-entry it charred and burned away, carrying off much of the heat generated by friction with the atmosphere.

Service Module (SM)

The Apollo service module was an aluminium alloy cylinder about 3.9m in diameter and 7.5m long. The main engine, used to place Apollo into orbit around the Moon and to begin the return to Earth, was mounted down the centre of the cylinder with its nozzle projecting from the rear. The 91kNewton thrust engine used a mixture of hydrazine, unsymmetrical dimethylhydrazine and nitrogen tetroxide as propellants. Every vital system of the engine was duplicated, for without it the astronauts could have been stranded in orbit around the Moon, and its propellants were hypergolic (self igniting) so that no ignition system was necessary. Six wedge-shaped segments surrounding the main engine provided equipment bays, four of which contained the propellant tanks. The others held the systems used to generate and regulate electrical power and some of the life-support systems, including the tanks which supplied most of the crew's oxygen. The last three Apollo lunar missions also carried a special scientific instrument module containing cameras and other sensors in one of the bays. This equipment was used to study the Moon from orbit.

At four positions around the outside of the cylinder the attitude control jets were mounted. Each unit consisted of four small (445Newton) thrusters pointing at right angles to each other. By firing various combinations of these, the entire Apollo spacecraft, including the lunar module if it was attached, could be pointed and stabilized in all three axes. The outer skin also carried radiators, used to dissipate waste heat, and both fixed and steerable radio antenna for communications with Earth.

Launch Escape System (LES)

Although it was jettisoned less than three minutes after launch, the LES was considered part of the Apollo craft. It comprised an open frame tower, 9.9m high, with a 653kNewton solid propellant rocket on top. In an emergency this rocket, together with a separate pitch control rocket, would whisk the entire command module upwards and sideways away from the launch vehicle. Once clear of the exploding launcher, the tower would be released and the command module would descend to safety under its parachutes. The LES could be used at any time from before lift-off until just after the first stage of the Saturn 5 launch vehicle had used up its fuel and been jetisoned. Once the second stage of the Saturn 5 was operating then other escape options were available and the tower was jettisoned. Although tested many times, the system was never used in action.

Apollo carried a supply of liquid oxygen and liquid hydrogen in tanks in the service module. Engineers estimated that these tanks were virtually leak free and that if a car tyre was built to the same standards it would take over 30 000 000 years to go flat.

The Apollo 15 command and service modules

Apollo programme

Apollo 7 1968, Apollo 17 1972

Men on the Moon

The Apollo programme sent 12 men to the lunar surface to fulfil the promise made by President John F Kennedy in 1961.

By 1961, stung by a series of Soviet space triumphs which had culminated in the flight of Yuri Gagarin a few weeks before the launch of the first US Mercury astronaut, US President John F Kennedy set out to define a major space project in which the USA could convincingly beat the USSR and thereby restore US prestige. He asked NASA if they could realistically expect to build the first orbiting space station, or make the first flight around the Moon or achieve the first manned landing on the Moon itself. NASA officials, knowing that it would take years to develop rockets more powerful than those already in use in the USSR, recommended the USA should try and beat the USSR to the lunar surface. So, at a joint session of the US Congress on 25 May 1961, President Kennedy announced 'I believe that this nation should commit itself to achieving the goal, before this decade is out, of landing a man on the Moon and returning him safely to the Earth'. The Moon race was on.

After much, sometimes acrimonious, debate NASA decided in 1962 to use the Lunar Orbit Rendezvous technique (see p161) for its Moon landing programme and set about the development of a lunar module (see p97) to accompany the basic Apollo spacecraft (see p3) to the Moon. While these new spacecraft and the huge Saturn rockets required to lift them were being developed, NASA instituted the two-man Gemini programme (see p65) to gain experience in rendezvous, docking, spacewalks and long-duration spaceflight. The Mercury and Gemini programmes went well and it seemed that the Apollo programme would easily reach its target, when disaster struck. On 26 January 1967 a fire swept through the Apollo 1 capsule during a routine countdown test, killing the three astronauts on board. An inquiry found that the Apollo spacecraft had a number of design and quality control problems and a major overhaul of the programme was called for, delaying the first flight by over a year and a half.

In the hiatus caused by the fatal fire, the Apollo programme was restructured. Some flights were cancelled (Apollo 2 and 3) and a number of unmanned missions (Apollo 4, 5 and 6) were performed to demonstrate that the lunar module and the huge Saturn 5 rocket were ready for a manned flight. It was October 1967 before NASA was ready to send men into space again.

Apollo 7 was a test flight of the Apollo command and service modules only and was launched into Earth orbit by a Saturn 1B rocket. The 11-day flight went well, although the three astronauts all caught colds and complained that mission control was overworking them. The next mission was intended to have been a test of the lunar module, but delays pushed the flight back into 1969 and instead Apollo 8 became the first manned flight of the Saturn 5 rocket. Originally intended as a flight in high Earth orbit, the objective was changed to a flight into lunar orbit when the US Central Intelligence Agency reported that the USSR planned to send a man around the Moon in a Zond capsule. Apollo 8 was launched in December 1968 and went into lunar orbit on Christmas Eve. During their 20 hours in lunar orbit the three astronauts returned television pictures of the Moon's surface and read extracts from the Bible. When Apollo 8 had splashed down in the Pacific Ocean on 27 December 1968 the road to the Moon was open; only the lunar module

Apollo mission summary

Mission	Dates	Crew CDR LMP CMP	Callsigns Command Module Lunar Module	Notes
Apollo 1	26 Jan 67	Grissom White Chaffee	-	Fatal fire during countdown test
Apollo 4	9 Nov 67	-	-	Unmanned test of first Saturn 5 rocket
Apollo 5	22 Jan 68	-	-	Unmanned lunar module test in Earth orbit
Apollo 6	4 Apr 68	-	-	Unmanned test of Saturn 5
Apollo 7	11 Oct 68 22 Oct 68	Schirra Cunningham Eisele	-	First Apollo flight; CSM only in Earth orbit
Apollo 8	21 Dec 68 27 Dec 68	Borman Anders Lovell	- -	First manned orbit of Moon (10 orbits)
Apollo 9	3 Mar 69 13 Mar 69	McDivitt Schweickart Scott	Gumdrop Spider	Manned test of lunar module in Earth orbit
Apollo 10	18 May 69 26 May 69	Stafford Cernan Young	Charlie Brown Snoopy	Lunar module tested in lunar orbit
Apollo 11	16 Jul 69 24 Jul 69	Armstrong Aldrin Collins	Columbia Eagle	First manned Moon landing 20 Jul Sea of Tranquillity
Apollo 12	14 Nov 69 24 Nov 69	Conrad Bean Gordon	Intrepid Yankee Clipper	Moon landing 19 Nov Ocean of Storms, near Surveyor 3
Apollo 13	11 Apr 70 17 Apr 70	Lovell Haise Swigert	Odyssey Aquarius	Mission aborted, ruptured oxygen tank, crew saved
Apollo 14	31 Jan 71 9 Feb 71	Shepard Mitchell Roosa	Kitty Hawk Antares	Moon landing 5 Feb Fra Mauro
Apollo 15	26 Jul 71 7 Aug 71	Scott Irwin Worden	Endeavour Falcon	Moon landing 30 Jul Hadley Rille, first lunar rover
Apollo 16	16 Apr 72 27Apr 72	Young Duke Mattingly	Casper Orion	Moon landing 20 Apr Descartes region
Apollo 17	7 Dec 72 19 Dec 72	Cernan Schmitt Evans	America Challenger	Moon landing 11 Dec Taurus Littrow, last and longest Apollo mission

Key:
CDR = Commander, LMP = Lunar Module Pilot, CMP = Command Module Pilot

remained unproven. This hurdle was passed in March 1969 when Apollo 9 tested the lunar module in Earth orbit during a sucessful 10-day mission which included a short spacewalk to test a new lunar spacesuit. Two months later the Apollo 10 lunar module descended to within 15km of the surface in the final dress rehearsal for the Moon landing.

The first Moon landing was accomplished on 20 July 1969 by the crew of Apollo 11 when Neil Armstrong and Edwin Aldrin steered their lunar module 'Eagle' to a safe touchdown on the Sea of Tranquillity while Michael Collins remained in lunar orbit aboard the command module 'Columbia'. Armstrong and Aldrin made a single moonwalk during which they planted a US flag, set up three experiments and collected samples of lunar soil. The men also conducted a brief telephone conversation with President Nixon and unveiled a plaque celebrating their landing. The astronauts returned to Earth on 24 July 1969, spending a period in biological isolation before rejoining their families.

Apollo 12, which was launched four months later, demonstrated the ability to make a pinpoint landing when the lunar module 'Intrepid' touched down close to the unmanned Surveyor 3 spacecraft. During two moonwalks the astronauts deployed several scientific instruments, and walked over to the Surveyor to examine it. The only major failure of the mission was damage to the lunar television camera which prevented viewers on Earth watching the astronauts as they went about their activities.

Apollo 13, launched in April 1970, was crippled by an explosion en route to the Moon (see p195) and further flights were delayed while the implications of this near catastrophe were examined and the Apollo spacecraft modified. Apollo 14, commanded by the USA's first man in space, Alan Shepard, landed in the Fra Mauro region of the Moon on 5 February 1971. The mission featured the first, and only, use of a lunar cart, which was used by the astronauts to carry extra tools and samples during their two moonwalks.

By this time there had been more changes in the Apollo programme. Financial cutbacks, together with the fear of another accident with fatal consequences, led to the cancellation of Apollo 20 and then of Apollo 18 and 19. The three remaining missions were spaced out at six-month intervals and used the improved 'J mission' lunar module which could carry more weight to the Moon and remain there longer. Apollo 15–17 all carried an electrically powered lunar roving vehicle (see p102) and each crew was to complete three moonwalks. For these final missions the service module was equipped with a battery of instruments so that scientific observations could be made from lunar orbit. Since these experiments included cameras which used conventional film, it was necessary for the command module pilot to make a spacewalk during the return to Earth to recover the film before the service module was jettisoned during re-entry.

All three of the final missions were successful, although the landing of Apollo 16's lunar module was delayed for a few hours while mission controllers resolved a problem with the main engine of the command module. Each mission deployed experiments on the lunar surface, explored interesting geological regions and returned large quantities of lunar samples. By working in conjuction with Earth-based scientists, and sending a geologist (Harrison Schmitt) on Apollo 17, the final missions were very productive and provided a fitting climax to the Apollo programme.

> Geologist Harrison Schmitt was originally assigned to Apollo 18. When this mission was cancelled he replaced Joe Engle on Apollo 17, denying Engle his chance of a lunar mission.

Apollo 15 astronaut Jim Irwin digging a sample trench

Apollo–Soyuz Test Project

Docking 17 July 1975

A handshake in orbit

The Apollo–Soyuz Test Project was the first international manned space mission.

By 1970 NASA had won the race to the Moon and the political relationship between the USA and the USSR was beginning to improve; for the first time since the 1950s, a joint US/USSR space flight seemed a possibility. Such a mission had attractions for both sides. It would keep NASA's astronaut team together during the hiatus between the end of the Skylab programme and the first flight of the Space Shuttle and for the USSR it offered a chance to be seen on equal terms with the Americans. Perhaps more importantly it would be a highly visible symbol of the detente established between US President Richard Nixon and Soviet Premier Leonid Brezhnev. To investigate the feasibility of such a mission a series of joint working groups was set up in October 1970 to consider the technical and safety issues involved in a docking between a US Apollo and a Soviet Soyuz spacecraft. No insuperable problems arose and an agreement to launch a joint mission in July 1975 was signed at a superpower summit in May 1972.

The technical meetings soon revealed that both parties would have to make compromises if the mission was to succeed. The most obvious need was a docking system in which either spacecraft could be the active partner. The systems then in use on both Apollo and Soyuz required one spacecraft to insert a probe into a receptacle on the other; naturally, these systems were not compatible and since one of the objectives of the flight was to demonstrate a capability for space rescue, a new, more flexible, system was needed. A second issue concerned the life-support systems; the Americans had always used a pure oxygen atmosphere at about one-third of normal atmospheric pressure, the Soviets preferred a more Earthlike mixture of oxygen and nitrogen at a higher pressure. This meant that a direct transfer from Soyuz to Apollo would have been impossible: the sudden reduction in pressure would have subjected the Soviet crew to an attack of 'the bends', a dangerous and painful condition familiar to deep sea divers who surface too quickly and find nitrogen gas bubbles forming in their body fluids. Other key technical issues included the standardization of radio links, docking techniques and flight control procedures. It was also necessary for both sides to agree which language they were going to use.

The solution to several of the problems was the development of a special docking module to be launched with the Apollo spacecraft. This had a standard Apollo docking system at one end and a specially developed ring-and-petal mutual docking system on the other. The docking module was carried in the US Saturn 1B launch vehicle and extracted from the second stage of the launcher by Apollo once in orbit. The docking module also acted as an airlock between the two spacecraft; astronauts transferring from one spacecraft to the other, would spend time in the docking module during which the atmosphere would be gradually changed from that of Apollo to Soyuz or vice versa. Scientific experiments for use by the American astronauts after the international space docking was over were also stored here. The language barrier was overcome by each side agreeing to speak in the other's language, and to listen in their own. Not surprisingly, the Americans would refer to the flight as Apollo–Soyuz, the Soviets as Soyuz–Apollo.

The US crew was commanded by Tom Stafford, a veteran of the Gemini 6, Gemini

9 and Apollo 10 missions. The other crew members were each making their first spaceflight: Vance Brand, who would go on to fly several Space Shuttle missions, and Donald K ('Deke') Slayton, one of the original Mercury astronauts who had been grounded for 15 years by a minor heart irregularity. The Soyuz commander was Alexei Leonov, who in 1965 had been the first man to walk in space, and his companion was Valeri Kubasov who had flown on the Soyuz 6 mission. In the years preceding their flight the two crews came to know each other well as joint meetings were held in both countries so that each side could become totally familiar with the other's equipment and procedures.

The USSR conducted a dress rehearsal for its part of the mission in December 1974 when Anatoli Filipchenko and Nikolai Rukhavishnikov flew the Soyuz 16 mission. During this 142-hour flight they checked radio procedures, adjusted the Soyuz cabin atmosphere and conducted simulated tests of the mutual docking system. All went well, increasing confidence that the historic mission could take place as planned. Just to be on the safe side, the Soviets prepared a complete back-up rocket, spacecraft and crew to be ready for launch should Leonov and Kubasov fail to reach orbit due to technical problems. NASA, short of money but not of confidence, took no such precautions.

Soyuz 19 had a perfect launch on the afternoon of 15 July 1975 and this was followed later in the day by the equally smooth departure of the Americans. By noon on 17 July a complex series of manoeuvres had taken place, moving Apollo and Soyuz into the same orbital plane ready for rendezvous, and Vance Brand could see Soyuz 19 about 50km away. A perfect docking was accomplished later that evening and Stafford and Slayton sealed themselves in the docking module and began to adjust the pressure to make it compatible with the atmosphere aboard Soyuz. After what seemed like an interminable delay the hatch between the two spacecraft was opened and a carefully choreographed sequence of handshakes and welcomes took place in front of a live television audience. The Americans entered Soyuz, where Leonov played the perfect host and messages were broadcast to the crew from Leonid Brezhnev and Gerald Ford (who by this time had replaced Richard Nixon as US president following the Watergate scandal).

Apollo and Soyuz remained docked together for two days during which there were four crew exchanges and various symbolic activities took place. On 19 July the two spacecraft separated, re-docked and then separated again.

Soyuz 19 landed on the morning of 21 July completing a perfect mission for the USSR. Apollo remained in space a further three days, carrying out a programme of Earth observations, astronomical studies and metallurgical experiments in what was the last US manned spaceflight for almost six years.

Apollo splashed down in the Pacific near the aircraft carrier USS *New Orleans*. From the outside the splashdown appeared normal but it was later revealed that, due to an error by one of the astronauts, excess fuel being dumped overboard as the capsule swung below its parachutes was accidentally sucked back into the capsule, filling it with toxic fumes. Brand lost consciousness for a time, but fortunately no serious harm was done to the three men.

The success of Apollo–Soyuz led to hopes that further joint missions might take place, perhaps with US astronauts visiting a Soviet space station, but these were not to be fulfilled. NASA needed all its available money to finance the faltering Space Shuttle programme and detente did not survive the Soviet occupation of Afghanistan and the policies of the Reagan administration.

The Apollo–Soyuz Test Project Apollo launch marked the last ever use of a Saturn rocket and the final flight of an Apollo capsule.

Donald K Slayton and Alexei Leonov meet during the Apollo–Soyuz mission in 1975.

Ariane

First flight 24 December 1979

European launcher

After a shaky start in the 1960s, Europe developed its own satellite launcher.

Although never on the scale of the projects undertaken by the USA and the USSR, a number of European countries were involved in the development of satellite launch vehicles in the 1960s. National programmes concentrated on small rockets, such as the British Black Arrow and French Diamant vehicles, intended to lift lightweight scientific satellites into orbit. The development of a larger rocket, able to lift communications and weather satellites was left in the hands of the multinational European Launcher Development Organisation (ELDO) which was set up in 1962.

ELDO's main project was the Europa rocket which used a modified British Blue Streak missile combined with upper stages provided by other European nations. The first tests were made at the Woomera rocket range in Australia before the programme was moved to the French launch site at Kourou, in French Guiana. Unfortunately, the Europa rocket never placed a satellite in orbit and, when Britain announced that it would end its involvement in space launchers after 1971, it sounded the death knell for Europa.

ESA decides on Ariane

The problems with ELDO's Europa project, and worries about dividing the limited amount of European resources between ELDO and its sister ESRO, the European Space Research Organisation, lead to the merging of the two organisations into the European Space Agency, ESA, in 1975. At the urging of the French, who offered to provide over 60% of the necessary finance in return for an appropriate share of the industrial contracts, ESA decided that it would develop a new launcher able to meet the demands of Europe in the 1980s. The decision to develop the French LIIIS proposal, which became the first member of the Ariane family, was doubtless partly due to the fear that if Europe did not have its own launcher it would be at the mercy of the USA in the increasingly important world market of satellite communications.

Low technology, high reliability

It was decided that the programme would be controlled by the French space agency CNES (Centre National d'Etudies Spatiales) and that to improve reliability and speed development, Ariane would make the maximum possible use of technologies already familiar to European industry. The result was a three-stage rocket able to launch a satellite weighing about 1 835kg into the geostationary transfer orbit demanded by most commercial communication satellite operators. The French-built first stage, was 3.8m in diameter and 18.4m high and was powered by four French-developed, Viking-V engines. These burned a mixture of unsymmetrical dimethylhydrazine and dinitrogen tetroxide, and together produced 2 462kNewton of thrust at lift-off. Stage two, diameter 2.6m and height 10.4m, was built in Germany and used a single Viking engine. The first and second stages would each burn for about 140 seconds during a normal launch. The third stage was 9.1m high and was topped by an equipment bay holding computers and other electronics, adding another 1.15m to the rocket. The third stage had a single HM-7 engine of 61kNewton thrust and burned liquid hydrogen and liquid oxygen, a more powerful combination than that used in stages 1 and 2. Problems with the third stage engine would eventually cause several failures later in the Ariane programme. In a typical flight the third stage burned for about 560 seconds to place the satellite into geostationary transfer orbit.

The first Ariane was launched successfully on Christmas Eve 1979 but the second rocket plunged into the Atlantic Ocean soon after launch. Two flights in 1981 were successful and Ariane was then declared ready for use as an operational launcher. In 1980 the Arianespace organization was founded to market launches commercially, although it was agreed that ESA would continue to develop new versions of Ariane. Once proven, operation of these newer versions would also pass to Arianespace.

Improvements to Ariane

The first change to the basic Ariane was relatively simple, with the operating pressures of the main engines being increased to provide extra thrust and the third-stage propellant tanks being lengthened to carry 10 tonnes of fuel instead of 8. The Ariane 2 rocket was introduced in 1983 and the eleventh and last Ariane 1 was used to launch the ESA Giotto probe in 1985. The final Ariane 2 launch was on 2 April 1989 when the Swedish Tele-X satellite was placed in orbit.

The Ariane 3, introduced in 1984, was similar to the Ariane 2 except for the addition of two solid rocket boosters, each carrying 7 tonnes of propellant, to the first stage. These boosters were separated about 35 seconds after launch, but their extra thrust raised the lifting capability of the Ariane 3 to about 2 600kg into geostationary transfer orbit.

Apart from a disastrous period in which failures of the third-stage motor led to the loss of two Ariane launches in rapid succession, Ariane soon became accepted as an important part of the world space industry. Furthermore, when continuing problems with the Space Shuttle showed that NASA would never be able to compete in the commercial launch vehicle business, the Ariane programme went from strength to strength. Following a carefully laid plan, ESA soon began the development of a still more powerful version called Ariane 4.

The Ariane 4 was a logical outgrowth of the earlier versions and had a lengthened first stage which increased its propellant capacity from 145 to 220 tonnes. The second and third stages of Ariane 3 were retained. An interesting feature of the Ariane 4 is that to provide extra thrust at lift-off various combinations of booster rockets can be fitted to the first stage. These boosters can use either solid or liquid fuels depending on the requirements of the particular mission, making Ariane 4 a very versatile launcher. Ariane 4 versions are identified by the code letters L (*liquide*) and P (*poudre* or solid), so an Ariane 44L has four liquid boosters, a 44LP has two liquid and two solid boosters and a 40 is the basic launcher without extra boosters. The payload which can be injected into geostationary transfer orbit ranges from 2 600kg for an Ariane 42P to 4 200kg for an Ariane 44L.

The first Ariane 4, an Ariane 44LP, was launched in 1988 and marked the beginning of the end for the earlier versions, which were soon phased out of service. All Ariane launches now use members of the Ariane 4 family and this will continue until the much heavier, and rather different, Ariane 5 comes into service in the mid 1990s.

Ariane 5

The Ariane 5 is a completely new rocket designed for use both as an unmanned satellite launcher and as the carrier for the European mini-shuttle Hermes (see p71). The objective of the Ariane 5 programme is to reduce the cost per kilogram of launching a satellite into geostationary orbit by about 40% compared with the costs of an Ariane 4 launch. An Ariane 5 should be able to launch a satellite weighing 6 900kg into a geostationary transfer orbit or launch a space station module weighing 18 000kg into an orbit 550km high or place 12 000kg into an orbit 800km high.

The first stage of the Ariane 5 is called the H155 and has a single Vulcan engine which burns liquid hydrogen and liquid oxygen propellants and has a thrust of 800kNewton at sea level. In a typical flight this motor will fire for about 615 seconds. To provide additional thrust at lift-off there are two large

P230 solid rocket boosters fitted to opposite sides of the H155 stage. Each P230 booster contains 230 tonnes of solid propellant and will burn for about 120 seconds. If the payload is to be one or more satellites then the H155 core stage is fitted with an upper stage to place the payload into the required orbit. If the rocket is being used to launch the Hermes spaceplane into low Earth orbit then no upper stage is required.

Ariane 5 launches will take place from the latest launch pad (ELA-3) at Kourou in French Guiana. It is expected that, after the first flight in 1995, production of Ariane 5 will soon increase to a level at which up to 10 launches per year should be possible.

A Geostationary Transfer Orbit (GTO) is an elliptical (egg-shaped) orbit which has a low point only a few hundred kilometres above the Earth's surface and a high point of about 36 000km. Satellites destined for geostationary orbit are first placed into GTO by their launch vehicle and then, when they are at the top of the ellipse, they fire another motor (often called the Apogee Boost Motor or ABM) to circularize their orbit 36 000km high. Once in a circular orbit at this altitude the satellite completes one revolution in exactly the same time it takes the Earth to turn once on its axis. As a result the satellite appears to hover over a point on the Earth below and is said to be in a geostationary orbit. The final mass of satellite injected into geostationary orbit is about half of the mass placed in the transfer orbit.

As well as providing a range of different versions of the basic Ariane launcher, Arianespace can offer additional flexibility in the form of special systems which allow more than one satellite to be carried on a single launch. These systems, known by the French acronyms SPELDA and SYLDA allow the owners of two medium-sized satellites intended for similar orbits to share the cost of launch. The Ariane 5 rocket will have a similar system called SPELTRA which will allow double launches and a device called SPILMA which, used in conjunction with SPELTRA, will allow three satellites to be carried on a single launch.

First Ariane 4 launch

Britain in space

Prospero 1971

Missed opportunities

After an encouraging start, the British national space programme faded away.

At the beginning of the space age, the UK was involved in a wide range of space activities, including the development of launch vehicles and both scientific and technological satellites. Despite this encouraging start, many of these programmes faltered or were merged into European programmes during the 1970s. Although Britain is a major contributor to the European Space Agency (ESA), and plays a pivotal role in many ESA projects, it is no longer a major space power and while national space activities continue, they are mostly in the areas of commercial satellite development or junior partnerships in scientific projects led by other nations.

Rocket development

In the late 1950s, Britain led Europe in the field of large liquid-fuelled rockets with its Blue Streak ballistic missile. The missile programme was stopped in 1960, but Blue Streak was chosen as the first stage of the international Europa space launch vehicle being developed by the European Launcher Development Organisation (ELDO). Although Blue Streak performed satisfactorily, the Europa rocket never placed a satellite in orbit and the project, and with it Britain's interest in rocket development, was abandoned in the early 1970s. The remaining Blue Streak rockets now languish in museums.

In a national space launcher programme running in parallel with the development of Blue Streak, Britain also developed the Black Arrow satellite launcher. Black Arrow, which was based on the small Black Knight research rocket, was a three-stage vehicle 13.1m tall. The first two stages were liquid fuelled and the final stage used a single solid rocket motor. The first test flight, which had a dummy third stage and was not intended to reach orbit, was launched from Woomera in Australia on 27 June 1969. Unfortunately, the rocket veered off course less than a minute after launch and was destroyed. The second test, on 4 March 1970, was a successful sub-orbital flight which cleared the way for an attempt to place a small satellite in orbit. The launch took place on 2 September 1970, but failed because the second stage shut down 13 seconds early, preventing the third stage from building up enough velocity to place the X-2 satellite in orbit.

In July 1971, development of Black Arrow was cancelled, but permission was granted to fire the remaining rocket and so, on 28 October 1971, the last Black Arrow was launched. Everything went smoothly and the X-3 satellite, also known as Prospero, was placed into a 547km × 1 582km orbit. The success of Black Arrow made Britain the sixth nation to place a satellite into orbit using its own rocket.

The Ariel satellites

The Ariel programme comprised a series of scientific satellites built as part of a collaborative agreement between Britain and the USA. The series began with four satellites devoted to the study of the Earth's ionosphere and to galactic radio astronomy; these were launched between 1962 and 1971. Ariel 1 and 2 were built in the USA and fitted with British experiments, but Ariel 3 and 4 were built in the UK by the British Aircraft Corporation (now British Aerospace). The Ariel 5 satellite was devoted to X-ray astronomy and was a particular success, remaining in operation for five years and producing a large number of important scientific results. The final mission, Ariel 6, was also devoted to astronomy. The main experiment on the satellite was designed to detect cosmic rays and smaller X-ray detectors were also carried. All six satellites were launched by NASA, the first five as part of the international collaboration. The final launch was a national mission and the UK reimbursed NASA for the cost of the launch.

Prospero and Miranda

The X-3 satellite, which was named Prospero once in orbit, was a pumpkin-shaped, spin-stabilized satellite weighing about 70kg. Prospero was intended to test basic systems, such as solar cells, telemetry systems and thermal coatings, which were being developed for future satellites. There was also a single scientific experiment. Prospero was the first and last satellite orbited by a British-developed launcher.

Miranda, also known as X-4, was another satellite devoted to the development of new space technology and was the first British-built satellite to be stabilized in all three axes. Miranda was launched into a polar orbit on 28 February 1974 by a US Scout rocket. Although quite successful, Miranda was the last of the X satellite series.

Uosat

The Uosats are small satellites built by the University of Surrey and are intended for use by amateur radio operators and for a variety of educational purposes. The Uosats are launched as supplementary payloads when a major space agency has some room to spare on a rocket being used to launch a large satellite. Uosat 1 (also known as OSCAR 9) was launched on 6 October 1981 by the Delta rocket used for the NASA Solar Mesospheric Explorer satellite. Uosat 2 (OSCAR 11) shared a Delta rocket with the Landsat 5 satellite on 1 March 1984. Uosat 3 and 4 were launched on the thirty-fifth Ariane rocket along with the SPOT-2 Earth observation mission and four other small payloads. Uosat 5, which weighs just 50kg and features a CCD camera able to relay pictures of the Earth back to schools equipped with suitable receivers, was launched with the European ERS-1 satellite on 15 August 1991.

International satellites

When the Ariel series ended with Ariel 6, British scientists were forced to work in cooperation with other space agencies to place experiments in orbit. This is usually done through participation in ESA missions, but the UK has also taken part in a number of other international collaborations. In 1983, the UK joined with the USA and the Netherlands in the highly successful tri-national Infrared Astronomical Satellite IRAS and in 1984 British scientists provided a small satellite as part of the US/German/UK AMPTE mission to explore the interaction between the solar wind and the Earth's magnetic field. X-ray telescopes developed at Birmingham University Department of Space Research have flown on the US Space Shuttle and on the Soviet Mir space station, and a British-built Extreme Ultraviolet telescope was carried by the West German ROSAT X-ray astronomy mission.

British astronauts

Unlike other major European countries such as France, Germany and Italy, who have negotiated flights on the Space Shuttle or aboard Soviet spacecraft, Britain has no national astronaut programme. A British astronaut, Nigel Wood, was scheduled to fly on a 1986 Space Shuttle mission to observe the deployment of a military SKYNET satellite, but the carrying of passengers on the Space Shuttle was suspended after the explosion of the Challenger and Wood's flight was cancelled.

The only Briton in space to date has been the result of Project Juno, a private venture in which it was hoped that British companies would sponsor the flight of a British astronaut to the Soviet Mir space station. Over 12 000 people applied for selection as the British astronaut and the candidate finally chosen was Helen Sharman, a 26-year-old researcher with the Mars® confectionary company. The project came close to collapse when it became clear that British industry would not sponsor the flight, and any attempt to have a major British science programme was abandoned for lack of money. Despite the failure to raise the necessary money, the USSR allowed the mission to go ahead. It took place during one of the regular flights to exchange the crew on Mir; Helen Sharman flew to Mir aboard Soyuz TM12 on 18 May 1991 and returned safely to Earth eight days later. There are no plans for another Juno mission.

> The British national satellites were all named after characters from the works of Shakespeare.

The first stage of Britain's Black Arrow launcher was powered by the Rolls–Royce Gamma 8 rocket engine

Buran

First flight 15 November 1988

Soviet space shuttle

The USSR developed a space shuttle which bore a remarkable similarity to its US equivalent.

The story of the Soviet space shuttle, the first of which has been named Buran, can be traced back to the 1960s when the Soviet chief designer Sergei Korolov considered the idea of a reusable spaceplane. The first design studies of such a craft were made by Artem Mikovan, the designer of the supersonic Mig fighter planes used by the Soviet air force. The project is said to have been called 50/50 and consisted of a hypersonic carrier aircraft which would fly at about six times the speed of sound and release a small spaceplane which would continue into orbit under its own power. As part of this programme a small hypersonic research aircraft was built and, like early US research aircraft, was carried to high altitude by a converted bomber and then released to glide back to the ground. Little came of this work and the project was abandoned in 1969 after three subsonic test flights.

By the early 1970s the USA was developing its Space Shuttle and the USSR decided that it could not afford to be left behind in such an important area of technology and so began its own shuttle programme. The project was concentrated at the Ramenskoye airfield near Moscow, site of the USSR's best aviation research facilities, but the work was spread between all the major Soviet space design groups. Wind tunnel tests were conducted to determine the best aerodynamic layout for the vehicle and scale models were launched to verify that new materials developed for the shuttle would be able to survive the heat of re-entry into the atmosphere. The first of these test flights, Cosmos 1374, was launched in 1982. In all, four such orbital tests were made, together with a number of sub-orbital flights. By 1985 a full-sized version of Buran had been built and fitted with ordinary jet engines so that its flying characteristics could be tested. Atmospheric test flights continued until 1988 and had the same objectives as the tests performed by NASA in 1977 when the Space Shuttle Enterprise was dropped from a Boeing 747 and allowed to glide down to a landing. Once the Soviet test flights were completed, attention switched to the Baikanour cosmodrome where the final assembly of Buran was underway.

Buran

At first sight the Soviet shuttle appears remarkably similar to its US counterpart. Buran has a double-delta wing layout, a large cargo bay and a two-level crew compartment near the nose. It is designed to carry two pilots plus some passengers. Like the US Space Shuttle, Buran uses special tiles to protect itself against the heat generated during re-entry and, like the Space Shuttle, it glides to a landing ready to be used again. The main dimensions of the two craft are very similar and they even have the same number of windows on the flight deck. There are however two important differences. Firstly, Buran can be flown automatically and secondly, Buran itself does not have reusable rocket engines. Buran is launched by the huge Energia rocket which carries it almost into orbit before releasing it. At this point small rockets, similar to the orbital manoeuvring system on the US Space Shuttle, take over and give Buran the final push into orbit. The same motors are used to slow down Buran to begin re-entry. Other smaller rockets mounted in clusters near the tail and in the nose are used to control the attitude of Buran when it is in orbit. The touchdown speed of Buran is about 340km per hour and its landing run is braked by three parachutes that are released from the tail section after landing.

The huge Energia rockets are assembled in buildings originally used during the development of the Soviet moonrocket. These include a facility in which the main Energia core is assembled before it is moved to another building for the attachment of its four strap-on boosters. Buran, and any sister ships that may eventually be constructed, undergo final assembly and testing in purpose-built buildings. Both Buran and the Energia are then taken to yet another facility where the two are joined together and loaded on to a rail transporter to be moved to a launch pad several kilometres away.

There are two Energia launch pads which can be used in conjunction with Buran and a third is available if Energia is carrying other cargo. They are located close to the centre of the Baikanour cosmodrome, not far from the pads used to launch the manned Soyuz spacecraft. The Buran pads each comprise a concrete base with fixed service towers on either side. These towers support bridges leading to Buran and Energia as well as pipes which are used for loading propellants. One of the towers also provides a means for cosmonauts to enter and leave Buran while it is on the launch pad. There is also a movable structure which runs on rails and can be swung into place around Buran to allow technicians access to the space shuttle itself while it is on the launch pad. To escape in an emergency the cosmonauts must leave Buran and jump on to trolleys which slide down a pipe into an underground bunker. The pads have high towers which serve to protect them from lightning strikes.

The runway on which Buran lands when returning from space is located about 12km from the launch complex and is equipped with an automatic landing system.

First flight

The first flight of Buran was unmanned and took place on 15 November 1988 after an attempt on 29 October was cancelled 51 seconds before lift-off because of a minor technical problem. The main engines in the Energia core stage ignited about eight seconds before lift-off, followed a few seconds later by the ignition of the four strap-on boosters clustered around the core. The Energia/Buran combination soon vanished into low clouds. About 2.5 minutes later the strap-on boosters were jettisoned, but the Energia core continued to fire until 8 minutes after launch when, at a height of 110km, Buran was released. When Buran reached an altitude of 160km it fired its own motors for 67 seconds. A second firing of the Buran motors a few minutes later placed the shuttle in a circular orbit 250km high.

Buran completed two orbits of the Earth before returning. During its flight through the atmosphere Buran conducted a series of turns, both to place itself on course for Baikanour and to help it slow itself down from Mach 25. A Mig 25 fighter plane intercepted Buran during its final approach and beamed television pictures of the last few moments of the flight back to mission control. The automatic landing was successful and Buran rolled to a stop after a flight lasting 3 hours and 25 minutes.

The future

Following the political turmoil in the USSR during 1991 the future of Buran is rather uncertain. It had been expected that a second flight, involving sending an unmanned Buran-type shuttle (probably the second example built and the first to be fully equipped with life-support systems) to dock with the Mir space station, would occur in 1990. This flight was then delayed until 1991 and had not taken place by the end of that year. Since it is far from clear that the Soviet space programme has any need for a Buran-type vehicle, it remains possible that the entire project may be cancelled in the near future.

Buran means 'snowstorm'.

	Buran	US Space Shuttle
Length (metres)	36	37.2
Wingspan (metres)	24	23.8
Cargo Bay Width (metres)	4.7	4.6
Cargo Bay Length (metres)	18.3	18.3

If viewed upside down the cyrillic letters for *Buran* appear to spell Hedgehog.

The Energia multipurpose space transport system, incorporating the new powerful booster rocket and the Buran (Snowstorm) reusable orbiter

Canada in space

Alouette 1 28 September 1962, Marc Garneau October 1984

Third nation in orbit

Canada has been involved in space development for 30 years, but has chosen projects with national applications and not aimed for space spectaculars.

With the launch of Alouette 1, by a US Thor-Agena B rocket, on 28 September 1962, Canada became the third country to have a national satellite in space. The satellite was built in Canada and carried four experiments, each designed to probe the ionosphere (the region between about 60km and 500km from the Earth's surface where atoms and molecules are ionized by the Sun's radiation) as part of a joint US/Canadian research programme. Alouette 1 was so successful that it led to an agreement between the two countries to develop a series of similar satellites under the name ISIS, the International Satellites for Ionospheric Studies. The first in the series, Alouette 2 was launched on 28 November 1965 and, like its predecessor, it operated for almost 10 years. Two more advanced satellites were launched in 1969 and 1971 to continue the programme.

The ISIS programme continued throughout an entire 11-year solar cycle in order to determine how the ionosphere reacted to changes in the Sun's radiation, an important area of study because long-range radio systems use the ionosphere as a reflecting mirror to bounce signals over the horizon. The ISIS satellites were also able to study the Aurora Borealis (The Northern Lights) by looking down on them from space. One further ISIS satellite was planned but this was cancelled when Canada decided to redirect its space activities from scientific research to programmes with more direct applications to the nation's economy.

Anik

In 1969 Canada decided that it should give priority to establishing a nationwide satellite communications system, an obvious move for such a large and sparsely populated country. A national organization, Telesat Canada, was set up to manage this system and in 1971 Telesat ordered three satellites from the Hughes Aircraft Company in California. Two of these were to be placed in geostationary orbit from which they would hover over the equator in constant view from Canada, the third would be kept ready for launch as a spare. Telesat insisted that some of the building of the satellites be subcontracted to Canadian firms and about 13% of this first series were of Canadian manufacture. The first satellite, named Anik (the Eskimo word for brother) was launched by a US Delta rocket on 9 November 1972 and went into full operation early the next year. The other two, including the spare, were launched in 1973 and 1975.

The first satellites, known as the Anik-A series, were followed by more and more powerful satellites as the system started to grow. Hermes, launched in 1976, and Anik-B were experimental satellites which could broadcast direct to rooftop dishes on individual buildings. These were used for a variety of medical and educational experiments in which doctors or teachers could consult with or lecture to groups thousands of kilometres away. The Anik-C series began with a US Space Shuttle launch in 1982 and can be used for television and electronic data transfer throughout most of Canada. The Anik-D series is used for telephone communications, replacing the old Anik-A satellites. Throughout the series, Canadian industry has become more and more involved in development and the Anik-D series were assembled by a Canadian company, Spar Aerospace, under a licence from the manufacturer, Hughes of California.

The latest version is the Anik-E series, the first of which ran into serious problems soon after its launch on 4 April 1991. Ground controllers were unable to unfold the satellite's antennae and solar panels, rendering it useless and it was expected that the satellite would have to be abandoned. However, after several weeks of experimentation, in early July the controllers were able to free the stuck panels. The second Anik-E was launched by an Ariane 44P rocket on 26 September 1991.

Canadarm

In the early 1970s Canada became involved in the US Space Shuttle programme when it agreed to develop the robot arm, or Remote Manipulator System (RMS) for the Space Shuttle. Canada would develop the system and provide one unit free of charge provided that NASA agreed to buy a further three at a cost of about $25 million each. The robot arm is a complicated device since it must be both very strong and very light and must be able to move heavy objects about with great accuracy. Development took seven years and the first example was delivered to NASA in 1981. The project was a great success and the arm has become a vital part of the Space Shuttle system. It has been used in some of the Shuttle's most daring missions, such as the rescue of the Solar Maximum Mission satellite in 1984.

Canada has decided to build on its experience with the RMS and has agreed to develop a similar system for the international space station Freedom (see p55). As presently planned the system will consist of a mobile service centre, equipped with manipulators and other tools, that can move around on the outside of the space station. The service centre will be used to carry out assembly and repair tasks, either under remote control or with direct astronaut supervision. As with the RMS, this project will develop Canadian technology in the area of manipulators and has great industrial potential.

Canadian astronauts

The success of the RMS led NASA to invite Canada to nominate a number of astronauts to fly on future Space Shuttle missions. Over 4 300 people applied and from these six were chosen to form the nucleus of the Canadian astronaut programme. The first to fly was Dr Marc Garneau, a naval officer, who flew aboard the thirteenth Shuttle flight in October 1984. During the flight Dr Garneau carried out a number of experiments including some associated with vision systems that can locate and track objects close to the Shuttle with great accuracy. These devices will have obvious uses in the new, more advanced manipulator systems planned for the space station.

A second Canadian astronaut, laser physicist Dr Steven McLean, had already been nominated when the explosion of the Shuttle Challenger disrupted the programme and his flight was cancelled. The next flight of a Canadian astronaut (Roberta Bondar) eventually took place during the International Microgravity Laboratory Space Shuttle mission in January 1992.

Radarsat

Canada's next major space project will be Radarsat. This satellite will carry a Synthetic Aperture Radar (see p107) that can gather information on crop conditions, ice distribution and geological data to exploit Canada's great natural resources. From a circular orbit about 1 000km high Radarsat will regularly survey the whole of Canada and will be unaffected by cloud which would limit satellites using ordinary cameras. Radarsat is expected to be launched in 1994 or 1995 by a US rocket.

In 1991 an orbiting Anik satellite was sold to the US operator GE American Communications Inc who wanted it to replace their Satcom 4 satellite which had encountered technical problems. Buying a secondhand satellite already in orbit was much cheaper than launching a replacement.

Msat

Msat is a mobile telecommunications satellite intended to serve air, land and sea traffic. The satellite should be able to support up to 3 200 different radio channels covering the whole of North America. Msat users will communicate with each other, via the satellite, using small radio antennas which can be mounted on the cabs of lorries, on small boats etc. The satellite, which is being developed jointly by the Hughes Aircraft Company in the USA and Spar Aerospace in Canada, is expected to be launched in 1994.

The Canadian-built robot arm for the US Space Shuttle

Cassini

Launch 1996?

Mission to Saturn

The NASA/ESA (European Space Agency) Cassini mission will explore the ringed planet Saturn and send a probe deep into the atmosphere of its enigmatic moon, Titan.

The planet Saturn has received only three visitors from Earth. The Pioneer 11 spacecraft, launched in 1972, flew past Jupiter in 1974 and was redirected towards Saturn, arriving there in 1980. Pioneer 11 returned useful data, and a handful of pictures, but its main achievement was to pave the way for the more sophisticated Voyager 1 and 2 spacecraft. The two Voyagers flew past Saturn in 1981 and 1982 and sent back an enormous amount of information but, because they spent only a brief time close to Saturn, they left many questions outstanding. To answer these questions is the objective of the US/European Cassini mission, which will enter orbit around Saturn and study the planet, its rings and moons for an extended period. The highlight of the mission will be the release of a European-built probe that will parachute into the atmosphere of Titan, the largest moon in the solar system, and investigate what lies below the clouds which permanently hide its surface.

Cassini will follow a complicated mission profile which may change in detail as planning for the mission evolves. One option is to launch Cassini with a US Titan IV-Centaur rocket in 1996 and to place it on a trajectory which will use flybys of Venus and the Earth to gain sufficient energy to send it outwards towards Jupiter. During its flight past Jupiter, instruments aboard Cassini will observe the planet, possibly in conjunction with the Galileo spacecraft (see p58) which may still be operating in orbit around the planet. Unlike Galileo, Cassini will not remain at Jupiter but will swing past the planet, receiving a gravitational boost that will propel it towards Saturn.

In 2004, as the spacecraft arrives at Saturn, it will fire its main rocket motor for over two hours. This will slow the spacecraft down and place it in orbit around the ringed planet. A month after arrival the motor will again fire to move Cassini into a new orbit that will take it towards Titan, target for the European probe. The main Cassini spacecraft will aim itself at Titan, release the probe, and then manoeuvre on to a new course that will allow it to fly past Titan just as the probe enters the satellite's atmosphere. During the probe's three-hour descent the main Cassini spacecraft will act as a radio relay station, collecting data from the probe and broadcasting it back to Earth. When the probe ceases to transmit, Cassini will continue with its own exploration of Saturn and its satellites.

Mission planners have discovered that if Cassini encounters Titan repeatedly it can use Titan's gravity to send the craft on a complicated tour of Saturn's other satellites without using an unacceptably large amount of fuel. So Cassini will spend several years studying Saturn and its rings while gravitationally bouncing from satellite to satellite. During each flyby of the major moons, Cassini will take close-up pictures that will be vastly superior to the best Voyager images now available. In addition, the chemical and physical properties of the satellites will be probed by remote sensing instruments such as spectrometers. During the satellite tour the angle between Cassini's orbit and the equator of Saturn will gradually increase so that towards the end of the mission, some four years after arrival, Cassini will be in an orbit that takes it high above the plane of the ring system and allows it to obtain unprecedented views of the Saturn's polar regions.

Cassini: the Saturn orbiter

The main Cassini spacecraft, weighing about 5 000kg, will be built by the NASA Jet Propulsion Laboratory in Pasadena, California. It will be one of a new type of interplanetary spacecraft called the Mariner Mk 2 series. These are based on a more or less standardized spacecraft which can be adapted for a variety of deep space missions by adding equipment suitable for each particular project. Each Mariner Mk 2 will use nuclear generators for power and will carry movable platforms on which cameras can be mounted. Many of the systems of these new craft will be based on equipment designed for earlier missions such as Viking, Voyager and Galileo. In this way, NASA hopes to save money and allow more planetary missions to be flown within a given budget.

The spacecraft releases the European Huygens probe into Titan's atmosphere

Huygens: the European probe

The European contribution to the Cassini mission will be the Titan probe, Huygens. Twelve days after release from Cassini, the Huygens probe will enter Titan's upper atmosphere at about 7km per second. The 192kg probe will be slowed down by friction between the atmosphere and a shallow, 3.1m diameter, conical skirt which functions as an airbrake. The maximum deceleration will be about 25g. The skirt will be jettisoned at an altitude of about 170km, at which point a drouge parachute will be released to stabilize the capsule. Shortly after, the main parachute will be deployed and the probe will be lowered towards the surface. During the parachute descent the probe's 40kg of scientific instruments will investigate the chemical composition of the atmosphere, measure windspeeds and cloud structure and attempt to take pictures of the hitherto unknown surface. Huygens will reach the surface about three hours after entering the atmosphere, impacting at a speed of about 5m per second. It is hoped that the scientific instruments will continue to transmit for a time after landing, although, from the little known about Titan, it is possible that by then the probe may be slowly sinking in an ocean of liquid methane.

The choice of names for the mission follows a recent tradition of naming planetary spacecraft after individuals from history. Giovanni Cassini was an Italian astronomer who first noticed a dark line in what was then thought to be a single ring around Saturn. Cassini's Division, as it became known, was subsequently found to be an area where the ring material is so thinly spread as to be transparent. The European probe is named after the Dutch physicist and astronomer Christian Huygens who discovered Titan in 1656 and whose book, *Systema Saturnian*, appeared in 1659.

China in space

Tungfanghung 24 April 1970

Long March into space

After a late start, China quickly demonstrated considerable capability in space and by the end of 1990 had orbited 33 satellites.

China is the traditional home of the rocket, the Chinese having invented fireworks by medieval times, but it was not until 1956 that the country's first rocket research institute was founded. The first Chinese-built sounding rocket (a sub-orbital research rocket) was launched in 1960, the year which also saw the flight of a Chinese ballistic missile prototype. Serious planning for launching satellites began a few years later but it was not until 1970 that a satellite was orbited, although there may have been a failed attempt in November 1969.

The East is red

Tungfanghung, as China's first satellite was named, was launched on 24 April 1970 from the Jiuquan launch site in the west of the country (41°N, 100°E). The launcher was a three-stage rocket called Long March. The satellite is best remembered for its radio transmissions of the propaganda song 'The East is Red' although it also returned data about the condition of its on-board systems until it fell quiet in June. The launch of Tungfanghung made China the fifth nation to launch a satellite using its own rocket. A second satellite, known as Practice 1, was launched on 3 March 1971. Practice 1 had a solar cell power supply and was used for scientific purposes. It remained in space until 17 July 1979.

After this promising start there were no further launches until 26 July 1975 when a satellite weighing about 2 000kg was placed in a low Earth orbit. This remained in space for about seven weeks before decaying naturally and may have been a prototype for the series of recoverable satellites that were to begin later that year. A similar satellite followed a near identical mission profile after a launch on 26 November 1975. Both of these flights, and another, possibly scientific, mission on 30 August 1976, used the two-stage FB-1 rocket. The final use of the FB-1 rocket came on 19 September 1981 with the simultaneous launch of three small scientific satellites from a single rocket.

Chinese eyes in the sky

A long-running series of recoverable satellites began with a launch of SKW-4 on 16 November 1975, although this had been preceded by a launch failure almost exactly a year earlier in which the Long March 2 booster was destroyed after only 20 seconds of flight. The SKW-4 satellite weighed about 3 000kg and remained in space for six days at which time it separated into two sections, one of which, weighing just under 2 000kg, returned to Earth. This made China only the third nation, after the US and USSR, to develop the ability to recover payloads from orbit. There were further launches of this type of satellite in 1976 and 1978, since when they have been launched regularly every year. At first it was believed that these satellites were for military purposes and, while this may be at least partly true, it now seems that they also carried at least some experiments designed for civilian use. Later missions in the series may have been exclusively civilian in nature.

Communication satellites

China is a very large country and so the setting up of a satellite communications network was an obvious step for the Chinese space programme. The first attempt, using a Long March 3 booster in January 1984 failed when the third-stage motor failed to re-ignite at the required moment and the satellite was stranded in low Earth orbit. The satellite separated from the errant booster and was used for tests while a second attempt was readied. This mission was successful, placing a satellite into a geostationary orbit over 125°E longitude. A third launch in 1986 positioned another satellite over 103°E longitude. Since then missions have continued from time to time, a recent example being a launch in February 1990 using a Long March 3 rocket.

Weather satellites

Feng-Yun (Wind and Cloud) 1-1, China's first weather satellite was launched in September 1988 but operated for only 39 days. A second satellite, Feng-Yun 1-2 was launched by a Long March 4 rocket on 3 September 1990 into a 900km, circular orbit inclined at 99° to the equator. The satellite carries high resolution radiometers able to scan cloud patterns during both daylight and darkness.

Launches for sale

In 1985 China announced that she would commercialize her launch vehicle industry and offer to launch satellites for other countries, in competition with the European Ariane and similar US rockets. They also offered to fly experiments in their recoverable capsules, an offer taken up by at least two European organizations. The first of these commercial recoverable flights carried a payload for the French company Matra in 1987. Commercial launches are based on the latest versions of the Long March family of two- and three-stage boosters and several different versions are available.

Launches can take place from one of three Chinese launch sites, depending on the version of Long March to be used and the requirements of the mission. The Xichang Satellite Launch Centre is near the city of that name in Sichuan province (102°E, 28.2°N) and entered service in the early 1980s. Its facilities have recently been upgraded to make it suitable for the demanding requirements of foreign satellite operators. The Taiyuan centre is in Shanxi province, west of Beijing. It was first used in September 1988 for the launch of the first Long March 4 rocket. The third site available is the one used for the original Tungfanghong satellite at Jiuquan.

The first commercial launch took place on 7 April 1990 when a Long March 3 vehicle orbited Asiasat 1 from the Xichang centre. This satellite, originally known as Westar 6, was launched from the US Space Shuttle in February 1984, failed to enter the correct orbit, was rescued by the Space Shuttle in November 1984, refurbished and resold to become Asiasat 1. Despite its success with Asiasat, China's commercial programme has been set back by the reluctance of the USA to allow further satellites which incorporate American technology to be launched from China.

Chinese astronauts?

China is also said to be planning a manned space programme, although details are scarce. Photographs have been released which claim to show Chinese astronauts in training and rumours of possible Chinese flights with the USA and the USSR circulated

Chinese launch vehicles

The Chinese launch vehicles are known by CZ (Cang Zheng or Long March) numbers. The CZ-1 was a two-stage, CSS-3 ballistic missile to which a third stage was added. The CZ-1 was 29.5m tall and weighed 81.5 tonnes. The CZ-2 was a two-stage rocket 33m tall and weighing about 191 tonnes. The CZ-2E is a larger version, 50m tall, which has four strap-on boosters added to the first stage and a more powerful second stage. The FB-1 was a modified CZ-2 and was retired in 1981. The CZ-3 comprises the CZ-2 stages plus a third stage using liquid hydrogen/liquid oxygen propellants and weighs 202 tonnes. The CZ-4 is an improved CZ-3 with an enlarged first stage and a new third stage. The CZ-4 stands just over 50m high.

at one time. The opportunity to fly on the Space Shuttle probably vanished following the the explosion of the Challenger in 1986, and the events of 1989 and the new commercial attitude of the USSR space programme since 1990 makes the possibility of a Sino-Soviet flight less likely than it once was. It is not clear how close China has come to developing its own manned spaceflight capability.

China's Long March 2E

Columbus

Approved 1987

Towards a European space station

Columbus is a European project which involves the international space station Freedom and various free-flying satellites.

In 1987 the European Space Agency (ESA) received provisional approval to start three large and expensive projects intended to provide Europe with its own space infrastructure. The objective of these three projects was to make Europe autonomous in all the important areas of manned spaceflight and reduce her dependence on the superpowers. The three projects concerned were the Ariane 5 rocket (see p13), the Hermes mini-shuttle (see p71) and a multi-element space station project called Columbus. The Columbus system comprised three different components: a pressurized module which would be joined to the international space station, a free-flying space laboratory which could be serviced by astronauts and an unmanned satellite, called the polar platform, for Earth observations.

Following the provisional approval of these plans, development of the Columbus system went ahead in collaboration with NASA and the other partners in the space station programme. It was expected that, following the studies and cost estimates made during the preparatory phase, approval to continue with the development and manufacture of Columbus and Hermes would be given by European space ministers in 1991. However, delays in the NASA programme, cost increases for both Columbus and Hermes and the political and economic turmoil in Europe during 1990-1 caused the ministers to delay a final decision when they met in November 1991, leaving some doubt over the long-term future of the original ambitious plans for European independence in space.

The Attached Laboratory

When President Reagan directed NASA to develop a permanently manned space station he invited other nations to join the US programme. Europe's response was to propose that ESA build a module, similar to those being provided by NASA, which could form part of the core of the space station. This module was originally known as the the Attached Pressurised Module and later renamed the Columbus Attached Laboratory.

The Columbus Attached Laboratory is intended to be launched in the payload bay of the US Space Shuttle and joined to the space station in orbit. Its main purpose will be research into the effects of weightlessness and it will be designed for operation by a crew of two to three astronauts. In keeping with the international nature of the space station, the facilities of the European laboratory will be shared by all the nations participating in the programme and the European astronauts will not live in the laboratory, but will share space in the living quarters used by the rest of the crew.

The original design for the Columbus Attached Laboratory was for a four segment module, similar in concept to the European Spacelab, about 12.7m long and 4m in diameter. The module would be launched partly outfitted and extra equipment, delivered during later Space Shuttle missions, would be installed once the module was docked to the space station. However, the design of the space station was cut back several times (see p55) and in consequence the length of the European module was reduced. One of the reasons for the reduction in length of the module is that it will simplify the internal structure and reduce the demands which the module places on the space station as a whole. The reduction in size, and hence mass, of the module also means that it will be possible to install more scientific equipment before launch without overloading the Space Shuttle that will carry it into orbit. This will reduce the amount of assembly work which has to be done once the laboratory is docked to the space station and enable the astronauts to start work in the module as soon as it has arrived.

The launch of the Columbus Attached Laboratory is dependent on the progress achieved on the space station but is expected to occur around the year 2000.

The Columbus Free Flyer

The Columbus Free Flyer will be an autonomous laboratory module designed for research which requires the weightless conditions of space. Since astronauts moving about inside a space vehicle cause tiny disturbances which make such research difficult, the Free Flyer will be designed to operate automatically for long periods. From time to time astronauts will visit the Free Flyer to replace or repair equipment and to collect samples of materials produced during the module's experiments. Once such a visit is over the Free Flyer will be left unmanned for another period of independent flight. In space jargon this mode of operation is called 'man-tended' and so the module was originally known as the Man-Tended Free Flyer or MTFF.

The main laboratory of the Free Flyer will be a cylindrical pressurized section resembling one of the manned modules of the space station. One end will have a docking system so that astronauts can enter the module to carry out servicing missions, and the other end will comprise a resource module. The resource module, which will probably be boxed shaped with two large solar panels, will provide the systems for attitude control, communications and power needed to operate the laboratory.

The Columbus Free Flyer was originally intended to be serviced at the space station and so the preliminary design included systems to allow it to rendezvous and dock with the space station. However, in an attempt to reduce costs and to avoid dependence on NASA, the mission was changed to allow servicing by the European Hermes mini-shuttle. This change made it possible to remove the complicated and expensive automatic docking systems and reduce the amount of fuel which was needed. The decision not to use the space station for servicing also means that the module can be placed in a higher orbit, in which the effects of atmospheric drag will be reduced. However, the reductions in the programme mean that the expected lifetime of the module will be limited to 10 years instead of the original goal of 30 years and that servicing will only take place once per year instead of the twice originally planned. The Columbus Free Flyer is expected to be launched by an Ariane 5 rocket during the early years of the next century.

The polar platform

The polar platform is an unmanned satellite intended to carry instruments to observe the Earth and provide data for both meteorology and Earth resources research. The platform will be large, about 10m long and 3m wide and will have a mass of about 8 000kg. Electricity will be supplied by a single solar wing about 16m long and 5m wide which should generate several kilowatts of power. The polar platform will be launched by an Ariane 5 rocket towards the end of the 1990s and will be placed in a Sun-synchronous, circular orbit about 800km high.

The polar platform was originally intended to be serviced by astronauts who would visit it using the Space Shuttle. However, following the changes to the Space Shuttle programme after the explosion of the Challenger in 1986, this requirement has been abandoned and the present plan is for the polar platform to operate entirely independently for at least six years.

Comet exploration

Vega and Giotto 1986

Into the coma

Spaceprobes provide almost the only means of studying the icy nucleus at the centre of a comet.

Comets have been known since antiquity but it is only this century that astronomers have gained a clearer understanding of what comets really are. In 1949 Fred Whipple proposed that a comet was a snowball of ice, gas and dust which had been in deep freeze since the formation of the solar system. When the snowball, more properly called the nucleus, approaches the Sun, material boils off the nucleus surrounding it with a cloud a gas and dust called the coma. Material from the coma is then blown behind the comet by the solar wind to form a tail.

A contemporary of Whipple, Dutch astronomer Jan Oort, suggested that thousands of millions of comets orbit the Sun far beyond the planet Pluto. From time to time the gravity of nearby stars stir up this cloud causing comets to fall towards the Sun. Most comets pass briefly by the Sun, are visible for a few weeks then return to deep space never to be seen again. These are called 'long period' comets. However, a few comets pass close to the planet Jupiter and are gravitationally captured into orbits in which they take from a few years to a few decades to circle the Sun. These are said to be 'short period' comets and reappear at predictable intervals.

A challenging mission

Comets are interesting because they consist of material which has been frozen since the solar system was formed and so they carry clues about conditions in the very distant past. A scientifically ideal target for a space mission would be a long period comet making its first visit to the Sun but unfortunately the orbits of such objects cannot be predicted sufficiently far in advance. A short period comet, with a well-known orbit, is an easier target but, since it will have made frequent passages close to the Sun, much of its volatile material will already have boiled away making it fainter and less interesting. A good compromise is Halley's comet, whose period of 76 years is longer than most short period comets and which still retains considerable amounts of gas and dust. In fact Halley is the only comet with a predictable orbit that shows all the properties of a typical comet: the development of separate gas and dust tails, the formation of a bright coma and occasional outbursts of fresh material from the central nucleus into the coma. For this reason its 1985/86 appearance was chosen as the target for a series of comet intercepts launched by Europe, the USSR and Japan. It was, however, a hazardous undertaking; Halley's comet goes around the Sun in the opposite direction to that of the Earth and during the flyby the spacecraft and comet met head-on at a relative speed of about 70km per second.

The Halley armada

First to reach the comet was the Soviet Vega 1, launched on 15 December 1985. Vega 1 (the name is a contraction of Venus–Halley using Cyrillic letters) was one of a pair of spacecraft which visited Venus (see p244) before swinging past the planet en route for the comet. Vega 1 passed within 8 889km of Halley's nucleus on 6 March 1986 and, during a three-hour period, returned 500 images and various other kinds of scientific data. Vega 2 (launched 21 December 1985) passed about 8 030km from the nucleus on 9 March and returned 700 pictures. Both craft were struck by dust from the comet's coma, but neither was put out of action.

Next to visit the comet was the small Japanese probe 'Suisei' which had been launched on 18 August 1985. Suisei was not intended to fly close to the nucleus but rather to observe the comet from a safe distance. This it did successfully, passing 152 400km from the comet on 8 March 1986 and taking a series of images in ultraviolet light. These pictures did not show the icy nucleus, but mapped out a huge cloud of hydrogen gas extending many millions of kilometres from the comet's centre. Suisei was preceded into deep space by a test craft named Sakigake. Sakigake was not expected to intercept the comet but did carry equipment that was able to measure the interaction between the solar wind and the tail of the comet from a distance of 11 million kilometres.

The most ambitious probe was the European Giotto, named after the painter Giotto Ambrogio di Bondone, who included a comet in his representation of the nativity. This 600kg spacecraft was launched on 2 July 1985 on a mission to fly deep into the coma of the comet and to try and photograph the nucleus. To protect the craft as it smashed into the comet's dust cloud Giotto was equipped with a two-layer dust shield that, it was hoped, would absorb the energy of the small dust particles as the spacecraft crashed into them. There was no way to protect the craft from larger impacts; Giotto scientists could only hope that the spacecraft would not suffer such an impact until after the probe had achieved its main objectives.

After last-minute course corrections using information from the Vega spacecraft, Giotto plunged into the inner coma of Halley's comet in the early morning of 14 March 1986. Live pictures were beamed back to the European Space Agency (ESA) control station at Darmstadt, Germany and as the nucleus approached it seemed that Giotto might survive. Suddenly, only a few seconds before closest approach, contact with the probe was lost. Giotto had been hit by a dust grain weighing about 1gm and this had knocked the probe out of alignment with Earth, cutting the radio link and stopping the flow of scientific data. Contact was resumed within an hour and some data, but no more pictures, were taken as the probe flew out the other side of the coma.

Giotto returned a number of pictures of the hitherto unseen nucleus and showed that it was a potato-shaped body about 16km long and 8km wide. The nucleus was very dark, with a few small active zones from which a mixture gas and dust was jetting into space. Analysis of the material in the coma showed that much of it was inorganic dust as expected, but that there were also particles of organic material containing just carbon, hydrogen, oxygen and nitrogen.

After the flyby Giotto was placed in hibernation with most of its systems switched off. The craft was reactivated in 1990 and was directed towards the comet Grigg–Skellerup which it encountered in 1992. Unfortunately some of its instruments, including the camera, were damaged during the Halley flyby so no pictures were possible when Giotto encountered comet Grigg–Skellerup.

In 1985 the International Sun–Earth Explorer (ISEE-3) satellite, originally launched to study conditions in the space around the Earth, intercepted comet Giacobini–Zinner. The satellite had never been designed to study a comet, so could not return any photographs of the nucleus, but did provide scientists with information on the interaction between the comet and the solar wind. The satellite was directed towards the comet by means of several close approaches to the Earth and the Moon during which gravity modified the spacecraft's orbit so that it flew off towards the comet. It was renamed the International Cometary Explorer for this phase of its mission.

The next comet mission was expected to be the US Comet Rendezvous and Asteroid Flyby (CRAF) project. Unfortunately, this project ran into finance problems and was cancelled in 1992. It is possible, but not likely, that the mission may be reinstated in the next few years.

An impression of the European Giotto spacecraft approaching Halley's comet

Communications satellites

SCORE 18 December 1958, Telstar 10 July 1962,

Syncom 14 February 1963

An electronic ring around the world

From a speculative article in a 1945 electronics magazine, communications satellites have become the biggest business of the space age.

The idea of using satellites as giant relay towers to bounce radio signals between continents predates the space age. In May 1945 Arthur C Clarke drew the attention of the British Interplanetary Society to the idea of placing a satellite into an orbit 35 880km above the equator. Such a satellite would complete a single orbit in 24 hours and would appear to hover motionless over a single point on the Earth. A system of three such satellites, suitably placed, could provide radio relays between any points on Earth (except regions very close to the poles). He promoted his idea further in the October 1945 issue of the magazine *Wireless World* and for this work has been called 'The father of the communications satellite'. However, despite Clarke's prophecy, the first satellites used for communications were not placed into the 35 880km geostationary, or 'Clarke' orbit; early space launchers were not powerful enough to lift suitable payloads this high, and lower orbits were preferred.

Early experiments

The first attempt to broadcast messages from space came with the US project SCORE (Signal Communications Orbit Relay Experiment), launched on 18 December 1958. SCORE used tape recorders onboard an Atlas B rocket to broadcast a pre-recorded Christmas message from President Eisenhower. SCORE could also record messages and play them back later, but it could not actually relay messages directly, and it operated for only 13 days before its batteries ran down. A rather more sophisticated satellite using this 'store and retransmit' system was US Army Courier 1B. This satellite was launched on 4 October 1960 and could receive and re-broadcast up to 68 000 words a minute. Courier 1B operated for 17 days.

An alternative approach, which avoided placing complicated electronic systems in space, was taken by the US Echo programme. On 12 August 1960 NASA launched Echo 1, a spherical balloon with a metalized skin. Once in orbit the balloon was inflated until it reached its intended diameter of 30m and it was then used as a reflector to bounce radio signals across the oceans. Echo was simply a radio mirror in the sky, and its efficiency decreased as the gas which inflated the balloon escaped and it lost its shape. Echo 2, a larger and thicker balloon satellite, was launched in January 1964 and used for a further series of tests.

Telstar and Relay

After these early experiments came the world's first real communications satellite, the American Telephone & Telegraph Company's Telstar. The first commercially developed satellite, Telstar was a multifaceted sphere just under 1m in diameter and weighing about 77kg. Unlike its predecessors, Telstar was able to receive messages from the ground, amplify them and then re-transmit them immediately. This made it possible to send high-quality data, for instance television pictures, from place to place. Telstar was placed into a low (952km

× 5 632km) orbit on 10 July 1962 and only 15 hours after launch was used in an historic exchange of live television pictures between Andover in the USA, Goonhilly Downs (UK) and Pleumeur-Bodou (France). Although crude by today's standards, the Telstar images were the beginning of a revolution in international television broadcasting. Telstar 1 ceased to function at the beginning of March 1963 and Telstar 2 was launched on 13 May 1963 to continue the experimental programme.

Contemporaries of Telstar were Relay 1 and 2, launched on 13 December 1962 and 21 January 1964. These were also placed in relatively low orbits, 1 323km × 7 436 km and 2 019 × 7 482km respectively, and were used for communications experiments. Like Telstar, the two Relays showed that satellite communications held great promise, but that satellites in low orbit were not really suitable since much time was lost locking on to each satellite as it came over the horizon, and continuous contact between different continents was not possible because the satellites soon drifted out of view of one or other ground station. There was no doubt that the geostationary orbit was the best place for communications satellites, just as Clarke had predicted.

Syncom

The first attempt to operate from geostationary orbit was the aptly, if unoriginally, named Syncom 1, launched on 14 February 1963. Syncom was a cylindrical satellite with a built-in rocket motor. After a perfect launch Syncom 1 was put into a highly elliptical orbit with a high point of 35 880km. When it reached this height Syncom was to fire its own motor to circularize the orbit at that height and become geostationary. Sadly, contact was lost just 20 seconds after the Syncom's rocket ignited and no experiments could be made. NASA tried again with Syncom 2 in July 1963. This satellite reached the correct height, but was not placed directly above the equator and so was not strictly geostationary, appearing to slowly wander around a figure-of-eight pattern as seen from the ground. None the less Syncom 2 was able to demonstrate the principles of a geostationary communications satellite and it was followed by Syncom 3 on 19 August 1964. Syncom 3 was truly geostationary and was used with great effect to broadcast live coverage of the 1964 Olympic games from Japan.

At the end of their experimental missions both Syncom 2 and 3 were transferred to the control of the US Department of Defense for military use, but by then they had achieved their main objectives. The value of the geostationary concept had been proved and the decision was made to set up an international organisation, INTELSAT, to manage a global satellite communications system. Starting with Early Bird the INTELSAT organization (see p81) has managed an ever growing network of communications satellites, which now form a key element of the world's communications network.

Molniya

The USSR, while not ignoring geostationary communication satellites completely, put considerable effort into an alternative type of system based on satellites called Molniya. This is for two main reasons. Firstly, the launch sites used are at high latitudes and this makes it difficult to undertake launches into a geostationary orbit which must, by definition, be over the equator. Secondly, from the extreme north of the country satellites hovering over the equator appear very close to the horizon where their signals are weakened by their passage through the atmosphere. To avoid these disadvantages the USSR launched several generations of Molniya satellites into highly elliptical (500km × 40 000km) orbits which make an angle of about 65° to the equator.

The orbits of these Molniya satellites have their high point, or apogee, over the territory of the former USSR and from the ground they appear to rise slowly to a point well above the horizon. Since this movement takes place over a period of several hours the satellites are relatively easy to track using large steerable radio dishes. The satellites then begin to sink again, heading for a low point, or perigee, on the other side of the Earth. By placing several satellites in similar orbits, but spacing them out, it is possible to arrange tracking equipment so that as each satellite begins to sink towards the horizon, the ground stations can re-tune to the next in the series as it appears. Although not as convenient as the fixed position of a geostationary satellite, the Molniya system has been in use since 1965.

The Molniya satellites are cone shaped with six petal-like solar panels radiating outward from their base. The first Molniya was launched on 23 April 1965 and by 1987 no less than 70 Molniya satellites had been launched. An improved version, Molniya-2, went into service in 1971 and remained in use until 1981 when the series was phased out by the still more capable Molniya-3, the first of which had been launched in 1974.

The USSR also used geostationary satellites known as Ekran, Raduga and Gorizont. The first of this stable was orbited in 1975. The Raduga and Gorizont systems are used to relay telephone, facsimile and television transmissions between fixed ground terminals, and several dozen have been launched. The Ekran series are used to broadcast television programmes from central studios in Moscow direct to individual homes in outlying regions.

Amateur radio satellites

A series of satellites called OSCARs, Orbiting Satellites for Communication by Amateur Radio, have been built by radio 'hams' around the world. The OSCARs are to no fixed design but are light (typically about 30kg) and are launched, usually free of charge, as passengers on rockets carrying other satellites. OSCAR 1 was launched by the US Department of Defense on 12 December 1961. In 1969 the Amateur Radio Satellite Corporation (AMSAT) was set up in Washington to coordinate work on the OSCAR series. OSCAR 10 was built in Germany and OSCAR 9 and 11 were also known as Uosat 1 and 2 respectively.

The USSR has also launched amateur radio satellites called Iskra. Iskra 1 was orbited on 10 July 1981 and Iskra 2 and 3 were deployed from the Salyut 7 space station in 1982. The Japanese amateur radio satellite JAS was launched during the first flight of Japan's new H-1 booster on 12 August 1986.

National satellites

Not all communications satellites are owned by large international agencies such as INTELSAT. For example, the AUSSAT system is an Australian national programme and the Palapa satellites are owned by the government of Indonesia. Canada operates a domestic communications satellite system under the name Anik and Mexico uses its Morelos satellites. Various other systems, such as the international Arabsat and Scandinavian Tele-X, also exist.

There are also communications satellites owned by individual companies. Examples of these are Western Union's WESTAR satellites, the ASTRA direct broadcasting satellite which is used for satellite television within Europe, the commercial Japanese JCSat system and the US Satellite Business System (SBS) series.

The amateur-built OSCAR 10 satellite, launched in 1983

The European Olympus multipurpose communications satellite

Cosmos

Cosmos 1 16 March 1962, Cosmos 2000 10 February 1989

Jack of all trades

The Soviet Cosmos programme included scientific missions, military projects, prototypes, and unadmitted failures.

The first Cosmos was a lightweight scientific satellite launched on 16 March 1962 from the Soviet launch site at Kapustin Yar. It was launched by a B-1 rocket and was followed by two similar satellites in the next few weeks. At first it seemed that Cosmos was a programme of small scientific satellites similar to the US Explorer series, but Cosmos 4 was heavier than its predecessors and was launched from Baikanour. After three days, a capsule from Cosmos 4 returned to Earth suggesting that Cosmos might be a more wide-ranging programme than originally thought.

Since 1962 over 2500 Cosmos satellites have been launched and the series continues today. Some are launched from Kapustin Yar, some from Baikanour and others from the northern cosmodrome at Plesetsk. Some return capsules to Earth, others carry out bizarre manoeuvres in orbit and explode. Some simply circle the Earth quietly and then re-enter. In most cases the Soviet authorities made only vague statements about 'conducting research' and 'completing their mission as planned', but despite the paucity of official information, various patterns have emerged. By comparing the orbits, transmission frequencies and other characteristics of the satellites it has become possible to identify several dozen sub-programmes within the Cosmos cover name.

Military Cosmos

The most common use of the Cosmos label has been for military satellites. Several families of Cosmos craft use modified versions of manned spacecraft as spy satellites, photographing foreign territory then returning the exposed film in the re-entry module that would otherwise be occupied by the cosmonauts. Ocean reconnaissance, often by satellites equipped with radar, is also carried out. A number of classes of electronic intelligence satellites (ferrets) have been identified, as have miltary communications satellites.

The most obviously military satellites have been those connected with tests of anti-satellite weapons during which one Cosmos has homed in on another and has then exploded. The first test was in late 1968 when Cosmos 248 was the target for Cosmos 249 and 252.

Cosmos satellites have also been used in tests of a 'Fractional Orbit Bombardment System' in which satellites are launched and then re-enter before completing an entire orbit. The objective of such a system is to attack from an unexpected direction, thus confusing any anti-missile defences. Since each satellite completes only part of an orbit the system does not infringe nuclear treaties which forbid placing nuclear weapons in orbit. Fifteen test flights of this system were carried out between 1967 and 1971.

Scientific missions

The original objective of the Cosmos programme, that of orbiting small scientific satellites, has continued although the number of missions has fallen over the years. This is not due to a reduction in scientific research, but reflects a trend for mounting individual experiments as 'hitchhikers' on larger satellites. None the less a wide range of programmes have been carried out under the Cosmos designation. A few examples are: astronomy (Cosmos 51, 215, 262, 731), solar physics (Cosmos 3,

230, 650), geophysics (Cosmos 1, 26, 426, 481, 721, 906), Earth observation and geodesy (Cosmos 243, 1076, 1410) and biology (Cosmos 936, 1129, 1514, 1887).

Prototype satellites

The Cosmos label is used for prototypes of new satellites. Only once the development programme is completed, and the design is ready for regular service, is the final name of the programme announced. For example, Cosmos 122 and 144 are now known to be early versions of the weather satellite series currently called Meteor.

Tests of manned spacecraft

Each new manned spacecraft design, even if it is only a modification of an existing model, is tested automatically before a cosmonaut is allowed to fly in one. The one-man Vostok capsules were called Sputniks when they underwent unmanned test flights, but the arrival of the Cosmos covername meant that subsequent test flights could be disguised within this programme. Cosmos 47 was a test of the Voskhod spacecraft which later carried the first three-man crew into space. Voskhod 2, which featured the world's first spacewalk, was preceded by a test flight. The test flight was not very successful since Cosmos 57 broke up soon after reaching orbit, but this did not prevent the launch of Voskhod 2 three weeks later.

The next Soviet spacecraft design was the three-man Soyuz. This design was tested at least twice, as Cosmos 133 and Cosmos 140, before Vladimir Komorov was launched aboard the ill-fated Soyuz 1. This crashed disastrously in April 1967 (see p192) and two further test flights were flown under the guise of Cosmos 186 and 188. These two craft were launched three days apart (27 and 30 October 1967) and carried out the world's first fully-automatic space docking. The mission profile was similar to that flown in January 1969 by the crews of Soyuz 4 and 5. Other Soyuz tests included Cosmos 496 in 1972 (which tested modifications made after the Soyuz 1 disaster), Cosmos 1001 (1978), and 1074 (1979), which tested the new Soyuz-T version.

Cosmos numbers were also assigned to launches connected with the Salyut space station programme. Cosmos 557 (launched 11 May 1973) would have been Salyut 3 had it not failed as soon as it got into orbit, and Cosmos 772 (launched 29 September 1975) was a test of equipment for the unmanned Progress re-supply spacecraft. Cosmos 929, 1267, 1443 and 1686 were modules designed to be docked with a Salyut space station; Cosmos 929 made a solo test flight in July 1977, and the other three did eventually dock with either Salyut 6 or 7. Cosmos 1669 was also connected with the Salyut programme and was described as a support satellite. It docked with the Salyut 7 space station, but was also capable of independent operation.

The Buran space shuttle was not tested under the Cosmos label, but early tests of a Soviet spaceplane were made as Cosmos 1374 and 1445. Other missions, including Cosmos 881, 882, 997, 998, 1100 and 1101 may also have been part of the Soviet shuttle programme.

Unannounced failures

In the early years of the space race, the Cosmos label was often used to hide the true objectives of missions which had failed. During the 1960s in particular, a number of Soviet lunar and planetary probes were stranded in low Earth orbit when the final rocket stage, intended to boost them into deep space, failed to ignite. Rather than admit to such a high failure rate, each of these missions was given a Cosmos designation. This deception was not limited to interplanetary probes; failed Earth satellites have also been given Cosmos numbers. Cosmos 1612 was probably the first flight of the improved Meteor 3 weather satellite which failed to reach its intended orbit and was hurriedly reclassified.

Four Cosmos missions, all orbited in 1970 and 1971, are believed to have been connected with the secret Soviet manned lunar programme. Cosmos 382 and 434 are suspected of having been tests of the Soviet lunar lander, with Cosmos 398 and 434 being related to tests of the lunar orbiting craft.

Daedalus

Study phase 1973–1978

The first practical starship design?

The Daedalus study was an attempt to produce a concept for an interstellar probe using credible technology.

In 1973 the British Interplanetary Society undertook a five-year study of an unmanned mission to another star. The project was named Daedalus and drew on the talents of 13 professional engineers and scientists from the society, who between them devoted 10 000 hours of study to the problem. The study was published in 1978 and soon became recognized as a major contribution to the subject of interstellar engineering. Daedalus was not intended to produce a definitive design for an interstellar probe, but rather to focus attention on what might be possible by the end of the next century. It was a bold initiative, as might be expected from a society which had published plans for a lunar spaceship as early as 1948.

The scenario

The Daedalus team assumed that by 2075 much of the world's energy would come from nuclear fusion powerplants that consumed deuterium (a form of hydrogen atom with one extra atomic particle) and helium 3 (an atom of helium with one atomic particle missing). The deuterium/helium 3 nuclear fusion reaction can be used to produce energy, but has few of the unpleasant side effects of other nuclear fusion reactions and so was thought likely to be in widespread use by the next century. Unfortunately helium 3 is very rare on the Earth and hundreds of tonnes a year would be required to provide the world's energy needs.

One means of obtaining sufficient helium 3 would be to extract it from the atmosphere of Jupiter, where it exists in large quantities, and ship it to Earth in tankers. Clearly this would require a culture which was both socially stable and had mastered the art of crossing interplanetary space on a regular basis. Such a society would have the resources, and probably the will, to mount an interstellar mission.

The chosen target for Daedalus was Barnard's star, a small red star six light years from Earth. This star is not the nearest to earth, but it is the nearest for which there is any evidence of a planetary system. The team also chose this target because a craft able to fly to Barnard's star would also be able to accomplish a number of other missions, whereas one designed just for the closest star (Proxima Centauri in the Alpha Centauri system) might not be useful for any other targets. It was originally decided that the probe should be designed to accomplish its mission in 40 years but, when preliminary studies showed that this would require a craft weighing over 150 000 tonnes, the allowable duration was increased to a little more than 50 years. This reduced the mass of the probe to about 50 000 tonnes.

The spacecraft

The key to the Daedalus concept was its propulsion system, a refined version of the nuclear pulse rocket designed for Project Orion (see p85). The Daedalus engine, which was based on research done at the Lawrence Livermore Laboratory in California, used the deuterium/helium 3 nuclear fusion reaction. Tiny pellets, comprising a frozen mixture of these gases, would be fired into a combustion chamber where they would be compressed by a sudden and very intense blast of electrons. This would cause the fuel pellet to implode to the point where a nuclear fusion reaction would occur. The result would be a superheated plasma (a mixture of atomic nuclei and electrons) that could be confined by powerful magnetic fields and ejected via a nozzle to produce thrust. Daedalus would ignite 250 pellets per second during a boost phase lasting several years. The probe would then coast towards its destination, making no attempt to slow down on arrival.

The final Daedalus design was for a two-stage spacecraft with a scientific payload of 450 tonnes. Each stage comprised a nuclear pulse motor above which were six spherical fuel tanks. The second stage was topped by a cylindrical payload module, 64m in diameter, which would have a shield to protect it from erosion by interstellar dust as the craft ploughed through space. Further protection would be provided by ejecting a small dust cloud ahead of the vehicle. Large objects encountering this cloud would be vaporized by collisions with the dust before they could strike the main Daedalus probe.

Since radio signals would take years to reach the probe once it was en route, Daedalus would be controlled by an intelligent computer system able to operate the ship during its cruising flight and to define the best possible scientific observations during the encounter with Barnard's star. To allow for the mechanical failures that were bound to occur during a 50-year voyage Daedalus would be equipped with 'Wardens', mobile computer-controlled robots able to make running repairs on the ship, and on each other. The payload module also carried a telescope system which would search for planets around Barnard's star so that they could be studied in detail during the encounter.

The mission

Daedalus would be assembled in space, either in lunar orbit or close to Jupiter. Once the assembly was complete and the main engine had been started the craft would reach solar system escape velocity in two days. The initial motor burn would continue for 250 days when, 0.0039 light years from the Sun, the first pair of empty fuel tanks would be jettisoned. A second pair would be released 0.0178 light years away and, just over two years after departure, the third pair would run dry and the first stage would have completed its mission. The second stage would then separate from the first and, dropping empty tanks from time to time, would continue to accelerate for a further 1.75 years. By then Daedalus would be travelling at 12% of the speed of light and on course for Barnard's star. The cruise phase would last for 45 years, during which time data on the interstellar medium would be returned regularly to Earth. If required, a mid-course correction might be made about 20 years after launch.

A few years before arrival the automatic telescopes on Daedalus would start to observe Barnard's star, searching for Jupiter-sized planets and determining their orbits. These planets would be the targets of automatic probes, which would be released as early as possible, in order to reduce the amount of manoeuvring required to place them on course. Earth-sized planets would be detected later and become targets for another set of probes.

The entire flyby would last only a few days as Daedalus rushed past Barnard's star, still travelling at 12% of the speed of light. Once the flyby was over, the first priority would be to return all the data, including that collected by the planetary probes, to Earth. This would probably be broadcast repeatedly to ensure that it was all received back on Earth. Finally, Daedalus would return to its cruise condition and continue to return data on interplanetary space until it either failed, or ran out of fuel for attitude control. Once the fuel had run out the craft would start to tumble, and, unprotected by its dust shield, would soon be destroyed by collisions with interstellar material.

> The mission duration of 40 years was originally chosen so that young scientists joining the project at launch might have a reasonable chance of still working on the project when the probes' data arrived at Earth. The longer flight eventually planned still would allow junior staff to have a chance of being alive at the conclusion of the mission.

ESA

Founded May 1975

European Space Agency

Formed by the merger of two earlier European space organizations called ESRO and ELDO, ESA (the European Space Agency) has made Europe a major space power.

The European Space Research Organisation (ESRO) was formed in 1964 by 10 European states in order to promote collaboration in space research for peaceful purposes. The countries which made the largest financial contributions to ESRO were France, Germany, Italy and the UK. ESRO was quite successful and developed seven scientific satellites, all of which were launched by US rockets. The ESRO-1A, ESRO-1B, ESRO-2B and ESRO-4 satellites studied the aurora and ionosphere, the TD-1A satellite was an ultraviolet astronomy mission and the two Highly Eccentric Orbit Satellites (HEOS) examined the solar wind and the Earth's magnetic field.

In parallel with ESRO, which concentrated on developing scientific satellites, Belgium, France, Germany, Italy, the Netherlands and the UK set up the European Launcher Development Organisation (ELDO) to build a satellite launcher. The rocket was named Europa and used the British Blue Streak missile as a first stage with upper stages built by France and Germany. Although the consortium built and launched a number of rockets, none succeeded in placing a satellite into orbit and ELDO collapsed in the early 1970s.

The failure of ELDO, and the desire for greater cooperation in European space activities, led to the setting up of a new organization, the European Space Agency (ESA). As part of the new arrangement it was agreed that ESA would assume responsibility for the programmes of ESRO and ELDO, which both ceased to exist when ESA came into being in May 1975.

ESA

The ESA convention is similar to that of ESRO, calling for European cooperation for peaceful scientific purposes, but with the additional objective of developing and operating applications programmes such as weather and communication satellites. The member states of ESA are Austria, Belgium, Denmark, France, Germany, Eire, Italy, the Netherlands, Norway, Spain, Sweden, Switzerland and the UK. In addition Finland is an associate member and Canada is a cooperating state.

At the highest level, ESA is managed by a council composed of representatives of each of the member states. This council generally meets four times a year and takes decisions on ESA policy towards scientific, technical and administrative matters. Day-to-day management is delegated downwards through a director general, who is appointed for a four-year term of office, and a variety of directorates. The total staff directly employed by ESA is about 2 000.

The annual ESA budget is about $2 000 million, although budgeting takes place in a carefully defined accounting unit which allows for the differing values of each member nation's currency. The funding of the organization is flexible and ESA's programmes are described as being either mandatory or optional. All member nations must contribute to the general budget and to the mandatory scientific programme. The level of each country's contribution is set by a formula based on its economic circumstances. Other programmes, such as the development of the Ariane rocket, the

Hermes spaceplane and the Columbus space station programme, are optional. Each member country must decide how much, if anything, they wish to contribute to optional programmes depending on their economic and political situation and the level of interest in the project within the country itself. For example, when ESA was first set up, Germany decided to pay more than half the cost of the Spacelab programme and France became the main contributor to the Ariane rocket. The British government, which was more interested in projects which might provide some commercial benefit, chose to invest in ESA's communications satellite programmes. To ensure a fair return for those countries investing heavily in the optional programmes, a system called '*jouste retour*' operates in which ESA places industrial contracts with European companies in proportion to each nation's contribution to the agency.

Scientific programmes

ESA has a range of scientific programmes which cover astronomy, solar physics and the exploration of small bodies within the solar system. Many of these are described elsewhere in this book. In 1985 ESA published a plan of scientific projects called Horizon 2000 which was intended to clarify the agency's long-term scientific programme and to help define which areas of new technology ESA should develop. Horizon 2000 proposed the development, over the next two decades, of four major projects referred to as 'cornerstones'. These projects are a fleet of solar physics probes called the SOHO/Cluster mission, a large X-ray telescope, an orbiting submillimetre telescope and a mission to return samples from the nucleus of a comet. Smaller, less expensive missions are expected to take place every few years in between each cornerstone. The small missions currently under development are the Infrared Space Observatory and the Huygens probe which forms part of the NASA/ESA Cassini mission to Saturn.

Applications programmes

The main thrust of ESA's applications satellites are in telecommunications, meteorology and Earth observation. Recent projects in these areas have included a large experimental communications satellite called Olympus and the Earth Resources Satellite ERS-1. As a general rule, ESA uses satellites like these to develop new technology and, once this has been done, transfers operation of the satellites to other organizations such as the International Maritime Communications Satellite Organisation (INMARSAT) and the European Telecommunications Satellite Organisation (EUTELSAT). This allows ESA to concentrate on developing the technology required for the next generation of satellites without becoming embroiled in each operational system.

ESA is also involved in a programme of research on the effects of microgravity, some of which may have commercial applications in, for example, the development of new materials. Microgravity experiments are also important for a wide range of medical research.

Space transportation systems

ESA is developing both manned and unmanned space transportation systems. The best known of these is the Ariane rocket (see p13) which was developed so that Europe did not have to rely on other countries to put its satellites into orbit. ESA is also developing a mini space shuttle called Hermes (see p71) so that European astronauts will not be dependent on other nations for trips into orbit. The ESA Columbus programme (see p30) is a series of orbiting laboratories which will operate either in conjunction with the international space station Freedom or independently as free-flying craft serviced by Hermes.

ESA establishments

ESA operates four main centres, each in a different European country and a fifth, a European astronaut centre, will soon be opened in Germany.

ESA headquarters, which house the main administrative departments of the organization are in Paris.

The European Space Research and Technology Centre (ESTEC) is at Noordwijk in the Netherlands. ESTEC is responsible for the study, development and testing of ESA spacecraft which are being built by European industry. ESTEC is also the home of technological development programmes which lay the groundwork for future missions.

The European Space Operations Centre (ESOC) at Darmstadt in Germany is responsible for controlling ESA satellites once they are in orbit. To do this they operate a network of tracking stations around the world.

ESRIN (European Space Research Institute), at Frascati near Rome, is responsible for ESA's programme of gathering data from Earth observation satellites and is the home of the Agency's information retrieval system.

ESA hopes to provide various elements of the International Space Station Freedom

Explorer programme

Explorer 1 1958

Science in space

The US Explorer programme is a series of low-cost satellites devoted to specific scientific objectives.

Explorer 1 was the the first US satellite (see p222) and when other similar satellites were launched, the original numbering series was extended to cover them. However, as the US space programme expanded, the name Explorer was assigned to a programme of small scientific satellites. At about the same time, the naming system was subtly modified so that satellites only received an Explorer number once in orbit, removing the embarrassing gaps in the series caused by launch failures.

Satellites within the Explorer programme generally belonged to one of a small number of families, each devoted to research into a specific scientific area. Many of them were known by names which reflected the aims of particular programmes. Examples of these are the Small Astronomy Satellites (SAS-1, 2 and 3) and the Interplanetary Monitoring Platforms (IMP-1 to 10), which were placed into high, eccentric Earth orbits to study solar and terrestrial magnetic fields and to observe the solar wind. One of these satellites was placed in lunar orbit and known as an 'anchored IMP'. Some of the satellites were simple in concept, for example balloon satellites like Explorer 9 which were tracked optically from the ground so that the gradual changes in their orbits could provide information on the density of the upper atmosphere. Others were complex scientific missions which operated for many years, often well beyond their expected design lives.

After Explorer 55 the use of individual numbers ceased and instead each satellite was known by a specific name. Several of these later missions were international in character, with European nations providing spacecraft for multi-satellite projects such as AMPTE and ISEE. The ISEE-3 spacecraft, which was originally placed in an orbit which kept it between the Sun and the Earth, was later commanded to make a series of close approaches to the Moon. These lunar flybys changed its orbit so that it intercepted comet Giacobini–Zinner in 1985, scoring a first over the fleet of spacecraft en route to Comet Halley. ISEE-3 was renamed the International Cometary Explorer during this phase of its mission.

At the time of the explosion of the Space Shuttle Challenger there were a number of Explorer-type missions either under development or still waiting for funds to become available. These included the Cosmic Background Explorer (eventually launched on 18 November 1989), the Extreme Ultraviolet Explorer, the X-ray Timing Explorer and the Far Ultraviolet and Spectroscopic Explorer. In the review of the US space programme which followed the Challenger disaster, it was recognized that although these missions were still of value, the rate at which Explorer spacecraft were being launched had become too slow.

In an attempt to break the cycle of increasing complexity and cost, NASA announced a new 'Small Explorer' programme with the aim of developing lightweight satellites, costing in the region of $30 million, which could be launched by a small and inexpensive Scout-type rocket. The objective of the programme was to develop and launch a satellite within a period of three years. In 1989 NASA announced the first missions in the series, which were the Solar, Anomalous and Magnetospheric Particles Explorer, the Submillimetre Wave Astronomy Satellite, the Fast Auroral Snapshot Explorer and the Total Ozone Mapping Spectrometer.

Explorer programme summary

Mission	Launch Date	Notes
Ex 1	31 Jan 58	First US satellite, discovered Van Allen belts
Ex 2	5 Mar 58	Launch failure
Ex 3	26 Mar 58	Similar to Explorer 1
Ex 4	26 Jul 58	Similar to Explorer 1, but with improved instruments
Ex 5	24 Aug 58	Launch failure
S-1	16 Jul 59	Launch failure
Ex 6	7 Aug 59	Irregular spheroid; studied magnetosphere and Van Allen belts
Ex 7	13 Oct 59	Double cone; studied solar X-rays, cosmic rays etc
S-46	23 Mar 60	Launch failure
Ex 8	3 Nov 60	Double cone, weight 41kg; for ionospheric research
Ex 9	16 Feb 61	Balloon satellite to monitor air drag
S-45	24 Feb 61	Launch failure
Ex 10	25 Mar 61	Cylinder plus sphere, studied Earth's magnetic field
Ex 11	27 Apr 61	Possible detection of cosmic gamma rays
S-45A	24 May 61	Launch failure
S-55	30 Jun 61	Launch failure
Ex 12	15 Aug 61	Studied relationship of Van Allen belts and solar wind
Ex 13	25 Aug 61	Technology development, research on meteoroids
Ex 14	2 Oct 62	Studied relationship of Van Allen belts and solar wind
Ex 15	27 Oct 62	Studied relationship of Van Allen belts and solar wind
Ex 16	16 Dec 62	Technology development, research on meteoroids
Ex 17	3 Apr 63	(AE-A) Atmospheric Explorer, 185kg sphere
Ex 18	26 Nov 63	(IMP-1) Studied magnetic fields and radiation
Ex 19	19 Dec 63	Balloon satellite to monitor air drag
S-66	19 Mar 64	(Beacon Explorer A) Launch failure
Ex 20	25 Aug 64	(IE-A) Double cone, weight 44kg; for ionospheric research
Ex 21	3 Oct 64	(IMP-2) Studied magnetic fields and radiation, in lunar orbit
Ex 22	9 Oct 64	(Beacon Explorer B) For ionospheric research
Ex 23	6 Nov 64	Micrometeoroid detection experiment
Ex 24	21 Nov 64	Balloon satellite to monitor air drag
Ex 25	21 Nov 64	(Injun-4) Studied magnetosphere and radiation belts
Ex 26	21 Dec 64	Studied Earth's radiation belts and solar wind
Ex 27	29 Apr 65	(Beacon Explorer C) For ionospheric research
Ex 28	29 May 65	(IMP-3) Studied magnetic fields and radiation
Ex 29	6 Nov 65	(Geos-1) Used for geodesy; flashing light and laser reflectors
Ex 30	18 Nov 65	(Solrad-1) measured solar X-rays in International Quiet Sun Year
Ex 31	28 Nov 65	(ISIS-X) Ionospheric research; Canadian involvement
Ex 32	25 May 66	(AE-B) Atmospheric Explorer, 225kg sphere
Ex 33	1 Jul 66	(IMP-4) Studied magnetic fields and radiation
Ex 34	24 May 67	(IMP-6) Studied magnetic fields, radiation and radio astronomy
Ex 35	19 Jul 67	(IMP 5) In lunar orbit, studied magnetic fields
Ex 36	11 Jan 68	(Geos-2) Used for geodesy, similar to Ex 29
Ex 37	5 Mar 68	(Solrad-2) Observed solar X-rays
Ex 38	4 Jul 68	(RAE-1) Radio Astronomy Explorer, very large antennae
Ex 39	8 Aug 68	Balloon satellite to monitor air drag
Ex 40	8 Aug 68	(Injun-5) Studied magnetosphere and Earth radiation
Ex 41	21 Jun 69	(IMP-7) Studied magnetic fields and radiation
Ex 42	12 Dec 70	(SAS-1, Uhuru) First satellite for X-ray astronomy
Ex 43	13 Mar 71	(IMP-8) Studied magnetic fields and radiation
Ex 44	8 Jul 71	(Solrad-3) Solar X-rays and ultraviolet radiation
Ex 45	15 Nov 71	Relationships between aurorae, magnetic fields and solar wind

Ex 46	13 Aug 72	(MTS) Meteoroid Technology Satellite
Ex 47	23 Sep 72	(IMP-9) Studied magnetic fields and radiation
Ex 48	15 Nov 72	(SAS-2) Gamma ray astronomy, failed after 7 months
Ex 49	10 Jun 73	(RAE-2) Radio astronomy from lunar orbit
Ex 50	26 Oct 73	(IMP-10) Studied magnetic fields and radiation
Ex 51	16 Dec 73	(AE-C) Atmospheric Explorer
Ex 52	3 Jun 74	(Hawkeye) Studied solar wind/geomagnetic field interaction
Ex 53	7 May 75	(SAS-3) X-ray astronomy
Ex 54	6 Oct 75	(AE-D) Atmospheric Explorer, 16-faceted cylinder
Ex 55	20 Nov 75	(AE-E) Atmospheric Explorer, 16-faceted cylinder
ISEE-1	22 Oct 77	International Sun–Earth Explorer, US-built spacecraft
ISEE-2	22 Oct 77	ESA-built spacecraft, joint launch with ISEE-1
IUE	26 Jan 78	International Ultraviolet Explorer (see p228)
AEM-1	26 Apr 78	(Heat Capacity Mapping Mission) Earth observation
ISEE-3	12 Aug 78	US built, diverted to Comet Giacobini-Zinner in 1985
AEM-2	18 Feb 79	(Stratospheric Aerosol and Gas Experiment) Earth observation
DE-1	3 Aug 81	Dynamics Explorer, solar wind and aurorae observations
DE-2	3 Aug 81	Joint launch with DE-1
SME	6 Oct 81	Solar Mesosphere Explorer, atmospheric research
AMPTE	16 Aug 84	Active Magnetospheric Particle Tracer Explorers, US/German/UK mission. Three satellites, triple launch

Key: AEM = Applications Explorer Mission

Explorer 53, an X-ray astronomy satellite weighing about 200kg

Food for space

Haute Cuisine?

The smuggled sandwich 1965

Considerable effort has gone into the development of astronauts' meals.

Eating and drinking is an essential activity in space just as it is on Earth. As spaceflights became longer it was necessary to provide astronauts with meals that were nutritious, easy to prepare, and psychologically acceptable. Today, Space Shuttle astronauts eat meals that are not too different from what they might eat at home, but it was not always so. Space cuisine has come a long way since the early days.

Before man first ventured into space there were serious doubts about whether or not it would even be possible to eat normally in weightlessness and the ability of astronauts to eat became a test objective of early missions. In retrospect these worries were groundless: swallowing food does not depend on gravity (as anyone who has eaten while standing on their head can confirm!), but it was found that the tendency of liquids to form spheres and float about did present a potential problem in the cramped cabin of a spacecraft.

First attempts

Early US spaceflights in the Mercury programme were so short that there was no need to provide the astronauts with meals, but with longer flights planned, the Mercury astronauts served as guinea pigs for a number of space foods. These included food cubes that could be gulped down in one bite, freeze-dried foods that needed to be mixed with water before being eaten and even semi-liquid foods that were squeezed, toothpaste like, from a tube. The results of these experiments showed that astronauts could eat fairly normally, but the taste and texture of these early meals left much to be desired.

The two-man Gemini programme, which started in 1965, dispensed with the toothpaste tubes because, apart from their unpopularity, the bulk of the weight came from the tube, not the food. Improved, bite-sized cubes were carried in clear plastic bags, and the cubes themselves were coated in gelatine to control crumbling. Rehydratable food was packed in clear bags with a valve at one end and a tube at the other. The astronaut squirted water in through the valve using a special water gun then kneaded the contents to form an edible paste that could be squeezed out of the tube. Having eaten, the astronaut popped a germicidal tablet, intended to stop the food residue spoiling, into the bag and resealed it. The used bags were stored in the cabin until after the landing.

Each Gemini astronaut could chose his own menu, and the meals were adjusted to provide a steady 2 800 calories per day. The meals were labelled with the astronaut's name and the day on which it was to be eaten and was then packed into the spacecraft ready for use during the flight. The menu choice was limited, and typically the same menu selection would come around every four days during the longer missions. Technically the meals were perfect, providing a balanced diet, with high energy and nutrient values and low residue. The astronauts, however, were not impressed, not least because the Gemini water dispenser produced only tepid water. In a now legendary incident on the first Gemini mission, astronaut John Young produced a sandwich and calmly passed it to his commander Gus Grissom. Grissom took a bite then put it away in case crumbs escaped into the cabin.

The Apollo moon flights used basically the same system as Gemini, with the minor improvement that hot and cold water were available to rehydrate the packaged meals. In addition some Apollo meals could be eaten with a spoon, rather than squeezed through a tube. These were rehydrated in the normal way, by squirting water into the bag and kneading the mixture, then a zip at the other end was opened and the food spooned out. Provided the food was moist, surface tension caused it to stick to the spoon and prevented it from floating about the cabin. Slices of bread, and sandwich spreads were also tried during Apollo and it was found that floating crumbs were not a serious problem.

Space stations

The advent of the Skylab space station in 1973 presented the food technicians with a new challenge. Skylab missions were planned to last for weeks and months, not days, and providing a varied and acceptable diet assumed much greater importance than it had during the brief Apollo missions. Skylab was equipped with a crew wardroom which featured a triangular table around which the crew could gather to eat. In weightlessness foot restraints replaced dining room chairs, but none the less Skylab offered the first chance for astronauts to eat in a normal social atmosphere. Precooked food was packed into aluminium cans with pull-off lids and the astronaut 'chef' would collect the appropriate cans for each meal and place them in a special tray. Each tray had eight cavities designed to hold the cans and served as both food warmer and placemat. Under some of the cavities were electric heaters which could warm the food to about 66°C. The trays were magnetized to hold the metal knives, forks and spoons in place when not required, and were clipped around the triangular table when the crew was ready to eat.

A total of 68 different food items made up the Skylab menu. To meet the requirements for crew nutrition and medical experiments, as well as satisfying personal preferences, preset menus were calculated which repeated every six days. During the longer missions, 56 and 84 days for Skylab 3 and 4, the crews found that this cycle became rather monotonous. They also found that after a time, all foods, even old favourites, began to taste bland and required extra seasoning to make them palatable.

Soviet cosmonauts, eating from cans aboard the Salyut space station, had similar experiences, and cravings for onions, garlic, horse radish and mustard became common. The use of unmanned Progress re-supply missions, and occasional visits by other crews, have gone some way to solve this problem for the cosmonauts assigned to long flights. Extra food items can be sent up in the Progress capsules and visiting cosmonauts from other countries usually bring along some of their national dishes. French 'spationaut' (space traveller) Jean-Loup Chretien carried crab soup, lobster, hare and cheese on his visit to Salyut–7 in 1982.

Space Shuttle

The US Space Shuttle features a galley which can be used to prepare meals for the crew. This galley has hot and cold water dispensers, food trays and an oven. The astronaut removes the prepackaged meals from their storage area, adds water to those foods that require it, and puts items to be warmed into the oven. The individual items are then placed on to trays which can be fastened to the wall, or held in the lap of the astronaut as he or she eats. Apart from the actual heating time, it takes about five minutes to prepare a meal for four astronauts. Unlike Skylab, where individual choice was allowed, most Space Shuttle meals are to a standard menu and each crewmember eats the same selection. A supply of other foods are available if individual astronauts do not like specific items on the regular menu. Snack foods, such as sweet bars, nuts, biscuits and fruit are also carried. The menu offers three meals a day, and repeats every six days on the longer flights.

France in space

A-1 (Asterix) 26 November 1965

First European spacewalker

The third nation to launch its own satellite, France is a major space power as well as the leading contributor to the European Space Agency (ESA).

When the first satellites were launched in the late 1950s France was no stranger to rocket development. The first version of the French liquid-fuelled Veronique sounding rocket was developed in 1949 and solid rockets, including the Antares and Berenice series, were being developed for research on re-entry vehicles. Other sounding rockets, such as the Centaure, Dragon, Pegasus and Dauphin vehicles were also in use or under development. In 1962, the French government decided that these activities should be brought together under the control of a single space agency to be called the Centre National d'Etudes Spatiales, or CNES. This new agency was charged with organizing French space policy, coordinating major space projects, assessing the value of research programmes in order to aid French industry and promoting French space science and technology. From an initial staff of 27 in 1962 CNES had grown to almost 100 times that number by 1991 and had made France a force to be reckoned with in space.

Today CNES has four main centres. Its headquarters are in Paris which is the base for the director general and his administrative teams. The French astronaut programme, and development of the Hermes spaceplane, are based at Toulouse. Development of unmanned launch vehicles is controlled from Evry and the French space launch centre, now used for the ESA Ariane rockets, is at Kourou, French Guiana.

The Diamant rocket

One of the first tasks facing CNES was the development of a national satellite launcher. This project was called Diamant (Diamond) and drew on a programme of smaller rockets named after precious stones: Agate, Topaze, Rubis, Emeraude and Saphir. Diamant was based on the two-stage Saphir rocket to which a new third stage was added, giving a capability to lift about 80kg into orbit. Diamant A stood 18.75m tall, was 1.4m in diameter and weighed 17 970kg. The first stage had a single Vexin liquid-fuelled motor, with stages two and three using solid rockets.

After a four-year development programme the first launch, on 26 November 1965, successfully orbited a satellite called A-1 (and nicknamed Asterix). This launch made France the first nation outside the superpowers to launch its own satellite. Further launches took place in 1965 and 1967 from Hammaguir in the Sahara, but in 1964 it was decided to develop a new launch site at Kourou and launches from Hammaguir were suspended. Launches from Kourou started in March 1970 using the improved Diamant B rocket which featured an enlarged first stage. The first launch orbited two technology satellites, one French, one German. Eight further Diamant launches were made, six successfully, the last one taking place in September 1975.

International collaboration

The French satellite programme was not restricted to its own launcher, France entered into launch agreements with both superpowers. The FR-1 scientific satellite was orbited in 1965 on a US Scout launcher and NASA Thor-Delta rockets were used to launch two Franco-German Symphonie communications satellites. The 237kg, three-axis stabilized Symphonie satellites were placed in geostationary orbit and were described as experimental craft to avoid conflict with the INTELSAT organization. Despite this compromise, political and technical problems were encountered which threatened to delay or even cancel the launch, and this may have been a factor in subsequent French insistence that Europe develop its own launch vehicle.

A long programme of collaboration with the USSR began in 1966 when General de Gaulle signed a Franco-Soviet accord in June 1966 and became the first westerner to visit the then secret Baikanour space centre. This accord led to various programmes of balloon and sounding rocket launches and to the placement of French experiments on a series of Soviet satellites. As well as missions in Earth orbit these collaborations extended to the Moon and planets, with French equipment being carried on the Lunakhod moon rover, the Mars 3 craft and several Venus probes. The collaboration included the launch of the SRET satellite in 1972. Several other French-built satellites were launched by the USSR but recently the emphasis of the collaboration switched to the French supplying large scientific experiments to be fitted to Soviet satellites. Recent examples of this are the payload for the Gamma-1 astronomy mission, the provision of the SIGMA gamma ray telescope for the Soviet Granat mission, and participation in the Phobos Mars probes.

Applications satellites

In parallel with its scientific programmes, and its contributions to the ESA programmes (see p43), France has also developed applications satellites for its own national use. The best known of these is the SPOT Earth observation programme (see p140) which will form the basis for the proposed Helios spy satellite programme. In addition, the Telecom 1A and 1B and TDF-1 communications satellites were launched in the 1980s. The Telecom series is used for telephone, facsimile and data transmission within France and for international telephone and television links with French overseas territories. TDF-1, which forms part of a joint project with Germany, is able to broadcast television pictures directly to homes equipped with a small rooftop receiving dish.

Manned space progammes

France has used its international programmes to secure joint manned flights with both NASA and the USSR and refers to its space travellers as 'spationauts' to avoid possible national sensitivity over the use of the word astronaut or cosmonaut. On 24 June 1982 spationaut Jean-Loup Chretien flew in the Soviet Soyuz T6 spacecraft to dock with the Salyut 6 space station. After a brief stay on Salyut 6 Chretien returned to Earth on 2 July.

Chretien's backup for the mission was Patrick Baudry who was soon nominated for a flight on the US Space Shuttle, giving him the unique distinction of having been trained by both superpowers. In 1985 he spent a week aboard the Space Shuttle Discovery on mission STS-51G. Baudry's role in the flight was a nominal one, he had no responsibilities for the main mission objectives but did carry out some small French experiments.

Chretien went on to join the crew of Soyuz TM7, on 28 November 1988 in a flight to the Salyut 7 space station. During his three-week visit he made a spacewalk, in conjunction with Soviet cosmonaut Alexander Volkov, becoming the first European to walk in space.

The future

The French space programme continues to look strong in the 1990s. In the area of manned flight France is heavily involved in the ESA programmes to develop a new, more powerful version of the Ariane rocket and is the driving force behind the European spaceplane Hermes (see p71). A further Franco-Soviet spaceflight is planned for 1992 with one spationaut flying to the Mir space station. Also, although there are no plans for another Frenchman to fly on the US Space Shuttle, France is continuing its cooperation with the USA, notably with the joint TOPEX/Poseidon satellite which is designed to study the oceans.

Spationaut Jean-Loup Chretien aboard a Soviet space station

Spationaut Patrick Baudry aboard the Space Shuttle

French satellites

Satellite	Launcher	Launch Site	Date	Mission
A-1 (Asterix)	Diamant A	Hammaguir	26 Nov 1965	Test Capsule
Fr-1	Scout	Vandenberg	6 Dec 1965	Scientific
D-1A(Diapason)	Diamant A	Hammaguir	17 Feb 1965	Geodesy
D-1C(Diademe)	Diamant A	Hammaguir	8 Feb 1967	Geodesy
D-1D(Diademe)	Diamant A	Hammaguir	15 Feb 1967	Geodesy
MIKA	Diamant B	Kourou	10 Mar 1970	Test Capsule
DIAL (German Satellite)		'	'	Dual launch
Peole	Diamant B	Kourou	12 Dec 1970	Meteorology
D-2A(Tournesol)	Diamant-B	Kourou	15 Apr 1971	Scientific
Eole	Scout	Wallops	16 Aug 1971	Meteorology
SRET-1	A-2-e	Plesetsk	4 Apr 1972	Technology
Symphonie-1	Thor-Delta	KSC	19 Dec 1974	Comsat
Starlette	Diamant BP4	Kourou	6 Feb 1975	Geodesy
D-5A(Pollux)	Diamant BP4	Kourou	12 May 1975	Scientific
D-5B(Castor)	"	"	"	Dual launch
SRET-2	A-2-e	Plesetsk	5 Jun 1975	Technology
Symphonie-2	Thor-Delta	KSC	27 Aug 1975	Comsat
D-2B(Aura)	Diamant BP4	Kourou	27 Sep 1975	Solar ultraviolet
Signe-3 (D-2Bγ)	C-1	Kapustin Yar	17 Jun 1977	Astronomy
Telecom 1A	Ariane 3	Kourou	4 Aug 1984	Comsat
Telecom 1B	Ariane 3	Kourou	8 May 1985	Comsat
SPOT-1	Ariane 1	Kourou	22 Feb 1986	Landsat
TDF-1	Ariane 2	Kourou	28 Oct 1988	Comsat
Telecom-1C	Ariane 3	Kourou	11 Mar 1988	Comsat
SPOT-2	Ariane 4	Kourou	22 Jan 1990	Landsat

Key:
Landsat: Earth Observation Satellite
Comsat: Communications Satellite
Technology: Technology Development Project
Wallops: Wallops Island, Virginia
Vandenberg: Vandenberg Air Force Base, California
KSC: Kennedy Space Center

During Chretien's six-hour spacewalk from Salyut in 1988 it was intended to deploy a French-designed lattice-work test structure which should have unfolded automatically from a small canister. At first the structure did not unfold as planned and it was later reported that one of the two spacewalkers had kicked the canister to speed up the deployment.

Freedom

Presidential commitment 1984

International space station

Freedom is the main US manned space programme for the 1990s and should establish an international space laboratory by the year 2000.

The idea of permanently manned space stations predates the space age; indeed, during the 1940s and 50s it was common to assume that a space station would be needed as a staging post for manned missions to the Moon and planets. This belief was encouraged by articles, often featuring huge wheel-shaped space stations, which appeared in science fiction stories and publications such as Colliers magazine. When the space age began in October 1957 attention was soon focused on the likely role of man in space and in 1959 the development of a US space station was identified as a logical step after the one-man Mercury programme. However, when President John F Kennedy committed the USA to landing a man on the Moon within a decade, it became clear that only programmes directly related to that goal were likely to be approved. Once the Lunar Orbit Rendezvous method (see p161) was chosen for the Apollo programme, hopes of building a US space station before 1970 all but vanished.

By the mid 1960s interest in space stations was rekindled since they were seen as the key to the long-term exploration of the Moon and Mars. With this in mind several studies of large space stations which might be built in the post-Apollo era were carried out by US aerospace companies. These studies were usually for 6–12 man space stations weighing about 150 tonnes which would be assembled on the ground and then launched by the lower two stages of a Saturn 5 rocket. In the interim, NASA decided to press ahead with a small space station in what was called the Apollo Applications Programme. This programme, which was based on using the upper stage of a Saturn rocket as an orbiting workshop, eventually matured into Skylab (see p171).

In early 1969 the new US President, Richard Nixon, asked his advisors for recommendations on the future of the US space programme, and later that year NASA released guidelines outlining its objectives for a permanently manned space station. Anticipating that this would soon receive presidential approval, work was begun on a programme which would lead to a 12-man space station in Earth orbit by about 1980. However, within a year it had become clear that the plans to develop the Space Shuttle would have a major impact on the space station. In particular, the Space Shuttle programme was so expensive that there would not be enough money to build a large space station at the same time and this realization led to the concept of a cheaper space station that could be assembled in orbit by joining together modules launched by the Space Shuttle.

However, with the US space budget falling throughout the 1970s, and with delays and cost overruns in the Space Shuttle programme consuming the bulk of NASA's resources, it was not possible to begin development of even a small modular space station during that decade. It was not until the 1980s, when the Space Shuttle had proved itself and the Soviet Salyut programme had established a significant lead in the field of long-duration spaceflight, that serious plans for a US space station re-emerged.

Freedom

The decision that a permanent presence in space would be the next goal of the US space programme was announced by President Ronald Reagan during his 1984 State of the Union message. The President directed NASA to develop the space station within a decade and invited America's allies to join the programme. Once this commitment had been made, NASA, in consultation with US industry, began to decide how best to meet this goal within the estimated $8.5 billion budget.

The proposed space station was soon named 'Freedom' and the first design concept was called the 'power tower'. This featured a long slender latticework mast pointing directly away from the centre of the Earth, a configuration which is naturally stable and avoids the need for regular thruster firings to maintain its attitude. The plan called for solar panels and pressurized modules to be joined alongside the tower. The 'power tower' would be accompanied in orbit by a free-flying laboratory module which would be used to conduct experiments away from any disturbances, such as caused by a Space Shuttle docking, that might affect the station itself. The free-flyer would be docked to the station from time to time for servicing but would usually operate automatically. The programme would also include an unmanned US platform for Earth observations which would be placed in polar orbit and serviced by the Space Shuttle but which would never be physically connected to Freedom.

In 1985 the 'power tower' concept was replaced by a design known as the 'dual keel' in which two vertical towers were joined and braced by three horizontal beams. In this configuration four pressurized cylindrical modules were to be mounted near the centre of the structure, with solar panels, radiators and other equipment attached to the central, horizontal beam. Scientific experiments, such as astronomical telescopes, were to be mounted on the upper and lower beams. By now the international nature of the project was well established and Europe and Japan had both agreed to provide a pressurized laboratory module for the core of the station (see p30 and p88) and Canada had offered a remote servicing system based on the Space Shuttle's robot arm (see p23). The other two modules, one to accommodate the crew of eight astronauts and the other for scientific research, would be provided by NASA. The modules would be linked together by smaller modules called nodes.

The explosion of the Space Shuttle Challenger, and the spectre of constantly rising costs, forced a re-evaluation of Freedom's design in 1986 and this led to a plan to build the station in two phases. Phase 1 would consist of just the central horizontal brace plus the four pressurized modules and the necessary solar panels. Phase 2, which need not be started until the initial station was already in operation, would add the vertical towers and upper and lower braces. The basic cost of the phase 1 version was expected to be about $16 billion, but when hidden extras such as the launch costs of Space Shuttle flights were included, the true cost rose to about $25 billion. Partly as a result of this tripling of the original $8.5 billion estimate, there were constant battles with the US Congress over the project's budget. Amongst some of the fiercest critics were space scientists who were concerned that Freedom would consume all of NASA's funds, leaving little or no money for other small scientific programmes.

Freedom's problems came to a head in October 1990 when the project's budget was cut by the US Congress and NASA was instructed to produce a new, cheaper design. NASA spent about five months on this process and in mid 1991 presented a design which was similar in layout to the phase 1 space station, but somewhat smaller. One major change was that the size of the pressurized modules which formed the core of the station had shrunk from 13.4m to 8.2m in length (although their diameter remained the same at 4.4m). This reduction in the size meant that the modules could be fully outfitted with equipment before launch without overloading the Space Shuttle and replaced the original plan to launch the modules only partly complete and then install extra equipment that would be delivered on later flights. Other changes were a reduction in the size of the central truss from 150m to 108m and the removal of one the four pairs of solar panels. A consequence of these reductions was that the permanent crew would be reduced to four. If all went well then an extra accommodation module could be added later to allow the crew to be increased to eight.

The present plans

The precise future of Freedom remains unclear. In 1991 an attempt was made by some members of the US Congress to remove all funds for Freedom and transfer the money to other programmes. This proposal was eventually overturned and funding for Freedom was restored, so it now seems likely that the programme will go ahead. Naturally the exact timetable is uncertain, but in mid 1991 NASA published a schedule which showed that assembly could begin in late 1995 and be complete by late 1999. To do this a total of 17 Space Shuttle flights would be required.

During the assembly phase astronauts will visit the station and while some of them are involved in the assembly itself, others will enter whichever pressurized modules are in place and work in them for a few days to gain experience of operating the station. These astronauts will then return to Earth in the Space Shuttle which carried them to the station. This is known as the 'man-tended phase' and will last until the phase 1 station is completed. Only then will the first crew of four board Freedom for a regular tour of duty. Once Freedom is operational each crew is expected to remain onboard for about 90 days, before being relieved by the next visiting crew, who will arrive by Space Shuttle.

> A critical issue on Freedom is that of crew safety. The Soviet Mir space station has always had a Soyuz ferry craft docked so that the crew can return to Earth at once if required. The plans for Freedom envisage that the Space Shuttle which delivers each crew will depart after a few days, leaving the astronauts marooned and so a special Assured Crew Return Vehicle is to be developed. This will be a capsule, probably similar to an Apollo command module, which will remain attached to Freedom whenever the station is permanently manned so that it is always available as a space lifeboat if needed.

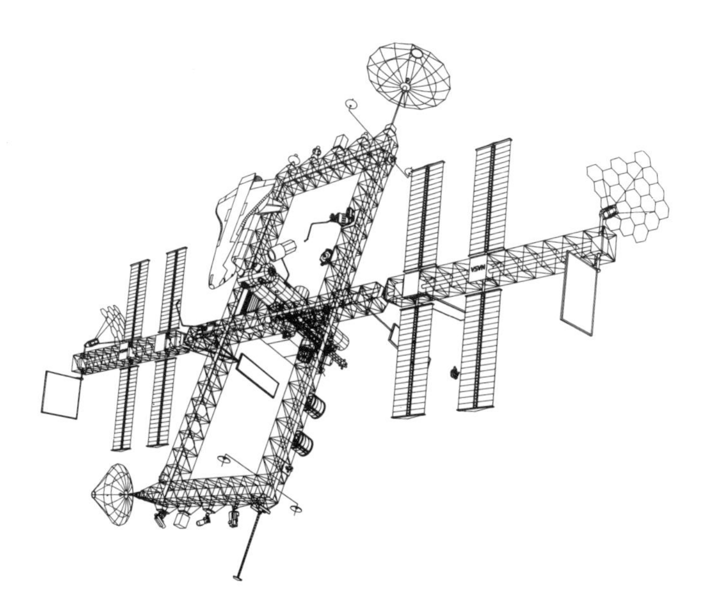

One of the many possible developments of the International space station Freedom

Galileo

Launched 10 October 1989

Mission to Jupiter

In 1995 the USA's Galileo spacecraft will arrive at Jupiter and release a probe into the planet's turbulent atmosphere.

Jupiter is the largest planet in the solar system and has a retinue of more than a dozen satellites and a complicated system of radiation belts. The planet was first visited by the US Pioneer 10 spacecraft in 1973 (see p151) and the success of this flyby focused attention on plans for a spacecraft to orbit Jupiter. This mission, originally called the Jupiter Orbiter/Probe and later named Galileo, envisaged a two-component spacecraft designed to explore the planet, its radiation belts and its major moons. The main spacecraft would go into orbit around Jupiter while a probe, released from the orbiter, would enter the atmosphere and study the atmospheric structure and composition as it descended by parachute. In 1977 Germany agreed to join the project and to supply the spacecraft's main propulsion system in return for participation in the mission operations and the scientific programme.

It was originally planned that Galileo would be lifted into Earth orbit by the US Space Shuttle in 1981 and then use a three-stage version of the Boeing Inertial Upper Stage (IUS) to boost itself towards Jupiter. However, as the Space Shuttle programme fell behind schedule, the launch was delayed until 1984. Galileo was delayed again when the three-stage IUS was cancelled and replaced by a liquid-fuelled Centaur G upper stage. The Centaur G stage was later cancelled, then reinstated, and Galileo was eventually scheduled for launch in May 1986 and arrival at Jupiter in 1988. Then, in January 1986, the Space Shuttle Challenger exploded and the Centaur stage was cancelled again in the safety review which followed. For a time the spectre of total cancellation hovered over the project, but mission planners developed a complicated flightpath that would use a two-stage IUS and flybys of the Earth and Venus to gain the energy needed to send the spacecraft to Jupiter. Known as VEEGA, the Venus–Earth–Earth Gravity Assist, this tortuous journey will take six years, instead of the two required for a direct Earth to Jupiter flight.

The Galileo spacecraft

The Galileo spacecraft weighs 2 500kg and comprises two sections. The larger section, which contains the power supply, computers, control electronics and the propulsion module, spins three times per minute to provide gyroscopic stability as Galileo cruises through space. The spinning section also houses a number of scientific instruments that benefit from sweeping regularly across the sky. Below this is a de-spun section, which actually rotates at the same speed but in the opposite direction to the main craft, so that it appears to remain stationary. This carries those instruments, such as cameras and spectrometers, that require a stable platform from which to operate. The de-spun section also carries the entry probe.

Galileo uses RTG nuclear generators (see p153) to produce electricity and these are carried on a boom that holds them clear of the rest of the spacecraft. To transmit scientific data back to Earth from Jupiter, Galileo requires a very large radio antenna and uses a type that can be folded up for launch and deployed once the spacecraft is on its way. The late change to the VEEGA trajectory meant that it was necessary to keep the antenna folded up for the first few months of the flight to protect it from overheating during Galileo's brief sojourn inside the

Earth's orbit. While the antenna is folded, communications with Galileo are only possible via small, low-gain antennas which are most effective when the spacecraft is close to Earth.

The Jupiter entry probe does not carry a propulsion system and so about five months before arrival at Jupiter Galileo will increase its spin rate and aim itself directly at the planet. The 336kg Jupiter probe will then be released and, stabilized by the spin imparted before release, will coast towards Jupiter. The main craft will then return to its normal spin rate and fire its thrusters to move on to a path that will place it on course to fly above Jupiter's clouds just as the probe enters the atmosphere below.

Six hours before the planned time of arrival, a clock will activate the probe and prepare it for atmospheric entry. The probe will slam into Jupiter's atmosphere at 185 000 km per hour and will decelerate at up to 350g, slowing down to about 200km per hour in two minutes. If all goes well a drouge parachute will then be released to pull away the rear heatshield which will in turn release the 2.5m-diameter main parachute and pull the probe clear of the front heatshield. As the probe begins to sample Jupiter's atmosphere the data it collects will be radioed to the main craft flying overhead. The data from the probe will be broadcast to Earth and, as a precaution, will also be recorded for re-transmission later if required. Data collection from the probe will stop about 75 minutes after entry, even if the probe is still operating, because the main craft will be passing out of range and will be busy preparing itself for the motor firing that will place it in orbit around Jupiter.

The first motor firing will place the main Galileo spacecraft into an orbit around Jupiter which has a period of about eight months. As the spacecraft climbs away from Jupiter and then falls back again, its cameras and other sensors will study the planet and its environment. As the first orbit finishes, Galileo will encounter the moon Ganymede and use this moon's gravity to divert itself so that it flies to encounter another moon a few months later. Final decisions on the details of the next phase of the mission will be made nearer to the time when Galileo is expected to arrive at Jupiter but each encounter will be designed to put Galileo on course for a flyby of another satellite without needing to use very much thruster fuel. This satellite tour will continue for at least two years, allowing Galileo to investigate each of the three outer Galilean moons in detail. Only innermost Io will not be included in the tour because mission planners have decided that the risk that the spacecraft will be crippled by the intense radiation close to Jupiter outweighs the scientific gains from repeated flybys. Accordingly, Io will be visited just once. Galileo will pass within 1 000km of Io whilst approaching the planet a few hours before the probe enters Jupiter's atmosphere.

On the way at last

Galileo was finally launched from the Space Shuttle Atlantis on 18 October 1989. After a careful checkout, the IUS was fired, sending Galileo to Venus, where it arrived on 10 February 1990. Since at this time the main antenna was still folded, most of the scientific data taken during the flyby was recorded on Galileo's tape recorder for later transmission to Earth. The encounter with Venus was a complete success and, as it finished, Galileo was on an almost perfect course back towards the Earth. The spacecraft returned to Earth on 8 December 1990 and executed another successful flyby, passing 960km above the Atlantic Ocean. The flyby of Earth added about 5km per second to Galileo's speed and sent the craft on a course that took it into the asteroid belt and towards a rendezvous with the asteroid Gaspra. During the flyby of the Earth, data recorded at Venus were transmitted to the ground via the low-gain antennas and a series of scientific observations of the Earth and Moon were made.

On 11 April 1990 ground controllers instructed Galileo to deploy its main radio antenna, but unfortunately the large dish did not unfurl completely. Analysis showed that a few of the ribs of the umbrella-like antenna had not disengaged from the central supporting mast and engineers immediately began to consider what to do. The best plan appeared to be to either warm or cool the antenna system in the hope that the mast and the stuck ribs would expand or contract by different amounts and this would provide enough clearance for the ribs to spring out. However, the first few attempts were not successful and the antenna remained useless throughout 1991 when further attempts to solve the problem were still underway. If the antenna cannot be freed by the time Galileo reaches Jupiter the scientific return from the mission will be drastically affected and contingency plans for this eventuality are being made.

On 30 October 1991 Galileo flew about 1 000km from the asteroid Gaspra, a potato-shaped body about 20km across. A number of images and other scientific data were taken and recorded on the spacecraft tape recorder for later playback. If the main antenna remains stuck the data will be returned via the low-gain antennas when Galileo returns to Earth for its final flyby in December 1992.

Galileo Galilei discovered the four largest moons of Jupiter in January 1610 when he observed the planet with his newly developed telescope. After a series of observations he realized that the four tiny stars he could see near the planet, and which seemed to move erratically from night to night, were not stars at all, but separate objects revolving around Jupiter. He announced his results in March and created a sensation since his discovery provided direct support for the Copernican theory that the Earth was not the centre of the Universe but, together with the other planets, moved around the Sun. In his honour, the four large moons of Jupiter are called the Galilean Satellites.

How the Galileo spacecraft will appear if all is well when it arrives at Jupiter

Gamma ray astronomy satellites

SAS-2 15 November 1972

Probing the most energetic objects in the Universe

Gamma ray astronomy is best done from space and a number of satellites have been launched to study these energetic messengers from beyond the Earth.

Gamma rays are parcels of electromagnetic waves similar to, but many thousands of times more energetic than, visible light. These parcels are called quanta, or photons, and are usually categorized by the amount of energy they carry rather than by wavelength as done for visible light. Gamma rays are of interest because they are produced in the most extreme physical conditions and allow astronomers to study exotic objects such as neutron stars (objects with the mass of the Sun squeezed into a sphere only 10km across), black holes and quasars.

Unfortunately, gamma rays cannot penetrate the atmosphere; those that try interact with the atoms in the air and are destroyed, so they are usually studied from high-flying balloons, or from satellites. To add to the difficulties gamma rays cannot be focused by lenses or mirrors and can only be detected when they interact directly with matter, so studying gamma rays is very difficult. Some instruments, called scintillation counters, use material that scintillates, or flashes, as a gamma ray passes through, others use sparks to reveal the trail of ionized atoms produced when a gamma ray passes through a chamber filled with an inert gas. With suitable ancillary equipment these types of instruments can reveal the energy of each gamma ray they detect, but they cannot provide accurate information on the direction from which the photon arrived. This means that they cannot be used to make detailed gamma ray pictures of the sky.

The first gamma ray experiments launched into space were very small since the rockets available were not powerful enough to lift large instruments into orbit. The US Explorer 11, launched on 27 April 1961, detected a handful of gamma rays but could not prove that they originated beyond the atmosphere. Other small experiments were carried on the US Orbiting Solar Observatory 3, Orbiting Geophysical Observatory 5 and a number of Soviet Cosmos satellites. These experiments confirmed that gamma rays could be detected from astronomical sources, but no individual gamma ray sources could be identified.

SAS-2 and COS-B

On 15 November 1972 NASA launched the first satellite dedicated to gamma ray astronomy. Officially designated Explorer 48, and also called Small Astronomy Satellite 2 (SAS-2), the satellite contained a spark chamber intended to map the sky and to try and identify individual gamma ray sources. Because the spark chamber was quite large, SAS-2 could estimate the position of a source to within about one square degree, which was quite good for a gamma ray experiment, but very bad by normal astronomical standards. SAS-2 detected gamma rays from the Crab Nebula and the Vela supernova remnant, both of which are the remains of stars which exploded hundreds of years ago. The existence of a few other sources was also hinted at, but after seven months a failure of the satellite's power supply ended the mission prematurely.

The European COS-B satellite, launched by a NASA Delta rocket on 9 August 1975, carried a spark chamber similar to that on SAS-2. COS-B surveyed large regions of the sky and it confirmed many of the results from SAS-2 as well as locating, at least approximately, the positions of 25 point-like sources of gamma rays. The satellite was deactivated after 6.5 years, more than three times its planned lifetime.

Gamma and Granat

Gamma-1 and Granat were both missions in which French-built gamma ray detectors were installed on Soviet satellites. Gamma-1 was based on an unmanned version of the Soyuz spacecraft, in which a French spark chamber was installed. Originally scheduled for a flight in about 1984, it was finally launched on 11 July 1990. Granat was a Soviet satellite which carried a large French experiment as well as a number of smaller instruments provided by other countries. The French experiment, called SIGMA, uses what is called a coded mask telescope to produce gamma ray images of the sky. At one end of the SIGMA telescope is a mask, a 30cm-square piece of tungsten which incorporates a number of metal blocks arranged in a special pattern. Some of the blocks are thin enough to allow gamma rays to pass through and some are too thick for the gamma rays to penetrate. Two and half metres behind the mask are a set of gamma ray detectors. Gamma rays passing through the mask cast a complicated shadow on the detectors and this pattern can be unscrambled by computer to produce an image of the sky. The system is rather like a pinhole camera, but with many pinholes instead of just one. Granat was launched on 1 December 1989 into an orbit which ranged from 1 957km to 210 693km. The mission was quite sucessful and the SIGMA telescope was used to make observations of the centre of our Galaxy.

The Gamma Ray Observatory

The largest and most powerful gamma ray satellite ever launched was the NASA Gamma Ray Observatory or GRO. This 15-tonne satellite was carried into orbit by the Space Shuttle Atlantis on 5 April 1991 and it is hoped that it will remain in operation for much of the decade. The GRO carries four experiments, three of which weigh 2 000kg each and are about the size of a small car. The fourth experiment consists of eight separate detector systems, one at each corner of the satellite. The objective of the GRO mission is to map the entire sky much more comprehensively than was done by earlier missions and then to spend a period concentrating on the most interesting objects which it has discovered. The GRO is in a relatively low orbit and atmospheric drag will cause it to spiral slowly towards the Earth. To prevent the satellite making an early re-entry, from time to time the GRO will use its thrusters to climb back into a higher, safer orbit. When fuel begins to run low NASA will have to decide either to send a Space Shuttle mission to refuel the satellite or to use what fuel remains to arrange for a controlled re-entry. This re-entry will be timed to occur over an uninhabited area where there would be no risk of people being injured by debris as the satellite falls to Earth.

Gamma ray bursters

In 1973 it was announced that a series of US military satellites called Vela, which had been launched to monitor clandestine nuclear explosions in space, had discovered a new kind of gamma ray source. These mysterious objects, which lie outside the solar system, suddenly erupt into a huge burst of gamma rays then rapidly fade away. These results have been confirmed by many other satellites, but the nature of the gamma ray bursters remains a mystery. The bursts probably originate in neutron stars. They may be caused either by 'starquakes' as a neutron star rearranges its internal structure, or by explosions of material at the surface of a neutron star or by comets crashing into a neutron star. As yet, astronomers just do not know. In 1991 it was announced that the GRO had detected many gamma ray bursts and that the results seemed to suggest that the bursts arise well beyond our galaxy. If confirmed this would mean that the bursters are not neutron stars but a hitherto unknown type of very energetic object.

Gemini

First flight 8 April 1964

Heavenly twins

The first spacecraft able to change its orbit and to dock with other satellites, the US Gemini bridged the gap between the Mercury and Apollo programmes.

When President John F Kennedy committed the USA to landing a man on the Moon, NASA had just 15 minutes experience of manned spaceflight. Even though the precise details of the Moon missions were unclear, it was obvious that NASA would soon require experience of long duration spaceflight, rendezvous and docking between spacecraft and precise control of re-entry and landing. The simple Mercury capsule could not provide this and so, on 7 December 1961, NASA announced its decision to develop a manoeuvrable two-man spacecraft capable of flights lasting up to 14 days. Initially called Mercury Mark II, the project was officially named Gemini, after the constellation of the heavenly twins, on 3 January 1962.

The 5.8m-long Gemini spacecraft was composed of three sections called the re-entry module, the retrograde section and the equipment module. The re-entry module was based on the truncated cone shape of the Mercury capsule but enlarged to provide a 50% increase in cabin volume. The two astronauts sat side by side in ejection seats and between them was a panel which housed the controls used to fire the small rockets to manoeuvre the spacecraft. In front of each crewmember was a small oval window below which a control panel stretched right across the spacecraft like the dashboard of a motor car. Each astronaut entered the spacecraft via a separate hatch, hinged along one side, which formed the roof of the crew compartment when closed. Since spacewalks were planned for the Gemini programme, the hatches were designed to be opened and closed in space. At the rear of the re-entry module was the heatshield. Although the use of a wing similar to that of a hang glider was considered for a time, it was eventually decided that the Gemini missions would use conventional parachutes and descend into the ocean and so the re-entry module was ballasted so that it would float upright. This allowed the astronauts to open their hatches while they waited for rescue after splashdown.

In front of the astronauts, outside the pressurized compartment in which they lived, was a cylindrical section extending forward about 1.5m. This contained the radar and other systems used during rendezvous and docking, and a set of 16 thrusters (in two independent sets of eight) for attitude control during re-entry. It also held the main parachute used to lower the spacecraft into the water for splashdown.

Behind the re-entry module was a gently tapering cylinder called the retrograde section. This housed eight thrusters for use during rendezvous and docking and four solid-fuelled retro-rockets. When it was time to return to Earth, and after the equipment module had been cast off, the retro-rockets were fired one after each other to slow the spacecraft down. The retrograde module was then jettisoned, exposing the heatshield.

The equipment module was an extension of the tapering cylinder of the retrograde module, and expanded outwards to match the diameter of the Titan 2 launch vehicle. The equipment module carried many of the support systems needed during the flight, such as fuel for the manoeuvring thrusters and most of the spacecraft's electronic systems. The equipment module also carried either batteries, for the short missions, or fuel cells which generated electricity by chemically combining liquid hydrogen and oxygen to produce water on the longer flights (see p153). Spaced 90° apart around the rear of the equipment module were four pairs of attitude control thrusters. The open end of the equipment section was covered by a gold foil blanket to help control the temperature of the equipment inside and to prevent the sun shining on to the tanks holding the liquid hydrogen and oxygen for the fuel cells.

Titan II

The rocket used to launch the Gemini spacecraft was a modified version of the US Air Force Titan II intercontinental ballistic missile. The Titan II was a liquid-fuelled, two-stage rocket, 3m in diameter and 33.2m tall with the Gemini spacecraft attached. The first stage had two motors which burned a mixture of unsymmetrical dimethylhydrazine and nitrogen tetroxide. This combination is hypergolic, that is, it burns spontaneously on contact, and requires no ignition system. The second stage was powered by a single motor using the same propellants. The motors were gimballed (tipped from side to side) to steer the rocket after launch.

Before the Titan II could be approved to carry astronauts, a number of extra safety systems were added. These included a malfunction detection system, to warn the crew if any of the rocket's vital systems were misbehaving, extra electrical and hydraulic systems which could take over if any of the main systems failed and additional instrumentation to assist in checking the rocket before launch and to provide monitoring from the ground during flight. In the event of an emergency being detected, the crew would use their ejection seats to escape.

Agena

The unmanned Agena D rocket was modified to serve as a rendezvous and docking target for the Gemini programme. The Agena, 1.5m in diameter and 7m long, was originally developed by the US Air Force as an upper stage for the Thor and Atlas rockets. In this role it was used in a number of military and civilian missions in Earth orbit, as well as for launching the Ranger Moon missions and Mariner probes to Mars and Venus. This record of achieving precise, predetermined orbits and its ability to be stabilized and controlled once in space made Agena an ideal choice for the Gemini programme.

To enable a Gemini capsule to dock with the Agena, a special collar was fitted to the front end. The collar included a hydraulic system to absorb the shock of docking, and latches and motors to pull the two spacecraft firmly together afterwards. Once this 'hard docking' had been achieved the two spacecraft effectively became one, with electric connections in the nose of the Gemini allowing the astronauts to send commands to the Agena. In particular the astronauts could restart the Agena engine, a procedure which could be carried out several times. The Agena engine could also be started by ground control.

Gemini 12 is launched by a Titan II rocket

Gemini programme

Gemini 1 8 April 1964, Gemini 12 11 November 1966

Stepping stone to Apollo

Through the 10 manned Gemini missions, the US developed many of the techniques required for a flight to the Moon.

The first two Gemini missions were unmanned and were intended to clear the way for the first manned flight. The objective of Gemini 1 was to confirm the compatibility of the Titan II launch vehicle and the Gemini spacecraft and no attempt was made to separate the Gemini from the booster. Gemini 2 was a sub-orbital test of the re-entry system. After climbing to about 160km the re-entry module splashed down in the Atlantic as planned. Both flights were successful.

Gemini 3 was commanded by Virgil 'Gus' Grissom who became the first man to fly into space twice. His co-pilot was John Young. It was a short flight, lasting about five hours, but for the first time a manned spacecraft moved from one orbit to another as the astronauts used the spacecraft's computers and rocket motors to plan and execute various manoeuvres simulating the rendezvous procedures that would be tested on later missions. Gemini 3 was the first of a series of spectacular missions that would see the USA leapfrog the USSR in the space race and set NASA on the road to the Moon.

Long duration flights

NASA used the Gemini programme to prove that men could survive the six to eight days required for a journey to the Moon. The Gemini 4 flight, best remembered for its spectacular spacewalk, lasted four days and was almost three times as long as the final Mercury mission. Gemini 5 lasted eight days, giving the Americans the space endurance record for the first time. This flight also marked the first use of fuel cells (see p153) to supply electrical power for the spacecraft and, although problems with the fuel cells threatened to terminate the flight, these were eventually overcome and the mission was allowed to proceed. The astronauts found that storage was a major problem, with loose items floating all around the cabin, causing one of them to describe the mission as 'eight days in a garbage can'.

Gemini 7 was the longest flight of the programme, and the longest spaceflight for the next five years. The objectives of the mission were to study the biological effects of a long flight and to act as a target for Gemini 6, which would attempt to rendezvous with them. To make the crew more comfortable, special lightweight spacesuits were worn, with the understanding that one or other crewmember, but not both, could remove his suit for a time and relax in his underwear. Despite a number of minor problems the flight was a success, and the crew suffered no ill effects from their long exposure to weightlessness.

Rendezvous and docking

Developing techniques for rendezvous and docking was a major objective of the Gemini programme. The Gemini 5 astronauts released a special Radar Evaluation Pod which simulated the equipment being developed for the Agena target vehicle. Unfortunately, trouble with the fuel cells caused the bulk of the rendezvous exercise to be cancelled due to a shortage of electrical power. Undaunted, NASA planned that Gemini 6 would attempt a rendezvous with an Agena rocket, but that mission was postponed when the Agena failed to reach orbit. Since another Agena was not available, it was decided to delay Gemini 6 and use Gemini 7 as a target instead. Two Geminis could not normally be docked together, and Gemini 7 commander Frank Borman vetoed proposals to fit a temporary docking adaptor to his spacecraft. However, rendezvous and formation flying techniques could still be practiced during the joint flight.

Gemini 7 was launched normally, but a Gemini 6 launch attempt failed when the main engines of the Titan rocket shut down seconds after ignition. With commendable restraint the astronauts did not eject and another attempt three days later was successful. Gemini 6 made the first space rendezvous a little under six hours after launch. The two craft flew together for about 20 hours during which time Gemini 6 moved to within 2m of Gemini 7 several times, and the crews waved to each other through the spacecraft windows. Gemini 6 then returned to Earth, leaving the Gemini 7 crew to complete their lonely duration flight.

Gemini 8 successfully rendezvoused and docked with an Agena target, achieving the world's first space docking. All went well for about 20 minutes and then the two craft began to tumble violently. After a battle to regain control (see p195) commander Neil Armstrong was forced to make an emergency re-entry and land after only 10 hours in space. Gemini 9 was also unable to complete its planned docking mission. Its original Agena target fell into the Atlantic soon after launch and a temporary docking target was hastily assembled to replace it. Although this was launched successfully, the Agena's nose shroud, designed to protect the satellite during launch, failed to separate and no docking attempt could be made.

After this long series of setbacks, the final Gemini dockings were a complete success. Gemini 10 docked with an Agena and the Agena's propulsion system was then used to send the docked craft towards a rendezvous with the Gemini 8 Agena, which was still in orbit after being abandoned during the earlier emergency. Gemini 11 docked with another Agena and used the Agena's rocket motor to propel itself to a record altitude of 1 369km. Gemini 12 also docked with an Agena and although problems with the Agena's main engines prevented any major rocket firings, some small manoeuvres were made.

Gemini flight log

Mission	Launched	Crew	Orbits	Duration (h min)	Notes
Gemini 1	8 Apr 64	Unmanned	–	–	Compatibility test
Gemini 2	19 Jan 65	Unmanned	–	–	Sub-orbital re-entry test
Gemini 3	23 Mar 65	Virgil Grissom John Young	3	4 52	Test flight
Gemini 4	3 Jun 65	James McDivitt Edward White	62	97 56	1st US EVA
Gemini 5	21 Aug 65	Gordon Cooper Charles Conrad	120	190 55	Duration flight
Gemini 6	15 Dec 65	Walter Schirra Thomas Stafford	16	25 51	Rendezvous with Gemini 7
Gemini 7	4 Dec 65	Frank Borman James Lovell	206	330 35	Record duration
Gemini 8	16 Mar 66	Neil Armstrong David Scott	6	10 41	Docking, emergency landing
Gemini 9	3 Jun 66	Thomas Stafford Eugene Cernan	45	72 21	Rendezvous, 2-hour EVA
Gemini 10	18 Jul 66	John Young Michael Collins	43	70 47	Docking, EVA, 2 rendezvous
Gemini 11	12 Sep 66	Charles Conrad Richard Gordon	44	71 17	Docking, EVA
Gemini 12	11 Nov 66	James Lovell Edwin Aldrin	59	94 34	Docking, 3 EVAs

The EVAs

Extra Vehicular Activity (EVA), or spacewalking, was also developed during the Gemini programme. Although Soviet cosmonaut Alexi Leonov was the first to walk in space, Gemini 4 astronaut Ed White made a 21-minute EVA from Gemini 4 which, in retrospect, made EVA appear a lot easier than it was. The planned EVA on Gemini 8 was prevented by the abrupt end to the mission and it was not until Gemini 9 that a second US EVA took place. On this occasion astronaut Gene Cernan, who had a series of complex tasks to perform, became overheated and his spacesuit visor began to fog up; he was instructed to give up and return to the cabin. On Gemini 10 Michael Collins was able to float across and retrieve an experimental package from the Gemini 8 Agena rocket, but the Gemini 11 EVA was also abandoned when Richard Gordon became overheated.

These problems forced NASA to rethink its plans and the Gemini 12 EVAs were simplified, with planned rest periods built in between activities. The new strategy worked, and astronaut Edwin Aldrin was able to accomplish all his tasks without major problems during three separate EVAs, two spent taking photographs through the open hatch of the spacecraft and one involving a series of complex operations outside the spacecraft and at a special workstation on the docked Agena rocket.

An Agena target vehicle attached to a Gemini by a flexible tether

Germany in space

Azur 1 8 November 1969

European space power

Although it has never launched its own satellites, Germany is a major force in European space activities.

At present, Germany is the second largest contributor to the European Space Agency (ESA) and is involved in many of ESA's largest projects. In addition, Germany has its own national space programme which has featured the launch of a variety of satellites, usually in collaboration with other nations, and the flight of German astronauts on Space Shuttle and Soyuz missions.

Public funding for German space activities comes from the BMFT, the Federal Ministry for Research and Technology, and is channelled through DARA, the German space agency. The German Aerospace Research Organisation, the DLR, is also involved in government space activities. German space policy is aimed at furthering scientific research and development, applying space technology to Earth and fostering the competitiveness of German industry.

Scientific satellites

The history of rocket research in Germany goes back to before World War II but despite this, Germany has never built its own space launcher, although it was a member of the ill-fated European Launcher Development Organisation (ELDO) which collapsed in the early 1970s. Since then, Germany has arranged for its satellites to be launched by other space agencies.

The first German satellite was Azur 1, a 71kg conical cylinder which carried equipment to study the Earth's radiation belts. Azur 1 was launched by a US Scout rocket on 8 November 1969. The Azur satellite was followed by the Dial/Wika mission, which comprised a small octagonal satellite built jointly by France and Germany and launched by a French Diamant B rocket. The next German mission was Aeros 1, a cylindrical satellite weighing 125kg, which was launched by a Scout rocket on 16 December 1972. The virtually identical Aeros 2 satellite was launched on 16 July 1974.

A more ambitious project was the joint German/US Helios mission. This featured two spacecraft, each weighing about 370kg, which were designed to fly close to the Sun and study the solar wind, the interplanetary magnetic fields and cosmic radiation, in this hitherto unexplored region of space. Helios 1 was launched on 10 December 1974 by a NASA Titan IIIE (Titan-Centaur) rocket and flew within 46.4 million km of the Sun. Helios 2 followed on 15 January 1976 and approached the Sun to within 43.5 million km.

Following a policy of working closely with other space agencies, Germany participated in the AMPTE (Active Magnetospheric Particle Tracer Explorer) mission in 1984, providing one of the three spacecraft used in this US/German/UK mission to explore the magnetosphere. After AMPTE, the next major German scientific project was the ROSAT (Rontgensatellit) X-ray astronomy mission. ROSAT (see p264) was carried out in collaboration with NASA, which agreed to launch the satellite free of charge, and the UK, which provided an extra scientific instrument. ROSAT was originally expected to be launched by a US Space Shuttle in 1987, but the explosion of the Space Shuttle Challenger caused the launch to be changed to an unmanned Delta rocket. After suitable modifications, ROSAT was finally launched on 1 June 1990.

Germany also contributed to the US Galileo mission to Jupiter by supplying the propulsion system for mid-course corrections and for inserting the spacecraft into orbit around Jupiter.

Communications satellites

In the late 1970s Germany collaborated with France on the development of the Symphonie communications satellites. The project, which involved two satellites in geostationary orbit, was described as experimental in order to avoid conflicting with the interests of the international INTELSAT organization of which both France and Germany were members. Symphonie 1 was launched by a US Thor-Delta rocket on 19 December 1974. Symphonie 2 followed on 27 August 1975. In both cases NASA was reimbursed for the costs of the launches.

The late 1980s saw the establishment of two German communications satellite systems called TVSat and DFS Kopernikus. TVSat is the German component of a German/French direct broadcast system which uses satellites in geostationary orbit to beam television pictures directly to individual buildings equipped with a small receiving dish. TVSat 1 was launched by an Ariane rocket on 21 November 1987, but one of its solar panels failed to open properly and the satellite was unable to enter service. A replacement, TVSat 2, was launched on 8 August 1989. DFS Kopernikus was the first satellite launched as part of a programme to supplement the existing German telecommunications network. The satellite, which can relay television programmes and accommodate 1 800 telephone and 200 data transmission lines, was launched by an Ariane 44L rocket on 5 June 1989.

Man in space

Germany has maintained a considerable interest in manned spaceflight and has been closely involved with the US Space Shuttle programme. In 1983 the German SPAS (Shuttle Pallet Satellite) was the first object to be deployed and then retrieved by the Space Shuttle's robot arm. The 1 800kg SPAS was later modified to carry free-flying experiments on a Space Shuttle mission in 1991. German industry also led the development of the ESA Instrument Pointing System, which can be carried in the cargo bay of the Space Shuttle and used to aim scientific instruments such as astronomical telescopes with great precision.

Germany was the major financial contributor to, and hence the industrial leader of, the ESA Spacelab project. Spacelab (see p200) is a pressurized laboratory carried inside the cargo bay of the Space Shuttle, and the first European to fly in it was Ulf Merbold, a German citizen working for ESA. German interest in Spacelab did not end with the flight of Spacelab 1. In 1985 Germany managed and funded an ambitious Spacelab mission (Spacelab D-1) which was devoted to materials science, biology and space medicine. The mission, which carried two German astronauts (Reinhard Furrer and Ernst Messerschmid), one ESA astronaut and five Americans, was launched aboard the Space Shuttle Challenger on 30 October 1985 and landed seven days later. Although control of the flight remained with NASA, the scientific programme was managed from a German space centre at Oberpfaffenhofen.

Although the second planned German Spacelab flight has not yet taken place, Germany remains interested in manned spaceflight and in March 1992 German astronaut Klaus-Dietrich Flade visited the Mir space station. Germany has also used the experience it gained with Spacelab to take the lead in the ESA Columbus programme, which is intended to develop pressurized modules for use in conjunction with the international space station Freedom and the European spaceplane Hermes.

With an eye on the future, Germany is also investigating a reusable spaceplane called Sanger (see p207), but the high costs of reunification mean that rapid development of the project is very unlikely.

During their close approaches to the Sun, the Helios spacecraft were exposed to 11 times the solar radiation they experienced in the vicinity of the Earth.

German astronaut Dr Ulrich Walter in a three-axis training device

Hermes

Provisional approval 1987

European mini-shuttle

The Hermes mini-shuttle will carry three European astronauts into orbit early in the next century.

At a meeting of European ministers in 1985 it was agreed that Europe should become independent of the superpowers in the area of manned spaceflight. Accordingly, in 1987 the European Space Agency (ESA) proposed that it should develop a small space shuttle called Hermes, the Ariane 5 rocket (see p13) and the Columbus space station project (see p30). These three projects were intended to make Europe autonomous in space by about the year 2000.

The original idea for Hermes came from the French space agency CNES (Centre National d'Etudies Spatiales) who carried out feasibility studies before attempting to have Hermes adopted by ESA. This strategy proved successful and Hermes was selected as an ESA optional programme to which the French would make a large financial contribution. The UK, the fourth largest contributor to ESA, declined to become involved in Hermes, arguing that it would be an extremely expensive project which could never compete with the US Space Shuttle. In response, the proponents of Hermes argued that the new craft was not intended as a rival to the Space Shuttle, but rather that it was a step towards a completely independent European manned space programme. The proposal was given provisional approval in 1987 on condition that a searching review of the development costs should be conducted in 1991 before authorizing full-scale development.

The original Hermes proposal was for a delta-winged space shuttle weighing about 17 tonnes which would be launched by an Ariane 5 rocket. Hermes was to be able to accommodate up to six astronauts and to carry about 4 000kg of cargo in a 35cu m payload bay which could be opened in orbit. It was expected that up to six Hermes flights could take place per year and that its missions would include the servicing of various elements of the Columbus system and the ferrying of crews to the international space station Freedom. The first flight was expected in 1995 or 1996. Unfortunately, as more detailed studies of the proposal were carried out, these plans were found to be too optimistic. In particular, the expected weight of the craft kept increasing and this problem was exacerbated by the decision to fit an escape system to protect the astronauts if the launch vehicle were to explode. Eventually the expected weight of Hermes increased to over 24 tonnes and in consequence the design was scaled back in an attempt to reduce weight and costs.

By 1991 the proposed Hermes design had been reduced to a three-seater spacecraft able to carry 1 000kg into orbit. The opening payload bay had been deleted and the number of flights restricted to one per year. Another significant change was that some of the Hermes systems, including such vital components as the docking mechanism, had been relocated in a disposable resource module fitted to the rear of the craft. The resource module, which also acts as an adaptor to connect Hermes to its Ariane 5 launch vehicle, is intended to be cast off just before re-entry. This drastic step is necessary because the performance of Hermes when it is gliding back to land after re-entry is set by the relationship of its mass and the area of its wings, a ratio known as the 'wing loading'. Since the size of Hermes's wings must be limited to prevent it imposing excessive loads on the Ariane 5 rocket during launch, the only way to achieve an acceptable wing loading is to restrict the mass of Hermes during the final phase of its mission. The disposable resource module achieves this mass reduction, but at the expense of increasing the overall costs, since the resource module and all the systems within it must be replaced for each flight.

The result of these changes is that Hermes now looks rather different from its original design. The basic delta wing shape has been retained, but the end of each wing tip is now folded upwards to provide stability. The crew of three sit in ejection seats in a cabin near the nose of the craft. Behind the cabin is a payload/living section with a volume of about 25cu m which provides storage for cargo and facilities for eating, sleeping and hygiene. Behind that is the conical resource module, which includes a pressurized volume of about 28cu m and which can function as an airlock and docking system if required. The resource module can also accommodate cargo and scientific experiments and its outer shell is equipped with radiators to dispose of waste heat from the entire Hermes craft.

The first flight of Hermes, which may take place without a crew, is now expected early in the next century. Once the system is operational a typical Hermes mission will begin with a launch from the European launch centre at Kourou and end at an airfield in France. The craft will then be serviced and returned to Kourou for its next flight.

In November 1991 the new plans for Hermes were presented to the European ministers in the expectation that the project would receive final approval. However, because of the political and financial situation in Europe and increasing worries about the usefulness of the scaled-down Hermes design, the ministers could not agree to begin development and the decision was postponed for another year.

An artist's concept of the Hermes mini-shuttle

Hubble Space Telescope

Launched 24 April 1990

A mirror ground to precisely the wrong shape

The Hubble Space Telescope was intended to fulfil the dreams of astronomers for a large optical telescope above the distorting atmosphere.

The first NASA studies of a large optical telescope in orbit were made in the 1960s, and by the early 1970s a project known as the Large Space Telescope, with a main mirror 3m in diameter, was under discussion. Between 1973 and 1976 further studies led to a proposal for a scaled-down telescope, with a mirror 2.4m in diameter, known simply as the 'Space Telescope'. In 1977 it was agreed that Europe would join the project, providing one of the five scientific instruments and a set of solar panels in exchange for 15% of the total observing time. Work on the Space Telescope began in 1977 and in 1983 NASA announced that it would name the satellite the Edwin P Hubble Space Telescope, a name so clumsy that it is invariably shortened to Hubble Telescope or HST.

After a long period of assembly and testing, during which a number of launch delays occurred, the HST was finally declared ready in late 1985. However, the fatal explosion of the Space Shuttle Challenger on 28 January 1986 totally disrupted the schedule and the launch of the HST was further delayed until 1990. Finally, the HST was launched by the Space Shuttle Discovery on 24 April 1990. Two days later the $1.5 billion satellite was carefully lifted from the payload bay of the Shuttle and released into space. After 30 years of studies, development and delays it looked as if the long wait for the HST was finally over. It was not; the HST's troubles were only just beginning.

HST description

The Hubble Telescope is a reflecting telescope with a main mirror 2.4m in diameter and a 0.3m secondary mirror arranged in a Ritchey–Chretien optical system. The mirrors were ground, polished with great precision, and then coated with a layer of reflective aluminium. Finally an ultra-thin protective layer of magnesium fluoride was added. The mirrors are supported in a graphite epoxy structure which is both light and rigid. Since the mirror was polished under Earth's gravity, and might change shape when placed in the weightless environment of space, special motors were installed behind the mirror to allow the shape to be adjusted slightly. A collar around the telescope tube, about level with the main mirror, contains the systems needed to operate the satellite, such as communications, power regulation and attitude control. The attitude control system, the most precise developed for a satellite, uses a combination of gyroscopes, star trackers and special sensors in the telescope itself to find and lock on to guide stars. Power is supplied by two solar panels which are unfurled once in orbit and can be retracted again if necessary.

Behind the telescope are five scientific instruments, each about the size of a telephone box. They comprise two cameras, two spectrometers, devices which split up the light from a star or galaxy and examine it in great detail, and a photometer, which measures the brightness of objects with very great precision. Since the HST is planned to operate for 15 years, each instrument is designed to be replaced should it fail or become obsolete.

It is planned that from time to time a

Space Shuttle will rendezvous with the HST and grapple it with the Space Shuttle's robot arm. The HST can then be held in position in the Space Shuttle's payload bay while spacewalking astronauts replace one or more instruments and carry out repairs. Once this servicing is over, the HST can then be released to continue its mission. Earlier plans to return the HST to Earth for major overhauls have been dropped; all servicing will now be done in space.

Hubble's troubles

Astronomers always recognized that a telescope as complicated as the HST would have some teething troubles and so a period was allowed for testing and adjustment before scientific observations would begin. At first, apart from a few minor hitches such as wobbles caused by flexing of the solar panels every time the HST moved into or out of the sunlight, all seemed to be going well. Then, in late June 1990, NASA revealed that the HST could not be focused properly. Preliminary studies, since confirmed, showed that the telescope suffered from a condition known as spherical aberration. This means that light from the centre of the mirror focuses at a different place to light reflected from near the edges. Depending where the secondary mirror is placed it is possible to get one sharp spot with a very out of focus halo around it, or to get a sort of average spot with almost all of the image slightly out of focus. There is no possible position of the secondary mirror which will produce a completely sharp image. Although much science could still be done, and NASA tried to limit the political damage by releasing a series of spectacular pictures, the mirror flaw was a disaster for many of the scientific programmes planned for the HST.

An investigation soon discovered what had happened: a measuring device used to check the shape of the main mirror during polishing had behaved in an unexpected way, reflecting back light from the wrong place during optical testing. This misled the mirror team into thinking the shape was correct when it was not. With the problem identified, NASA began to consider how the situation could be rectified.

It was already known that the adjusting motors could not bend the mirror enough to solve the problem and so the next option seemed to be to build new instruments with correcting lenses that could remove the spherical aberration. This was too expensive for most of the instruments, but a spare camera NASA was already building for the telescope will be fitted with correcting lenses and installed during a servicing mission in 1993. To correct the remaining instruments is more difficult and an ingenious solution has been chosen. One instrument, the photometer, will be removed completely during the first servicing mission. It will be replaced by a special gadget that will position correcting optics in front of the other three instruments to cancel out the effects of the aberration. If successful this device, called COSTAR (Corrective Optics Space Telescope Axial Replacement unit), should restore the full focusing capability of the telescope and allow the postponed scientific projects to be completed. It is a daring gamble, but without it the HST will remain partially blind.

Edwin P Hubble

Born in 1889 Edwin Hubble was one of the greatest astronomers of the twentieth century. In 1924 he was the first to prove that the spiral nebula in Andromeda was another galaxy. He also discovered that the galaxies were all receding from ours and that the more distant ones were receding fastest. This is known as Hubble's Law and it shows that the Universe is expanding, as predicted by theories of the Big Bang.

The HST's problems are not restricted to the defective primary mirror. The satellite has six gyroscopes in its attitude control system of which any three are required for normal operation. Gyroscope No. 6 failed in December 1990 and No. 4 failed in July 1991. By September 1991 another gyroscope, No. 5, was also showing signs of failing. Fortunately, the gyroscopes are designed to be replaced in orbit and repairs will probably be made during the first mission to service the satellite.

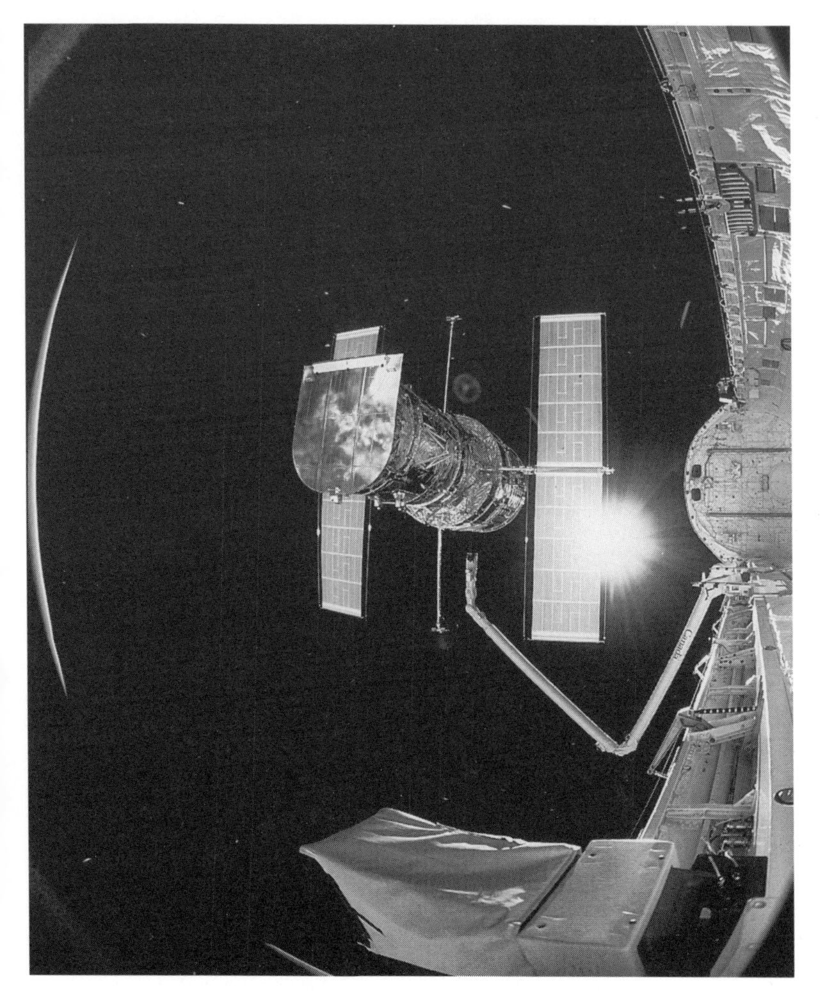

Deployment of the HST from the Space Shuttle in 1990

India in space

Aryabhata 19 April 1975

Bringing space down to Earth

India is using space technology to deal with the problems faced by a developing country.

The development of space activities in India began in 1963 with the setting up of a rocket range at Thumba which was used to launch sub-orbital rockets for scientific purposes. From this modest beginning, India's space activities have expanded to encompass satellites, launch vehicles and the flight of an Indian cosmonaut.

The Indian space programme is managed by the Indian Space Research Organisation (ISRO) which has several facilities spread around the country. The main centre for space technology development is the Vikram Sarabhai Space Centre at Trivandrum near the southern tip of the sub-continent. The main satellite launching site is at the SHAR (Sriharikota High Altitude Range) centre which lies on an island off India's east coast.

The first Indian satellite was Aryabhata, which was launched by a Soviet rocket on 19 April 1975. Although mainly intended to test the technology required to build and operate a national satellite, Aryabhata also carried some astronomical experiments. Unfortunately, few scientific data were received because the satellite failed after only four days in orbit. Since then, although some scientific programmes have been carried out, the main thrust of India's space programme has been the application of space technology to meet national needs. Accordingly, ISRO's major projects are concerned with the development of Earth observation and telecommunications satellites and with the rockets required to launch them.

Earth observation

India's first experimental Earth observation satellite was Bhaskara 1 which was launched by a Soviet rocket on 7 June 1979. Bhaskara 1 was equipped with visible and infrared cameras and a microwave radiometer designed to observe the oceans. The satellite operated until March 1981. On 20 November 1981 a slightly improved model, Bhaskara 2, was launched.

India's first operational Earth observation satellite was the Indian Remote Sensing (IRS) satellite, IRS 1A. This 850kg satellite was launched into a 900km polar orbit on 17 March 1988 by a Soviet rocket. IRS 1A exceeded its planned three-year lifetime and was still operating on 29 August 1991, when the IRS 1B satellite was launched from the Soviet Baikanour cosmodrome. India paid the USSR about $8.5 million to launch IRS 1B. IRS 1C is expected to be launched in 1993.

Communications

India has long recognized the potential of communications satellites and in 1975–76 participated in a Satellite Instructional Television Experiment (SITE) in collaboration with NASA. The SITE project used NASA's ATS-6 satellite to beam educational television programmes directly to remote areas that lay beyond the range of terrestrial television stations. During the SITE project, thousands of villages were equipped with a television set and a small receiving dish able to pick up the programmes broadcast from ATS-6.

India's first step in developing a national communications satellite came with the launch of the Ariane Passenger Payload Experiment (APPLE). This experimental communications satellite was launched on 19 June 1981 during the third test flight of the European Space Agency's Ariane rocket.

APPLE was a three-axis stabilized satellite which was placed in geostationary orbit and, despite problems with a stuck solar panel, operated until 19 September 1983.

India's first operational communications satellite was INSAT 1A, which was launched into geostationary orbit by a NASA Delta rocket on 10 April 1982. The INSAT 1 series, which were built by the US Ford Aerospace corporation to a specification laid down by ISRO, provide both telecommunication and meteorological services. The telecommunications payload can be used for telephone, data and facsimile lines and to relay television programmes. The meteorology instrument can observe a region from Egypt to the China Sea. INSAT 1A ran out of attitude control fuel for an unknown reason after only 147 days and was replaced by INSAT 1B, which was launched by the US Space Shuttle Challenger on 30 August 1983. INSAT 1C was orbited by an Ariane 3 rocket on 21 July 1988 and INSAT 1D was launched by a Delta rocket on 12 June 1990. The INSAT 1 satellites will be replaced by the Indian-built INSAT II series during the mid 1990s.

Rocket development

So as not to be totally dependent on other countries for satellite launches, ISRO embarked upon an ambitious, but not always successful, Satellite Launch Vehicle (SLV) development programme. The first result of this was the SLV-3, a four-stage, solid-fuelled rocket, 22.7m high. The first successful launch of the SLV-3 was on 18 July 1980 when a 35kg satellite called Rohini was placed in low Earth orbit. This flight made India the seventh nation to orbit a satellite with a national launch vehicle. The next launch, on 30 May 1981, was less successful and the RS-D1 satellite was placed into an orbit so low that it re-entered and burned up after only nine days. A further flight, on 17 April 1983, placed the RS-D2 satellite in orbit.

The SLV-3 formed the basis for the Advanced Satellite Launch Vehicle (ASLV). This new rocket used the SLV-3 as a core to which two boosters were attached. The first two flights of the new rocket, on 24 March 1987 and 13 July 1988, were intended to orbit the improved Rohini satellites designated SROSS, but both failed.

Despite the problems with the ASLV, ISRO is already developing its next rocket, to be known as the Polar Satellite Launch Vehicle (PSLV). The PSLV will use liquid propellants for two of its stages and is intended to place satellites weighing about 1 000kg into polar orbits. If all goes well, the PSLV will be able to launch versions of the IRS Earth observation satellites. Plans are also underway for a fourth-generation rocket called the GSLV (Geostationary Satellite Launch Vehicle), which will be able to launch payloads intended for geostationary orbit.

Indian cosmonaut

In 1984 an Indian cosmonaut, Squadron Leader Rakesh Sharma, flew to the Salyut 7 space station. The mission, launched aboard Soyuz T11 on 3 April 1984, lasted eight days and featured a series of experiments in biology and materials science. During the flight Salyut 7 made 11 passes across India and during these the cosmonauts used Salyut 7's MKF-6M and KATE-140 cameras to conduct Earth resources observations. The photographs were used to reveal information related to geology, hydrology and land use.

It was expected that an Indian astronaut would fly on the US Space Shuttle to observe the deployment of the INSAT 1C spacecraft, but this opportunity was lost when the launch of INSAT 1C was transferred to an Ariane rocket following the explosion of the Challenger in 1986.

> The Soviet–Indian spaceflight featured specially developed samples of Indian cuisine including mango bars and special curry.

Infrared astronomy satellites

IRAS 1983, ISO 1994

Supercold satellites

Telescopes cooled by superfluid helium can make important astronomical observations which would be completely impossible from the ground.

Infrared energy is a form of electromagnetic radiation rather like ordinary light but with longer wavelengths. It is usually regarded as covering the wavelength range from 1 micron to 100 microns. This region is of interest because infrared energy, often thought of as heat radiation, is produced by relatively cool material, for instance objects colder than 1 000°C radiate most of their energy in the infrared. This makes infrared astronomy a key tool in studies of asteroids and comets, dust around stars and in the space between the stars. What is more, infrared wavelengths penetrate the clouds of dust and gas in space better than the shorter wavelengths of visible light and so allow astronomers to look inside the dust clouds where stars are being formed. In a similar but more general way, star formation in other galaxies can also be studied. Unfortunately, much of the infrared radiation that reaches the Earth from space is absorbed by the atmosphere and to take full advantage of the potential of infrared astronomy it is necessary to place telescopes in orbit.

An infrared telescope orbiting above the atmosphere has another important advantage over its ground-based competitor, it can be cooled to very low temperatures. Ordinary telescopes, and the atmosphere above them, are warm and so produce their own infrared radiation and this background is so bright that it makes ground-based infrared astronomy very difficult. A space telescope does not have to observe through a warm atmosphere and it can be cooled to reduce its own, unwanted, infrared emission, which means that satellites are powerful tools for infrared astronomy.

IRAS

The first spaceborne infrared telescopes were launched on US sub-orbital rockets and spent only a few minutes above the atmosphere. None the less the results that they produced showed that it would be well worth launching a satellite to survey the whole sky in the infrared. The objective of this joint US/Netherlands/UK mission, which became known as the Infrared Astronomical Satellite (IRAS), was to map the whole sky at a number of wavelengths and highlight the regions worthy of further study.

The design of IRAS set the standard for subsequent missions. At the bottom was a service module which housed systems to control the satellite, point it in the required direction, to receive commands and to transmit its results back to Earth. Computers controlled the satellite for about 12 hours at a time without the need for human intervention and new sets of instructions were radioed up twice a day. The top section, or payload module, was built around a large dewar vessel, or vacuum flask, filled with superfluid helium. This is a form of liquid helium which boils at about 2° above absolute zero. A telescope, with a mirror 60cm in diameter, and an array of 64 individual infrared detectors was installed within the dewar. Because it was surrounded by superfluid helium, the whole telescope was kept very cold; any heat which leaked in just boiled away some of the superfluid helium. As long as the dewar contained some helium, the IRAS telescope was kept at a temperature below 10 degrees absolute

(10K or –263°C). The outer wall of the dewar was fully insulated to stop the heat from the Sun penetrating and boiling off the helium, and a sunshade prevented sunlight shining into the open end of the telescope and warming it up. To prevent gasses freezing on to the cold surfaces of the telescope either before launch or during the first few days in orbit (satellites tend to release vapours from various materials for a short time after launch) the telescope was sealed with an ejectable cover.

IRAS was launched on 26 January 1983 (25 January at the launch site in California) by a Delta rocket and was placed into a polar orbit at a height of 900km. During each orbit, which lasted about 100 minutes, the IRAS telescope scanned a strip of sky 0.5° wide and recorded all the infrared sources it could detect. Each orbit overlapped the one before by 0.25°, which allowed sources found on one orbit to be confirmed on another, and for moving objects to be identified. Over a period of 300 days IRAS surveyed almost the entire sky (only about 2% was missed out) and recorded almost a quarter of a million sources of infrared radiation. Amongst these were six new comets, unsuspected discs of dust around nearby stars, and thousands of galaxies, some of which were much brighter in the infrared that anyone suspected before launch. Astronomers using IRAS also discovered huge dust trails behind comets, bands of dust in the solar system and huge, wispy patterns of cold dust in interstellar space. IRAS was, by any reckoning, a huge success and a decade after the mission the catalogues it produced are being used by astronomers all over the world.

ISO

Although the Spacelab-2 mission carried an infrared telescope into orbit for a week, the next step after IRAS will be the European Infrared Space Observatory (ISO) which will be launched in 1994. Like IRAS, the ISO satellite has a service module and a payload module and like IRAS the 60cm telescope is enclosed in a dewar of superfluid helium. The ISO mission is, however, totally different from IRAS. ISO will not survey large areas of sky but will zoom in on small, selected, regions and study them in great detail. To do this ISO will carry four instruments which between them operate throughout the 2–200 micron wavelength band. The instruments are an infrared camera operating from 2.5–17 microns, a photometer and two spectrometers which between them cover the range 2.5–200 microns .

ISO will be launched by an Ariane 4 rocket into a highly elliptical Earth orbit stretching between about 1 000 and 70 000km. For much of the 24-hour orbital period the satellite will be visible from a control station in Spain and it is from here that ISO will be operated. When the satellite is out of contact it will be switched into a standby mode and no observations will be made. It is expected that a comprehensive scientific programme will be possible during the 18–20 months that it will take the helium to boil away. Once the helium is exhausted the telescope will begin to warm up and the mission will be over; it will not be possible to refill the dewar in orbit.

SIRTF

There will be a short Japanese mission called Infrared Telescope in Space that will be launched in 1994, but the next major infrared astronomy mission is the NASA Space Infrared Telescope Facility (SIRTF) which is expected to be launched in about the year 2000. This long-delayed project was originally conceived as a telescope attached to a Space Shuttle. From this it evolved into a free-flying satellite which would be placed in low orbit, and would receive frequent

Solar system astronomers took advantage of IRAS's sky survey to develop computer software which could detect moving objects in the data. This led to the discovery of six new comets. However, under international rules regarding discoveries using satellites the comets were not named after the members of the comet team (John Davies, Simon Green and Brian Stewart), but were all named IRAS.

top-ups of liquid helium from the Space Shuttle, and then changed again into a completely independent satellite in a 100 000km-high circular orbit. The SIRTF will be equipped with a 0.9-metre telescope and three sophisticated infrared instruments. It will carry enough superfluid helium to operate for about six years.

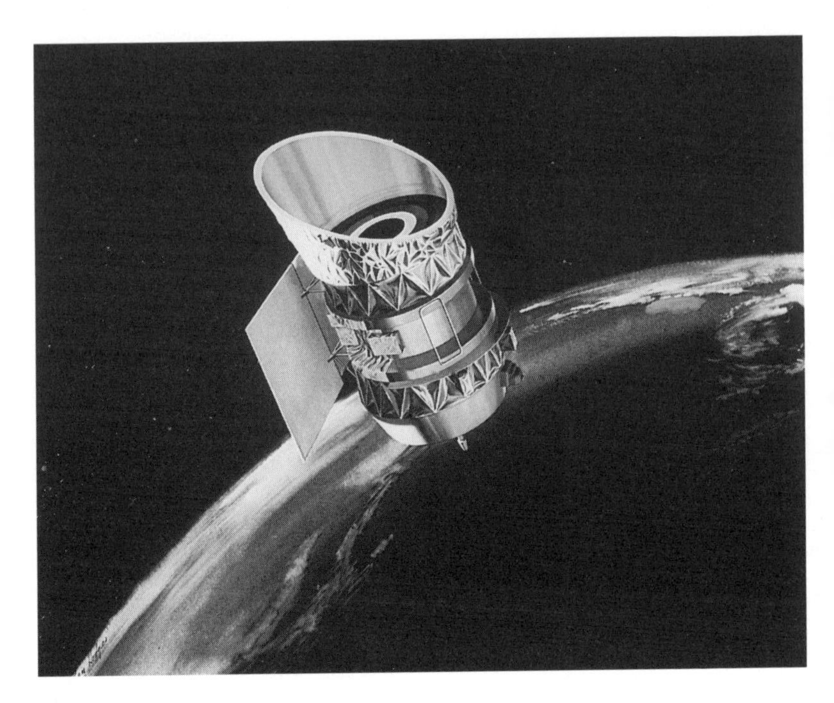

The IRAS satellite in orbit

INTELSAT

Founded 20 August 1964

International Telecommunications Satellite Organisation

Since 1964 INTELSAT has established a world-wide system of satellites which provide most of the world's international communications.

The international nature of the communications satellite revolution which began with Telstar and Syncom (see p35) showed the need to form a single organization to develop a worldwide communications satellite system. The desirability of such an entity was realized by US President John F Kennedy who, in 1961, called 'for the nations of the world to participate in a satellite system in the interests of world peace and closer brotherhood among people throughout the world'. This dream took its first step towards reality in August 1964 with the formation, by 11 countries, of INTELSAT, the International Telecommunications Satellite Organisation. INTELSAT, which is owned jointly by all its member nations, set out to develop and operate a series of communications satellites while leaving the provision (and ownership) of the necessary ground stations in the hands of individual countries. Since its foundation in 1964 INTELSAT has grown steadily and by 1990 had 119 members. The major users, such as the USA and the UK, retain quite large shares in the organization (24% and 12% respectively) but some of the smaller nations have registered their interest with a share as small as 0.05%. The headquarters of INTELSAT are in Washington DC.

INTELSAT's first satellite, INTELSAT 1 or Early Bird, was launched on 6 April 1965 by a Thrust Augmented Delta rocket and was placed in geostationary orbit over the Atlantic Ocean. It went into service on 28 June of that year and achieved the distinction of being the world's first commercial communications satellite. Early Bird was a cylinder, 0.72m wide and 0.52m high, with solar cells around its circumference. The satellite was spun around the axis of the cylinder to provide stability in space. Early Bird could relay 240 telephone lines or one television channel and remained in service for three and a half years before being withdrawn from use.

Early Bird was followed in 1967 by the INTELSAT 2 series. These were somewhat larger than Early Bird and weighed 86kg. The first craft was destroyed in a launch failure but three subsequent launches were successful and allowed INTELSAT to broaden its coverage by providing links across the Pacific, as well as the Atlantic, Ocean.

The continuing increase in demand for its services forced INTELSAT to develop bigger and bigger satellites. The INTELSAT 3 series weighed 150kg and could carry 1 200 telephone circuits or four television channels or a combination of both. Unlike the earlier craft which used a fixed antenna, INTELSAT 3 used an antenna that was spun in the opposite direction to the spin of the satellite, enabling the antenna to point continuously towards the Earth. The first INTELSAT 3 was launched in 1968, although several in the series of eight failed to operate as planned.

The INTELSAT 4 series marked a huge increase in size and capability. The satellite weight jumped to 720kg, and the overall height, including extended antennas, to about 5m. These new satellites were able to carry 6 000 telephone circuits and colour television channels. They also marked the shift to the more powerful Atlas-Centaur launcher. The first in the series was launched in January 1971. In 1975 the even more powerful INTELSAT 4A series was introduced.

The satellites in the INTELSAT 2, 3 and 4 series retained the spinning-drum shape of Early Bird but the INTELSAT 5 satellites, which were able to transmit 12 000 telephone channels plus colour television, adopted a different design. They were kept stable in all three axes by gas jets, and had a pair of large solar wings to generate electricity. This system allows all the solar panels to be exposed to sunlight all the time and produces more power, which in turn increased the capability of the satellite. This capacity was further increased to 15 000 telephone channels by the INTELSAT 5A version. The INTELSAT 5 series weighed 1 000kg each.

The first INTELSAT 5 was launched in December 1980 and a total of 15 were built. Of these 13 were orbited successfully, with the last craft being placed in orbit in January 1989. The remaining two satellites were destroyed during launch failures, one involving an Atlas-Centaur rocket and the other an Ariane launcher.

INTELSAT 6, the next in the series, returned to the traditional spinning design for a series of five satellites to be used in the busy Atlantic service. These huge craft, 11.7m high and weighing 2 500kg, can relay up to 24 000 telephone lines plus three television channels. The first in the series was launched by an Ariane 44L on 27 October 1989 and went into full operation in April 1990. On 14 March 1990 the second in the series was lifted into orbit by a commercial Titan rocket but, due to a wiring error in the second stage, the satellite failed to separate from the rocket and was stranded in the wrong orbit. Discussions between NASA and INTELSAT led to a rescue mission in May 1992 during which the Space Shuttle Endeavour rendezvoused with the satellite and attached a motor to boost the satellite into its originally planned orbit. The fifth and final INTELSAT 6 was launched on 29 October 1991 by an Ariane 44L rocket.

The next generation of INTELSATs, the INTELSAT 7 series, is already being designed and these will replace the aging satellites currently serving the Pacific and then other oceans. A contract for five satellites was placed with Ford Aerospace (who also built the INTELSAT 5 series) in 1988. If all goes well it is likely that many more than five will eventually be built and the contract includes an option for up to another nine similar craft. Like its earlier design, Ford has opted for a three-axis stabilized platform with two large solar wings. Initial craft in the series will carry three solar panels in each wing but the design allows for another panel to be added to each side if extra power is required as the series develops. Each INTELSAT 7 satellite will be designed to operate for about 11 years, but will carry sufficient fuel to operate for up to 15 years if the satellite exceeds it design life. The INTELSAT 7 series will be able to relay 18 000 telephone lines and three television channels.

As well as ordering craft in the INTELSAT 7 series, the INTELSAT organization decided that it needed to obtain an additional satellite to serve the North Atlantic region and Latin America. Accordingly, in 1989, they placed an order for a partially completed satellite originally intended for another company. This satellite, which would have been known as SATCOM K-4, was modified to meet INTELSAT's requirements and became known as INTELSAT-K. It is due to be launched in early 1992.

A 3.6m-diameter INTELSAT 6 satellite in geostationary orbit over the Atlantic Ocean

Interstellar travel

Interstellar ark proposed 1929

Arkships, hibernation and time dilation

Interstellar travel is where science meets science fiction, and sometimes the borderline is none too clear.

When considering interstellar travel it is necessary to come to terms with the scale of the Universe and with some of the consequences of Einstein's theory of relativity. Our nearest stellar neighbour, the triple star Alpha Centauri, is 42 000 000 000 000km away and one of the fastest of mankind's spacecraft, the automatic Pioneer 10 (which left the solar system at about 40 000km per hour) could not get there in less than 110 000 years. Light, which travels at 300 000km per second, takes more than four years to cross this gulf and so astronomers sometimes refer to the distance to Alpha Centauri as about four light years (1 light year is about 9 460 000 000 000km). There are about 60 stars within 20 light years of the Sun, mostly dull red stars unlikely to have planets suitable for human life. These nearby neighbours represent only a tiny fraction of the 100 000 000 000 stars in our galaxy, which is a system of stars about 100 000 light years across. The nearest major galaxy, the great spiral nebula in Andromeda, is about 2 500 000 light years away.

The scale of the Universe is not the only obstacle facing interstellar travellers. Technology has developed means of crossing great distances and while the Pacific Ocean would have seemed uncrossable to a caveman, today jets fly over it regularly. Unfortunately, crossing interstellar distances presents problems which no foreseeable technology can overcome; Einstein's theory of relativity has shown that it is impossible to travel at the speed of light. What is more, even if new technologies produce a craft that can approach the speed of light, then relativity predicts that strange things will start to happen to time itself. Einstein's equations show that, as seen by an observer back on Earth, time on a speeding spaceship will pass more slowly as perceived by the crew, an effect known as time dilation. This means that the crew of a spaceship travelling at almost the speed of light will be able to visit another star, but when they return their friends will have aged much more than they have and may have died of old age.

Of course, just as scientists once 'proved' that craft heavier than air could not fly, relativity theory might, one day, be found to be wanting. Science fiction writers often use shortcuts through other dimensions, space warps and the like, to move their characters around the galaxy. We cannot say they are wrong, but serious consideration of interstellar travel must remain within the bounds of physics as we understand it today.

Three solutions

Three different strategies for interstellar journeys have been suggested: arkships, suspended animation, and small, near light-speed spaceships.

Arkships are huge, self-sustaining spaceships which would be designed to carry thousands of travellers on a journey that would last centuries. As the journey progressed, the original crew would pass on knowledge of the ship and their mission to their children and only the far descendants of the original crew would live to see their destination. Possibly the first to consider this idea was the British physicist J D Bernal who used it in his book *The World, The Flesh and the Devil* in 1929. The technical problems of building such a vessel would be huge—Bernal suggested using a hollowed-out asteroid as a starship — but these pale into insignificance compared with the social, moral and philosophical problems of sending a large human population on a mission that none of them can live to see finished.

Suspended animation is another trick suggested to allow humans to undertake long space voyages. If drugs and refrigeration techniques could be developed that would slow the rate of human metabolism so that a human being aged only a day every year, then a 100-year voyage would not seem unduly long. It is possible to imagine that, having left the solar system and set course for another star, the astronauts might leave their spaceship under the control of computers and hibernate until it was time to emerge at their destination. As yet, no medical techniques able to induce such hibernation have been found, and there are serious concerns that the damage caused to individual cells by freezing and then thawing a human being might make revival impossible. An alternative, which demands even greater technological and philosophical advances, is to send frozen embryos across space, and incubate them in an artificial womb as the destination approaches.

The third, and probably most palatable alternative, is to accept that there will be explorers who are prepared to face the consequences of time dilation and to attempt journeys at speeds close to the speed of light. These astronauts will require new propulsion systems that far exceed those available today, but such systems are being considered already, even if the engineering reality has not yet caught up with the theoretical calculations.

New propulsion systems

One of the most promising propulsion systems suggested for starships is the nuclear pulse jet. This stems from Project Orion, a concept developed in the USA in the 1950s by Theodore Taylor and Freeman Dyson. Orion-type spaceships consist of a manned section attached by a system of shock absorbers to a giant pusher plate. Atomic bombs are exploded just behind the pusher plate every few seconds and the energy of the blast, smoothed out by the shock absorbers, pushes the craft forward. Tests of a small-scale craft powered by chemical explosives showed that the idea was feasible, but development was stopped because of lack of interest and by the signing of treaties banning nuclear explosions in the atmosphere or in space. Dyson's final design, proposed in 1968, was for a *Queen Mary*-sized starship launched on a 130-year journey to Alpha Centauri by 300 000 1-megaton bombs. A modified nuclear pulse system was proposed for the Daedalus starship study in 1978 (see p41).

Another idea, proposed by American Robert Brussard in 1960, was based on using interstellar hydrogen as fuel. The Brussard ram jet would use a nuclear fusion powerplant that would collect its fuel by using magnetic fields to funnel interstellar gas down into the engine. Unfortunately, this ingenious idea has been proved impractical because the density of interstellar hydrogen is so low that it would not be possible to collect enough without a magnetic scoop millions of kilometres in diameter.

A very efficient way of generating power for a starship might be the controlled annihilation of matter and anti-matter. This is not science fiction; individual atomic particles of anti-matter can be made in nuclear research laboratories and they destroy themselves, producing relatively large amounts of energy, when they encounter a particle of ordinary matter. So much energy can be produced in this way that only a few tonnes of anti-matter would be required to power a starship. The problem, yet unsolved, would be how to make and store the material without it coming into contact with ordinary matter before the crew were ready to use it.

All of these ideas, together with others such as using Earth-orbiting laser beams to beam energy at a sail on a starship, are far from being realizable yet. Even so, today's astronauts may be no further from starships than Leonardo da Vinci was from helicopters, a design concept that he sketched about 450 years before one was ever built.

Israel in space

Ofeq 1 12 September 1988

Middle Eastern space power

With the launch of Ofeq 1, Israel became the eighth nation to orbit a satellite using its own national launcher.

The Israeli space programme is managed by the Israeli Space Agency (ISA) which was founded in 1983. The role of ISA is to coordinate research into space studies within Israel and to encourage the development of space-related products by the country's industries.

The first Israeli satellite was Ofeq 1, which was launched from a site south of Tel Aviv using a rocket known as the Shavit. The Shavit is produced by Israel Aircraft Industries and is based on the Jericho medium-range ballistic missile. The launch was unusual in that, to avoid the political complications of flying over Arab countries to the east of Israel, the rocket was launched towards the west. This direction is opposite to the direction of the Earth's rotation and meant that instead of getting a small boost from centrifugal force, the rocket had to work against it, reducing the payload which could be carried.

Ofeq 1 was 2.3m long, basically cylindrical in shape and weighed 156kg. Power was provided by solar cells arranged all over the outside of the body. The satellite was placed in a low elliptical orbit and remained in space for 118 days before re-entering due to atmospheric drag. The purpose of the flight was to collect data on the space environment and to test such equipment as the solar cells and radio transmission systems.

Ofeq 1 was followed by Ofeq 2 on 3 April 1990. Although similar in size and shape to its predecessor, Ofeq 2 was able to respond to commands transmitted to the satellite from a ground station in Israel. The orbit of Ofeq 2 ranged from 210km to 1 500km and the satellite re-entered on 9 July 1990.

Future plans

It is expected that following the success of the first Ofeq satellites Israel will soon launch at least one scientific satellite. The satellite is likely to be relatively small and to have an intended life of about two years. One experiment which might be carried on the proposed satellite is an ultraviolet telescope weighing about 20kg. The satellite is tentatively known as the National Scientific Satellite or NSS.

Israel is also interested in developing a small geostationary communications satellite which would be built in Israel, but would have considerable foreign involvement. The satellite, which is expected to be called Amos, would form the basis of a regional system serving the Middle East, North Africa and the Mediterranean. The launch of such a satellite would be completely beyond the capabilities of the Shavit rocket and so Israel would need to arrange a commercial launch using a rocket such as Ariane.

Israel is naturally wary about discussing any plans which it might have for the development of military satellites. However, Israel's military and political leaders are known to be unhappy about depending on the USA to supply them with information from US spy satellites. It seems certain that Israel will attempt to launch a reconnaissance satellite as soon as it has developed the appropriate technology.

Ofeq is the Hebrew word for 'horizon' and *Shavit* means 'comet'.

Japan in Space

Land of the rising star

Osumi 11 February 1970

Japan has an active and highly successful space programme, in which scientific and technological missions are segregated.

Japan operates two national launch vehicle programmes and has orbited several dozen satellites. It is involved in manned spaceflight in collaboration with both the USSR and the USA and is providing an experimental module for the international space station Freedom. Programmes devoted to space technology are managed by NASDA, the National Space Development Agency of Japan, which was founded in 1969 (the National Space Development Centre, founded in 1964, was incorporated into NASDA) and consumes the lion's share of the Japanese space budget. Scientific missions are controlled by ISAS, the Institute of Space and Astronautical Sciences, which is an offshoot of the University of Tokyo.

NASDA launch vehicles

Rather than attempt to develop its own space launch vehicle from scratch, NASDA decided to import US technology and to base its first rocket, the N-1, on the US Thor-Delta rocket. The first and third stages of the Thor-Delta were built under licence in Japan and combined with a locally developed second stage to produce the N-1 rocket. The N-1 was a three-stage vehicle, 32.6m high, which could place 135kg into geostationary orbit. The first N-1 launch was the Kiku Engineering Test Satellite in 1975 and this was followed by six other launches. An improved version, the N-2 was in service from 1981 to 1987. The N-2 had a lengthened first stage and a new second stage and could place up to 350kg into geostationary orbit. There were eight successful N-2 launches.

The N-2 was replaced by the H-1 rocket which was able to place 550kg into geostationary orbit. The H-1 used the first stage of the N-2, but had a second stage which used liquid oxygen and liquid hydrogen propellants and a solid-fuelled third stage. The first H-1 launch, using only the first two stages, was on 13 August 1986 when a 685kg experimental geodetic satellite and a Japanese amateur radio satellite were placed in an orbit 1500km high. The H-1 rocket is now in regular service.

NASDA is now developing a new rocket, the H-2, which will be of entirely Japanese manufacture. The H-2 will be a two-stage rocket which will use liquid oxygen and liquid hydrogen propellants for both stages. Two large solid rocket boosters will be used to provide extra thrust at lift-off. The first flight of the H-2 is expected in 1993, but may be delayed due to a series of problems with the development of the LE-5 rocket motor used in the first stage.

NASDA satellites

The NASDA programme is concerned with the development of applications satellites, mostly in the areas of communications and meteorology. Satellites in the Engineering Test Satellite (ETS) series are launched from time to time to demonstrate new techniques and test new equipment. The first ETS launch was on 9 September 1975 and ETS 6 will be used to verify the performance of the first H-2 rocket.

The first step in developing communications satellites came with the launch, by a US Delta rocket on 15 December 1977, of an experimental geostationary satellite called CS-1 (Sakura). The next two satellites, Ayume 1 and 2, were both launched by Japanese N-1 rockets, but both failed to reach the correct orbit. The programme regained momentum after 1983 when two CS-2 (Sakura 2) communications satellites were placed in regular service relaying telephone and telex lines throughout Japan. The CS-2 satellites were superseded by the CS-3 series. CS-3a was launched on 19 February 1988 and CS-3b followed on 16 September 1988.

NASDA have also developed satellites for direct broadcasting, beginning with the experimental BSE satellite (Yuri) launched on 8 August 1978. This was followed by the operational BS-2 series, the first of which was launched in 1984. BS-2a suffered technical problems and BS-2b was launched in 1986 to replace it. BS-2b and the new generation BS-3a, which ran into difficulties soon after its launch on 28 August 1990, were replaced by BS-3b which was launched by an H-1 rocket on 25 August 1991.

JCSat 1, launched on 6 March 1989, and JCSat 2, launched 1 January 1990 form part of a commercial venture unconnected with the NASDA programme. They are used to supply telephone, television, facsimile and data transfer services for Japanese businesses.

Japanese Earth observation missions began with the launch of a Geostationary Meteorological Satellite (GMS-1 or Himawari) by a US Delta rocket on 14 July 1977. GMS-1 was manufactured in the USA by the Hughes Aircraft Company which acted as a subcontractor to the Nippon Electric Company. Since 1977 several other satellites in the GMS series have been orbited using Japanese launchers. The latest, GMS-4, was launched by an H-1 rocket on 5 September 1989. The Maritime Observation Satellite MOS-1, which was intended to develop Japanese technology in the area of remote sensing was launched by the last N-2 rocket on 19 February 1987. An identical MOS satellite, MOS-1b, was launched on 7 February 1990 by an H-1 rocket. The Japanese Earth Resources Satellite, JERS-1, was launched on 11 February 1992.

Manned spaceflight

Japan expressed an interest in taking part in a US Space Shuttle mission as early as 1980 and it was intended that a Japanese astronaut would fly on a Spacelab mission (Spacelab-J) in the late 1980's. However, the disruption to the Space Shuttle programme caused by the explosion of the Challenger delayed the mission until 1992 and the first Japanese in space was a 48-year-old journalist, Toyohiro Akiyama, who flew to the Mir space station in December 1990.

Despite the delay to the Spacelab-J flight, Japan has committed itself to providing a module for the international space station Freedom. The main Japanese Experiment Module (JEM) will be a 10m long, 4m-diameter pressurized module packed with equipment. One end of the JEM will be attached to the space station and the other end will feature a pallet on which experiments can be exposed directly to the vacuum of space. A robot arm on the pressurized module will be used to service the equipment on the pallet.

Japan has also expressed interest in developing a small reusable spaceplane called HOPE (H-2 Orbiting Plane). This 18m long unmanned craft may be launched at about the turn of the century.

The ISAS programme

The ISAS programme of scientific satellites are launched by solid-fuelled rockets which were developed from small rockets originally used for atmospheric research. Japan's first satellite (Oshumi) was launched by a four-stage rocket known as the L (Lambda) 4S before the M (or Mu) series came into use. The M-4S was a four-stage rocket, but later versions replaced the top two stages with a single, more powerful one. The M-3SII, which is the latest of the ISAS rockets, is 27.8m tall and uses two strap-on boosters to augment the first stage. The M-3SII can lift about 770kg into low Earth orbit. ISAS rockets are launched from the Kagoshima Space Centre which lies on the east coast of the Ohsumi peninsula in the south of Japan.

To prevent any conflict with the NASDA programme, the diameter of ISAS rockets is limited to 1.41m. This restriction has forced ISAS to concentrate on relatively small satellites and yet has produced a lively scientific programme. Since the satellites must be small, they can be developed and launched quite regularly, avoiding the enormous delays and cost escalations that have plagued

NASA space programmes like the Hubble Space Telescope. On average, ISAS launches a scientific satellite about every two years.

The first and second ISAS missions, launched on 11 February 1970 and 16 February 1971, were devoted to engineering tests of the ISAS launcher and of satellite technology. Test satellites are still launched from time to time, but most of ISAS's satellites now have specific scientific objectives.

ISAS has tended to concentrate on a few specific scientific disciplines and its first satellites, launched between 1971 and 1978, were all devoted to studies of the Earth's ionosphere, magnetosphere and radiation belts. Since then, ISAS has branched out into X-ray astronomy and solar physics, as well as continuing its studies of near-Earth space. Launches tend to alternate between scientific disciplines, so an X-ray satellite might be followed by one devoted to solar physics and so on. In 1985 ISAS launched its first deep space mission when two probes, a test vehicle and a scientific spacecraft, were sent towards Halley's comet. Another ambitious mission was launched on 24 January 1990 to gain experience in using what are called 'gravity assists' to send spacecraft from one orbit to another without using any fuel. As part of this project a small satellite called Hagoromo was released from a larger craft called Hiten and sent into lunar orbit. In early 1992, the Hiten spacecraft was also placed in lunar orbit. The Hiten/Hagoromo flight was the first mission to the Moon for 14 years. A solar physics mission, Solar-A, was launched on 30 August 1991.

Future ISAS missions are expected to include a fourth X-ray astronomy mission, an infrared or submillimetre satellite, an orbiting radio telescope and GEOTAIL, a project to explore the Earth's geomagnetic tail which will employ the gravity assist techniques developed during the Hiten mission.

> Traditionally each Japanese satellite has a technical designation during its development and is renamed once in orbit. Here are some examples: CS-1 = Sakura (Cherry Blossom) ASTRO-B = Tenma (Pegasus) PLANET-A = Suisei (Comet) MUSES-A = Hiten (A heavenly maiden) SOLAR-A = Yoko (Sunlight)

> NASDA rockets are launched from the Tanegashima space centre, but this site has launch restrictions placed on it by the powerful Japanese fishing industry. Discarded rocket stages and other debris from rockets launched from Tanegashima fall into rich fishing grounds to the east, so launches are restricted to the months of February and August to avoid disrupting the fishermen.

A CS-2 commnications satellite being prepared for launch

Kourou

From the stone age to the space age

Diamant B 1970, Ariane 1 1979

Kourou is the launch site for the European Ariane rocket and one of the world's most important space centres.

When France agreed to give up its launch site at Hammaguir in the Sahara, the French space agency CNES (Centre National d'Etudies Spatiales) was forced to look for a new base for its rocket programme. They chose a spot in French Guiana near a fishing village called Kourou and not far from the old convict settlement of Devil's Island. The location of Kourou, on the eastern coast of South America and only 5.3° north of the equator, has two important advantages as a launch site. Firstly, the open sea to the east and north means that jettisoned rocket stages can fall harmlessly into the Atlantic Ocean, allowing efficient flightpaths to be used for launches into both polar and equatorial orbits. Secondly, rockets launched eastwards from Kourou get a free boost from the Earth's rotation, increasing the payload that they can lift into orbit.

The Guiana Space Centre (French acronym CSG) was used to launch nine French Diamant B rockets (including two failures) between 1970 and 1975 before attention shifted to Kourou as a launch site for the Europa 2 rocket being built by the European Launcher Development Organisation (ELDO). Europa 2 was cancelled and replaced by Ariane when the European Space Agency (ESA) came into being in 1975 and the first Ariane launch from Kourou was on Christmas Eve 1979.

The space centre comprises two main parts. There is a technical centre near Pariacabo, about 5km west of Kourou town. This provides facilities for satellite preparation and checkout and for the administration of the centre. The launch complexes themselves are about 12km further to the north-west.

The first Ariane rockets were assembled on the launch pad, which was known as the Ensemble de Lancement Ariane (ELA) 1. To protect the rocket and the engineering team during the integration of the rocket and its payload, the pad was enclosed by a protective building which could be removed before the launch. ELA-1 was used to launch early versions of the Ariane, but was closed in 1989 when the larger and more powerful Ariane 4 series came into regular use.

The second launch complex (ELA-2) adopted a different approach more akin to the mobile launcher concept used at the Soviet Baikanour cosmodrome. At ELA-2 the Ariane rocket is assembled in a special building 950m from the launch pad itself. Once assembly is complete, the rocket is moved to the launch pad on a trolley that slides on rails. Final preparations, including fitting the payload, takes place at the pad, leaving space in the assembly building for work on the next rocket.

A third launch complex, ELA-3, will be used for launches of the Ariane 5 rocket which is due to enter service in the mid 1990s. ELA-3 will also be able to support the European spaceplane Hermes which will be launched by Ariane 5.

> One of the drawbacks of Kourou is the proximity of the jungle with its population of insects. A typical problem are the builder flies which make nests in the rockets, blocking up the air vents. To prevent this, the launch teams constantly blow air out from the launcher during its preparation.

Lifting bodies

Test phase 1963–1974, orbital flight 2000?

Pathfinders for the US Space Shuttle

Descending at 5 000m a minute, the M2-F2 was the fastest sinking aircraft ever flown.

Lifting bodies are wingless vehicles that obtain aerodynamic lift by virtue of their shape. A number of such craft were flown in the 1960s and 70s during a programme to investigate manned spacecraft designs which could glide to controlled landings rather than parachuting into the ocean. The lifting body concept was potentially valuable because it eliminated many of the problems associated with the design of wings able to withstand re-entry heating. They did, however, present numerous challenges in developing a shape which could be controlled over the enormous range of speeds involved in a vehicle returning from space and landing on a runway. In the end, the USA did not develop a small manned spaceplane and NASA chose a more-or-less conventional winged layout for the Space Shuttle. None the less, the lifting body programme produced useful data on the behaviour of these unusual craft and this may well be valuable for the future development of spaceplanes.

The US Air Force programme

The US Air Force began its research programme with the unmanned START (Spacecraft Technology and Advanced Re-Entry Test) programme in 1961. This was followed by a more advanced programme called PRIME (Precision Recovery Including Manoeuvring Entry) in which a craft known as the SV-5D (also called the X-23A) was developed. The SV-5D was a remotely controlled triangular wedge-shaped lifting body with a rounded top and a flat bottom. It was 2m long and 1m across at its rear, where two vertical fins provided stability. Three missions were flown, each with a test vehicle lofted into a sub-orbital trajectory by an Atlas rocket and then flown through a controlled re-entry. The first two flights (on 21 December 1966 and 5 March 1967) were successful, with control being exercised over the vehicle at hypersonic speeds, but in each case the craft was lost when recovery efforts failed. The third mission, on 18 April 1967, ended with the lifting body being recovered and this allowed a detailed study to be made of the effects of the flight on the craft's experimental heatshield.

The US Air Force, in conjunction with NASA, went on to develop the manned X-24A lifting body, which was used to explore the low-speed flight characteristics of the shape tested in the PRIME programme. This 7.5m long, 5 000kg craft was carried into the air under the wing of a B-52 bomber and released at an altitude of about 13 500m. The X-24A could then either simply glide down to Earth or could use its own rocket motor to accelerate to speeds of up to 1 600km per hour or climb to 21 500m to explore the characteristics of the design under different flight conditions. At the conclusion of its brief (2.5-minute) powered flight, the X-24A became a glider again and returned to Edwards Air Force Base in California, landing about 7 minutes after release from the B-52. A total of 28 flights, including 10 glides, were flown between April 1969 and April 1971 before the X-24A was converted to a delta-winged configuration and renamed the X-24B. In this configuration, which included a longer, more pointed nose and wider, more tapered rear surfaces that almost amounted to stubby wings, a further 12 glides and 24 powered flights were made. The X-24B flew for the first time in January 1973 and the programme ended in November 1975.

The NASA programme

The first manned lifting body flight-tested by NASA was known as the M2-F1 and was a simple craft constructed from plywood and tubular steel at a cost of only about $30 000. Like the X-24A, it was wedge-shaped with a rounded bottom and a flat top. Initial tests were made with the M2-F1 pulled along by a car at heights of a few metres and speeds of 200km per hour so that the pilot could experiment with the controls. In later flights, starting in August 1963, the M2-F1 was towed aloft by an R4D (a modified DC-3 Dakota) twin-engined aircraft to altitudes of 3 000m or more. The M2-F1 made over 100 flights and proved that lifting bodies could be controlled during the critical low-speed landing phase. The success of these tests led NASA to place contracts for two heavier lifting bodies known as the M2-F2 and HL-10.

Both craft were about 6.5m long, but the HL-10 was slightly higher and wider than the M2-F2. The two craft were designed to test quite different lifting body concepts: the M2-F2 was flat on the top and rounded on the bottom, the HL-10 was the opposite, flat on the bottom and rounded on top. Like the X-24, both craft were designed to be lifted to an altitude of about 13500m under the wing of a B-52 bomber before being released for either a simple glide flight or a more complex mission in which a rocket engine would be used to reach the speed or altitude required for a specific series of tests. In particular, the engineers and pilots wished to explore the performance of these unconventional craft at transonic speeds, that is, just above and just below the speed of sound, since this is where the most unpredictable behaviour usually occurs.

The first flight of the M2-F2 took place on 12 July 1966 and progressed quite well, developing techniques of energy management that would eventually be used to guide the Space Shuttle in to land. Then in May 1967 the programme suffered a severe setback when, on its sixteenth flight, the M2-F2 developed a rapid rolling motion during its approach to land. Although the pilot, Bruce Peterson, regained control after 11 seconds he found that he was heading at an angle from the normal landing run, and without the runway markings to guide him the M2-F2 struck the ground about half a second before the undercarriage could be fully lowered. The lifting body bounced and rolled over several times before coming to a stop upside down, and Peterson suffered severe facial injuries. However, despite the damage, the M2-F2 was rebuilt with an additional stabilizing fin and renamed the M2-F3. In this form it went on to make a number of additional flights.

The HL-10 programme began on December 1966 with a glide test. Stability problems were encountered during the flight and it took almost a year of wind tunnel testing and studies to identify and cure the problem before the HL-10 was allowed to fly again. The second flight was a complete success and from then on the HL-10 encountered no serious problems. The first successful powered flight of the HL-10 was made on 13 November 1968. Later flights of this vehicle reached altitudes of 27 785m and a top speed of just under 2000km per hour.

The HL-20

Lifting body research received a boost in the 1990s when NASA began to consider a small, eight-seater, spaceplane called the HL-20. The wedge-shaped HL-20 has three tail fins and a wingspan of about 7m. It would be much smaller than the Space Shuttle and could be used for missions to replace crews at a space station. A manned flight by the end of the century might be possible.

> The real-life crash of the M2-F2 featured in the opening credits of the television series, *The Six Million Dollar Man* in which fictional test pilot Steve Austin is seriously injured and his body rebuilt using tiny mechanisms which give him incredible powers.

Luna (Lunik) programme, phase I

First to the Moon

Lunar impact 1959, soft landing 1965, Moon orbit 1966

The unmanned Soviet Luna spacecraft scored many important space firsts.

The Soviet lunar programme began with the launch of Lunik 1, which was also referred to as Mechta (Dream), on 2 January 1959. This 361kg sphere was sent directly towards the Moon from the Soviet launch centre at Baikanour. Lunik 1 was probably intended to hit the Moon, but it missed by about 5500km. Lunik 1 returned data on radiation, magnetic fields and meteorites until its batteries ran down about three days after launch. It was the first spaceprobe to escape from the Earth's gravitation field and go into orbit around the Sun. A similar craft, the 390kg Lunik 2, was launched on 12 September 1959 and, as planned, crashed into the Moon two days later. Lunik 2 was the first man-made object to reach the Moon and it carried emblems of the USSR.

Lunik 3 achieved an even more impressive feat. Launched on 4 October 1959 it swung around the Moon and took photographs of the hidden far side. The film was developed onboard the satellite and the resulting pictures were sent back to Earth by a television-type system. Although crude by later standards, the nine Lunik 3 images were the first glimpse of the far side of the Moon and caused great excitement amongst astronomers and public alike.

The pathfinders

After the success of the first three Soviet probes there was a wait as the USSR developed new techniques necessary to send more complex craft to the Moon. Henceforth, each lunar probe would be placed into a temporary parking orbit around the Earth until another rocket was ignited to send it towards the Moon. This method is more accurate than a direct ascent straight into a lunar trajectory because errors introduced during the launch can be corrected during the second motor firing. There is, of course, also a risk that the motor will not start, marooning the craft in Earth orbit; several Soviet probes suffered this ignominious fate and were described as Cosmos satellites to disguise their true nature.

The precise purpose of this series of probes, which extended to Luna 14, is not known for certain, but it is quite likely that they were related to a planned Soviet manned lunar mission (see p182). Just as Apollo was preceded by a series of robot craft to scout out the lunar terrain, the Luna programme was intended to be the trailblazer for Soviet cosmonauts.

The first of the new series, launched on 4 January 1963, suffered a motor failure in Earth orbit and re-entered the next day. Another attempt broke up during launch and crashed into the Pacific Ocean. The first officially announced attempt, named Luna (rather than Lunik) 4 was launched on 2 April 1963 and passed the Moon a few days later. Although it was claimed at the time that a flyby was all that was intended, Luna 4 was probably the first of a number of unsuccessful attempts to make a soft landing on the Moon. The run of bad luck continued during 1965 when Luna 5 (launched on 9 May) crashed while attempting to land, Luna 6 (launched on 8 June) missed the Moon completely and Luna 7 and Luna 8 (launched on 4 October and 3 December) both crashed into the lunar surface.

The jinx was broken with Luna 9 which was launched on 31 January 1966 and landed in the Ocean of Storms four days later. The landing capsule was a sphere weighing about 100kg which was dropped rather roughly on to the Moon by a cylindrical mother craft which carried the main braking rocket. Its work done, the mother craft was allowed to fall to destruction and attention focused on the fate of the landing capsule bouncing along the surface. Once the sphere had rolled to a stop, four petal-like covers opened and four radio aerials extended, allowing Luna 9 to broadcast news of its landing back to Earth. Later, a television camera returned the first-ever pictures from the surface of the Moon, showing a 360° panorama around the landing site. Luna 13, launched on 21 December 1966, was similar and made a safe landing in the Ocean of Storms. As well as an improved television camera, Luna 13 was equipped with two spring-loaded arms. One of these was used to determine the density of the lunar surface by measuring the effect of the landing capsule's impact on the soil, an obvious piece of information needed by the designers of a lunar landing craft. The other arm probed the chemical composition of the surface. Luna 13's landing capsule operated for six days until its batteries ran down.

In lunar orbit

At the same time as the soft landing programme, a series of probes were placed into orbit around the Moon to make extended observations. The first of these was the 1 500kg Luna 10 which was launched on 31 March 1966. This craft was similar to the Luna 9 mother craft except that it carried a set of scientific instruments for studying conditions in space around the Moon instead of a landing capsule. Luna 10 was the first craft to go into orbit around the Moon and its data showed that the radiation field in lunar orbit would not be harmful to any human who might venture there. Luna 11 (launched 24 August 1966) went into lunar orbit but did not return much data and may have suffered a partial failure of its instruments. Luna 12 (launched on 22 October 1966) was more successful and returned a series of detailed lunar photographs. These may well have been intended for use as part of a manned landing programme, a speculation encouraged by the release of press photographs of cosmonauts examining the pictures as if looking for landing sites. One other important result emerged from Luna 12: the Moon's gravitational field was not completely regular and as a result Luna 12's orbit gradually twisted out of shape. Since a detailed understanding of the Moon's gravitational field is essential for complex rendezvous and docking operations in lunar orbit, it is quite likely that the mission of Luna 14 (launched on 7 April 1968) was to provide more data on these gravitational anomalies.

By the end of this phase of the Luna programme, the USSR had gathered a considerable amount of information about the Moon and, had the development of a new super booster been more promising, they might well have been preparing for a manned flight to the Moon. However, it seems that realizing that the Moon race was as good as lost, they diverted their efforts into a new series of spacecraft able to explore the Moon by remote control.

> The first pictures of the Moon's surface from Luna 9 were scooped by the British newspapers. Astronomers at the Jodrell Bank radio telescope were monitoring transmissions from Luna 9 and realized that the spacecraft was returning a picture. Hastily borrowing a facsimile machine from the Manchester offices of the *Daily Express* newspaper they were able to reconstruct the first pictures as if they were newspaper photographs and release them to the world.

Luna programme, phase II

Unmanned sample return 1970, unmanned Moonrover 1971

Moonscoopers and moonrovers

The Soviet Luna programme evolved to reduce the propaganda impact of losing the Moon race.

Although the details were shrouded in mystery at the time, it is now known that the USSR tried to develop an unmanned craft that could bring a sample of lunar material back to Earth before the USA succeeded in landing men on the Moon. This required a new, and much more complicated lander than used on previous missions and the result was a 5 500kg craft which had the same basic design as the Apollo lunar module. The descent to the Moon was carried out by a landing stage which had four footpads and which served as a platform for an upper, ascent stage. After touchdown, the design called for a remotely controlled drill to be lowered to the ground, dig out a sample and then load the precious lunar dust into a re-entry capsule on top of the ascent stage. This done, the upper stage would blast off directly back to Earth and the re-entry capsule would parachute down into the USSR where helicopter search teams would recover it.

The new heavier craft could only be launched by a version of the USSR's most powerful rocket, the Proton, which had already suffered a number of failures during attempts to send probes to the planets. Sometimes the Proton placed its payload into parking orbit, but its motor then failed to re-ignite to boost the probe away from Earth. This fate befell the first few attempts to launch the new generation of Luna probes and it was not until 13 July 1969 that Luna 15, the first of this series, was able to escape from Earth orbit. Luna 15 was just three days ahead of the American Apollo 11 moonlanding crew and since rumours of an unmanned Soviet 'moonscooper' had been circulating for some time, there was naturally a widespread suspicion that Luna 15 would try to return a sample of lunar dust before the Apollo astronauts. Luna 15 arrived in lunar orbit on 17 July and over the next few days it carried out a number of manoeuvres which moved its orbit lower and lower. It appears to have attempted to land on 21 July, just as the Apollo astronauts prepared for lift-off from the Moon. Something went wrong as Luna 15 approached the Moon and it crashed and was smashed to pieces in the Sea of Crises.

Although it failed to beat the Americans, the Luna 15 design was basically sound and Luna 16, which was launched on 12 September 1970, landed in the Sea of Fertility on 19 September. About an hour after touchdown its remotely controlled drill collected 101gm of lunar dust and placed it in the 0.5m-diameter return capsule. On 21 September the ascent stage was ignited and the return capsule placed on a course that would bring it back to Earth. Three days later the capsule parachuted to a landing in Khazakhstan and, at last, the USSR had its own sample of lunar dust. Luna 18, launched on 2 September 1971, landed on the Moon, but seems to have toppled over because radio contact was lost at the moment of touchdown. Luna 20, launched on 14 February 1972 was aimed at the same spot and was successful, returning another 100gm of lunar soil. The Luna 23, launched on 28 October 1974, landed safely, but it was unable to collect any samples because, for some unknown reason, its drill failed. The final sample return mission, and the last flight in the programme, was Luna 24. This mission, which carried an improved lunar drill, was launched on 9 August 1976 and recovered samples from 2m below the surface of the Sea of Crises.

Lunakhod

After the success of Luna 16 it was natural to expect another sample return mission, but once again, Soviet space designers had a new card to play. Luna 17 was launched on 10 November 1971 and landed in the Bay of Rainbows a week later. On command from Earth, ramps were lowered and a 756kg, remotely controlled roving vehicle called Lunakhod 1 rolled down to the ground. This eight-wheeled vehicle, 2.5m long and 1.6m wide, was shaped rather like a bathtub and had a lid covered with solar cells which was joined to the rear of the vehicle by hinges. During the lunar day, electricity generated by the solar cells was used to power the rover and to charge its batteries. As the two-week lunar night approached the lid was closed to help keep Lunakhod from getting too cold. If all went well the lid reopened and the mission continued when the Sun rose again.

Lunakhod 1 was controlled by a team of five engineers in the USSR who used television cameras to observe the terrain ahead. This ingenious mooncar operated for 321 Earth days, driving a distance of 10.5km around its landing site to take photographs and measure the chemical composition of the lunar soil. After one such excursion it returned to the landing stage, presumably as a test of its navigation system.

A second Lunakhod was launched towards the Moon on 8 January 1973 aboard Luna 21. Lunakhod 2 was 100kg heavier than its predecessor and could travel twice as fast. After a short period during which it relayed images of its landing site near the crater Lemmonier, Lunakhod 2 set off to explore more difficult mountainous terrain to the south. Over the next few weeks the craft covered a distance of 37km and relayed thousands of television pictures. Unfortunately, Lunakhod 2 could not be reactivated at the beginning of its sixth lunar day and was abandoned. There were no other similar missions, which is surprising in view of the success of the first two.

New orbiters

The third element in this series of advanced unmanned Soviet moonprobes was a new lunar orbiter. The first of these, Luna 19, was launched on 28 September 1971. During a mission lasting about a year, Luna 19 returned data from orbit, including a series of televised pictures of an area on the south-east of the Moon. A similar craft, Luna 22, was launched on 29 May 1974 and operated in lunar orbit for about a year and a half. During its mission, Luna 22 changed its orbit occasionally, swooping to within 25km of the surface at times for detailed observations of selected regions.

Lunakhod 1 was the first wheeled vehicle to operate on another world. The Apollo 14 astronauts used a hand-drawn cart in February 1971 and manned lunar roving vehicles were used on the three remaining Apollo missions.

Luna 16 blasts off from the Moon

Lunar Module (LM)

First manned flight March 1969, last manned flight December 1972

The Eagle has wings

The first true spaceship, the ungainly US Lunar Module carried the first men to the Moon.

The need for a spacecraft specifically designed to land on the Moon stemmed from NASA's decision in 1962 to adopt the Lunar Orbit Rendezvous technique (see p161) for the Apollo programme. Since the main Apollo spacecraft would remain in orbit around the Moon, a special lander would be required to carry the astronauts to and from the lunar surface. The resulting vehicle, called a lunar module, was a self-sufficient, two-stage spacecraft weighing about 15 000kg. Since the lunar module did not have to fly in the atmosphere (it was launched inside a special protective adaptor) its design made no concessions to aerodynamics and this produced its distinctive, and arguably beautiful, shape. To save weight, much of the lunar module's structure was very light, and it was sometimes referred to as a 'tissue-paper' spacecraft.

Lunar Module description

When in its landing configuration the lunar module stood about 7m high and was 9.5m across its landing legs. The upper, or ascent, stage consisted of a pressurized crew compartment, equipment areas, fuel tanks and the rocket engine that would lift the craft off the Moon. Not surprisingly, this engine was designed for absolute reliability and to reduce complexity it was decided that it would not be steerable. Since the thrust from the engine had to pass through the centre of gravity of the ascent stage, and the fuel was lighter than the oxidiser, the fuel tank was set further out than the oxidiser tank to keep things in balance and this produced the lunar module's well-known, asymmetric appearance. A variety of antennas for both communications and radar systems were mounted around the top of the ascent stage. During the landing the two astronauts, restrained in position by support harnesses, stood side by side either looking out from large triangular windows or glancing down at an instrument panel between them. There were few concessions to crew comfort in the cabin and in order to sleep while on the Moon the astronauts either curled up on the floor or used hammocks.

The lower, or descent, stage was octagonal with four large and four small sides. At the centre was the single rocket motor which was used to brake the craft so that it fell out of orbit, and was then restarted to allow the astronauts to slow down, hover above the surface to select a landing site and then to make a soft touchdown. Fuel tanks and other systems were clustered around the rocket motor. From each of the small sides a complex landing leg extended, ending in a circular footpad 1.94m in diameter. A probe, 1.7m long, extended below each footpad (except the one below the ladder used by the astronauts) and this triggered a light in the cabin when it struck the ground. At this point the astronauts, whose view was obscured by dust thrown up by the motor blast, stopped the rocket engine and allowed the lunar module to fall to the ground. Shock absorbers in each landing leg softened the landing. The large footpad was chosen before anyone knew how strong the lunar surface might be and the designers, fearing a very soft surface, selected large landing pads. In retrospect, simple skids such as those used on a helicopter would have been sufficient. The long sides of the descent stage contained bays that carried various equipment, including items that would be used on the Moon.

Lunar Module history

The lunar module was built by the Grumman Aircraft Company who had started to study the design problems involved as early as 1960. The first engineering mock-up was displayed in spring 1964 and featured five landing legs, seats for the crew and a circular forward hatch. As the design evolved, the seats and one of the legs were removed and, at the request of the astronauts, the hatch was changed to a square one which was easier to use when wearing a spacesuit with a clumsy backpack. In 1967 a test model was delivered to NASA for testing in a laboratory that could simulate the conditions that would be encountered during a flight to the Moon.

A test article, which radioed data, for example on, vibration to the ground to confirm that a real lunar module would not be damaged during launch, was carried aboard Apollo 4, an unmanned flight of the first Saturn 5 rocket. The first flight-qualified lunar module was launched, unmanned, by a Saturn 1B rocket during the Apollo 5 mission. During this eight-hour flight the Apollo 5 lunar module's propulsion systems were test fired in space under remote control from the ground. Although Apollo 5 was not a complete success, another unmanned test was not thought to be necessary. Further test articles were launched aboard the Saturn 5 rockets which launched Apollo 6 and Apollo 8.

The first manned flight of a lunar module was in March 1969 when the crew of Apollo 9 tested the fragile mooncraft in Earth orbit. In May 1969 Apollo 10 carried out a test in lunar orbit during which two astronauts swooped to within 15km of the surface in a lunar module codenamed Snoopy. The success of these missions led to the Moon landing by the Apollo 11 lunar module 'Eagle' in July 1969. The basic lunar module was used for the next three Apollo missions, including the near disastrous Apollo 13 flight in which the lunar module was used as a lifeboat when the main Apollo spacecraft suffered a major explosion. The last three missions, Apollo 15–17, used a specially modified lunar module version known as the J-series. These carried extra batteries and oxygen supplies so they could remain on the Moon twice as long as the early models. These versions were also equipped to carry the Lunar Roving Vehicle (LRV) (see p102) or Moon buggy. The longer stay on the Moon allowed the astronauts on these missions to carry out a much more detailed programme of exploration.

Several lunar modules were never used. These included Lunar Module 2, which was the vehicle planned for the second unmanned test and Lunar Module 9, which was the last of the early manned versions and was originally planned for use on Apollo 15 before the programme was shortened by budget cuts. Both of these spacecraft and other examples intended for later missions are now on display in US museums.

The original title of the lunar module was Lunar Excursion Module (LEM) but NASA felt the use of the word excursion was frivolous (perhaps suggesting a day at the beach), and it was dropped. However, in spoken English, the craft was still referred to as a 'lem'. The word bug, frequently used by the press, does not seem to have been popular with anyone else

The idea of using a two-stage spacecraft to land on the Moon, and then using the lower stage as a launch pad, probably originated with the moonship designs of the British Interplanetary Society published in 1947.

The Apollo 14 lunar module Antares

Lunar Orbiter

Orbiter 1 10 August 1966, Orbiter 5 1 August 1967

A camera circling the Moon

A series of five unmanned satellites mapped sites for the US Apollo landings and provided a wealth of other scientific data.

As planning for the Apollo programme got underway, scientists realized that they would need to examine large areas of the Moon to find safe landing sites. To do this NASA and the Boeing company developed a 380kg unmanned spacecraft able to go into orbit around the Moon and then transmit high-quality images of potential landing sites back to Earth. The project was given the rather prosaic name of Lunar Orbiter.

The spacecraft had a roughly conical body about 1.75m high, with four almost square solar panels mounted around the base of the cone. A low-gain radio antenna and a dish-shaped high-gain radio antenna projected from opposite sides of the body near the base. The propulsion system used to enter orbit around the Moon, and to adjust the orbit after arrival, was mounted in the upper section. The lower section contained the control, navigation and communications equipment as well as the 68kg photographic system developed by Kodak. The photographic system comprised a combination camera, processing lab and television station. Two cameras, one with a wide-angle lens and one with a telephoto lens, were mounted on an image compensation system which, in order to reduce blurring, allowed for the satellite's motion during each exposure. The photographs were taken on to ordinary film which was then passed through a processor where it was developed and dried. The film then passed on to a video scanner which examined the film line by line and prepared a video signal that could be radioed back to Earth. Stations on the ground recorded this signal and used it to reconstruct the images, resulting in the characteristic 'stripy' appearance of Lunar Orbiter pictures.

A typical mission for one of the first three spacecraft would begin with a launch by an Atlas-Agena D rocket into an Earth parking orbit. The Agena then re-ignited its motor to send the spacecraft towards the Moon. After a flight of about three days the Lunar Orbiter fired its own rocket motor to enter an elliptical orbit of about 200km × 2 000km around the Moon. This orbit was at an angle of about 10° to the lunar equator so that the satellite spent its time over the low latitude regions in which the Apollo missions were expected to land. After a few days the motor was used to lower the orbit until the satellite was dipping to within 40km of the surface so that very high-quality photographs could be taken of selected areas. Once the film in the cameras was finished, and all the images had been returned to Earth, the spacecraft continued to serve a useful function by returning information about the number of micrometeorites and the levels of radiation in lunar orbit. The Lunar Orbiters also gave ground stations invaluable practice in tracking spacecraft in lunar orbit, experience that would be useful for the manned Apollo programme. Once each Lunar Orbiter had outlived its usefulness it was deliberately crashed on to the Moon so that it could not interfere with subsequent missions.

Lunar Orbiter I provided large-area photographic coverage of nine potential Apollo landing sites as well as covering considerable areas of the still largely uncharted far side. The second mission concentrated on images of 13 potential Apollo landing sites as well as developing techniques for taking photographs at oblique angles. The purpose of these images was to assist in developing navigation techniques and in choosing landmarks that could be used by astronauts as they swept low over the Moon during their descent, but they also produced some stunning photographs of the large crater Copernicus. The third mission provided further photographs from which a shortlist of eight Apollo landing sites could be made, and scored a first by returning a picture showing the unmanned Surveyor 1 spacecraft sitting on the Moon where it had landed the year before.

The first three Lunar Orbiters effectively completed the original task of obtaining detailed photographs of potential landing sites, so it was decided to use the last two craft to photograph the entire surface of the Moon. To do this it would be necessary for the satellites to be placed in orbits over the Moon's poles so that they could cover regions at high latitudes not observed by the earlier craft. Lunar orbiter IV was the first spacecraft to go into a lunar polar orbit and before its mission ended it had covered 99% of the Moon's near side and 60% of the far side. Lunar Orbiter V covered most of the gaps left by earlier spacecraft and provided extra pictures of regions for which more detail was required. The satellite was also used to observe a lunar eclipse and for further tests of the Earth tracking stations being prepared for the Apollo programme.

The Lunar Orbiter project, now largely forgotten, was a tremendous success. When the photographic mission of Lunar Orbiter V was completed in August 1967 only 40 months had elapsed since a contract had been placed with Boeing. Five satellites had been flown in a period of just over a year, and every one had succeeded. Maps of the Moon showing detail 10 to 100 times greater than possible from the Earth had been made and anomalies in the Moon's gravitational field had been explored. Twenty-five years later, the images taken by these tiny craft are still the best available photographs of large areas of the Moon.

Lunar Orbiter programme summary

Mission	Launched	Spacecraft crashed	Pictures	Notes
I	10 Aug 1966	29 Oct 1966	211	First US Lunar Orbiter
II	6 Nov 1966	11 Oct 1967	184+	Imaged Ranger 8 site
III	4 Feb 1967	9 Oct 1967	182	Film jammed
IV	4 May 1967	Oct 1967	163	Contact lost July 24
V	1 Aug 1967	31 Jan 1968	213	Used as tracking target

The best images from the Lunar Orbiter series show features as small as 2m in diameter. One image shows the track left in the lunar dust by a boulder that had rolled down a hillside.

An artist's impression of a Lunar Orbiter circling the Moon

Lunar Roving Vehicle (LRV)

LRV 1 1971, LRV 3 1972

A spacecraft on wheels

Described by Apollo 17 commander Eugene Cernan as 'the finest machine I ever had the pleasure to drive', the LRV was a key element of the USA's last three Apollo missions.

During the planning for the Apollo missions it was clear that if the astronauts were to explore the Moon, rather than just land and pick up samples within walking distance of their lunar module, they would require a means of travelling some distance across the surface. Various methods of lunar transportation were studied including rocket-powered hoppers and different types of wheeled vehicles. The simplest of the wheeled vehicles were little more than motorized seats, but the most refined, such as Boeing's MOLAB, were literally mobile laboratories able to support astronauts during a two-week, 400km drive across the Moon. The large, MOLAB-type, rovers were too large to be carried by the Apollo lunar module and would have had to have been delivered to the Moon by a special unmanned carrier vehicle and checked by remote control. This would have been a complex and expensive business, so when it became clear that the Apollo programme would be cut back, NASA opted for the simpler, cheaper, seats-on-wheels variety which used the astronauts' spacesuits to provide life support and did not have any kind of pressurized cabin. After a design competition in which Boeing Aerospace and the Bendix Corporation were finalists, a contract to develop a manned Lunar Roving Vehicle (LRV) was awarded to Boeing in October 1969. In just two years a durable, reliable lunar vehicle that could carry two astronauts, their equipment and any samples they collected, over a wide variety of difficult terrain was developed, tested and driven across the Moon.

The LRV was just over 3m long, almost 2m wide and weighed about 220kg. It could carry more than twice its own weight, rather better than most terrestrial cars which only carry half their weight. It had four wheels, each driven by an individual electric motor, and had a cruising speed of about 13km per hour. The wheels had an aluminium central hub and tyres made of woven piano wire which could absorb the shock of driving over bumps or small rocks. Chevron treads were riveted to the wire mesh to give better traction in the lunar dust. This design saved 40kg compared to conventional wheels with rubber tyres. Fibreglass 'mudguards' were fitted to each wheel to prevent the LRV being covered in dust as it drove along. Power for the electric motors and other electronics was supplied by two, 36-volt, silver zinc batteries giving a total range of about 90km. Either battery was sufficient to power the LRV, but two were provided for safety reasons. As an additional precaution, the drives across the Moon were planned in such a way that should the LRV fail, the astronauts always had sufficient oxygen in their backpacks to walk back to the safety of the lunar module.

The two astronauts sat side by side with a control console between them. The console housed instruments to indicate the heading and distance travelled, as well as the direction and distance back towards the lunar module and a number of switches and displays connected with the electrical power system. The central console also supported a T-shaped hand controller, which combined the function of steering wheel, brake and accelerator. Pushing the controller forward moved the LRV forward, the further forward it was pushed the faster the

LRV went. Moving the controller left or right sent the LRV into a turn, which could be very tight since both front and rear wheels pivoted, and pulling it backwards (after pressing a safety switch) put the vehicle into reverse. The LRV could be driven by a single astronaut in either seat. Velcro strips, which stuck to strips on the astronauts backpacks, and seat belts were provided to prevent the astronaut parting company with the LRV in the low lunar gravity.

Since the Moon has no magnetic field a normal compass would have been useless and so the LRV carried its own inertial guidance system. This used a gyroscopic unit to sense direction and odometers which counted the number of turns of the wheels. With this information a small computer used trigonometry to calculate the distance and bearing travelled and provided information to the navigation displays on the instrument console. The LRV had a complex communications system which could transmit information on vehicle performance and navigational data to Earth so that mission controllers could advise the astronauts during their drive across the Moon. There was also a remotely controlled television camera that could broadcast live pictures of the astronauts back to Earth. The television system was not just for publicity purposes, it allowed scientists to describe features that they wanted the astronauts to investigate and permitted the astronauts to show samples as they were collected. This system used an unfurlable, umbrella-like antenna to transmit data. Live television transmissions were only possible when the LRV was stopped; a 16mm movie camera recorded the scene as the LRV bounced across the lunar surface.

The LRV was carried to the Moon in a storage bay in the descent stage of the lunar module. To save space, the frame was hinged so that the forward and rear sections folded over the central portion and the wheels folded inwards, over the chassis. Springs unfolded the vehicle and its wheels so that they locked into position automatically. The LRV could be deployed and activated in a matter of minutes by a single astronaut standing on the Moon and pulling on a series of lanyards. After a quick check of its vital electrical and navigational systems the LRV was then ready to be driven away.

A total of eight development versions of the LRV were built. These included full-scale mock-ups, a unit to check compatibility with the lunar module, two versions to test the deployment mechanism, models for vibration and environmental testing and a heavier, stronger model for use during training sessions on Earth. Three LRVs eventually went to the Moon, on Apollo 15, 16 and 17.

The first use of the LRV clearly demonstrated its advantages. Apollo 15 astronauts Dave Scott and Jim Irwin made three drives around the Hadley Rille area of the Moon, covering a total of 27.4km. This was more than six times the distance covered by the Apollo 14 astronauts a few months earlier. What is more, Scott and Irwin were able to carry more equipment, recover more samples and visit a wider range of scientifically interesting sites than any of their predecessors. Interestingly, when the astronauts prepared for the first drive the front wheel steering was inoperative but the problem apparently cured itself before the next day's drive.

On the Apollo 16 mission, which also featured three drives across the lunar surface, commander John Young spent a few minutes carrying out a series of trials to investigate the performance of the LRV at speed and in tight turns. This was officially an exercise to provide information for the designers of later Moon rovers, but also provided a chance to let off steam and produced some spectacular film footage. The Apollo 17 LRV drove up slopes of up to 25° and covered a total distance of 36km during its three excursions. The second of these, which covered 20km, was the longest sortie of any LRV.

> Although the spectacular cine film of the lunar speed trial carried out during the Apollo 16 mission is often featured on television, the lunar speed record was set by Eugene Cernan and Harrison Schmitt when they reached almost 18km per hour during one of their regular drives.

> The lunar 'mudguards' were not very robust and one of the Apollo 17 LRV mudguards broke. The astronauts repaired it using old maps and sticky tape.

Apollo 15 astronaut Jim Irwin walks away from the LRV which is parked close to Hadley Rille

Magellan

Launched 4 May 1989

Mapping the surface of Venus

Although not the first radar mission to Venus, the US Magellan spacecraft provided more detailed coverage of a larger area of the planet than ever before.

Astronomers have known for hundreds of years that Venus, a planet almost the same size as Earth, hides its surface behind a perpetual veil of cloud. To study the planet's surface requires the use of radar systems that can penetrate the clouds and provide data on the nature of the ground below. Radar observations have been made from Earth using radio telescopes, but much better results can be obtained from spacecraft orbiting the planet itself. The first such radar mission was the US Pioneer Venus Orbiter (see p152) which was launched on 20 May 1978. This spacecraft carried a small radar altimeter system that provided a general impression of the large-scale structure of Venus's surface, and found that the planet had three raised continents and perhaps some large volcanoes. The Pioneer radar could not, however, produce the photograph-like images required for detailed study of Venus's geological history. The Soviet Venera 15 and 16 craft, which were equipped with more powerful systems, returned detailed radar images of small regions of the planet in 1983, but the most significant radar mission to Venus was the US Magellan probe.

Like other recent US planetary missions, Magellan has undergone several changes during its development. As soon as the Pioneer Venus mission was underway, NASA began thinking about a follow-up project called the Venus Orbiting Imaging Radar (VOIR — the French verb meaning to see). VOIR was to use the technique of Synthetic Aperture Radar (SAR) to produce images which would show details as small as 100m across, a resolution a thousand times better than Pioneer. VOIR would also carry a package of six instruments designed to study Venus's atmosphere from orbit. The VOIR mission, which even in 1982 was expected to cost $360 million, was considered too expensive and was cancelled. Undeterred by the fate of the VOIR, scientists soon made a new proposal for a smaller spacecraft called the Venus Radar Mapper (VRM) which carried the SAR radar, but no other instruments. To save costs, the VRM could be assembled using spare parts left over from the Voyager, Galileo and Ulysses spacecraft. In this way, many of the scientific objectives of the VOIR mission could be achieved for about half the original cost and the VRM was authorized in the 1984 NASA budget. In 1986 the project was renamed Magellan.

The spacecraft

Magellan was designed and built by Martin-Marietta Astronautics of Denver, Colorado. The spacecraft stands 6m tall, 4.5m across, weighs over 3 000kg and comprises four main sections. At the top is the 3.5m-diameter dish-shaped antenna, a spare from the Voyager project, which serves both for communications with Earth and as the radar dish. The next section is a box containing the radar electronics, the reaction wheels used to point the spacecraft and various other sub-systems. Below this is a 10-sided spacecraft bus which supports two solar panels, and houses the star trackers and the command and data systems. The solar panels, which have a total area of 12.6sq m, generate 1200 watts to operate the spacecraft and to charge the two nickel cadmium batteries that provide electrical power when the solar panels are not illuminated. The final section is the solid-fuel Star 48 rocket motor that is fired to slow Magellan so that it can be captured into orbit around Venus. Once this is done, the spent motor casing is jettisoned. Any other manoeuvres, such as those to make small orbital adjustments, are carried out using some of the 24 small thrusters in the spacecraft bus.

The mission

Magellan was scheduled for launch in 1988, but was delayed for a year by the explosion of the Space Shuttle Challenger in 1986. The changes in safety rules after the Challenger accident also meant that the Centaur rocket, originally planned to boost Magellan from Earth orbit to Venus in just four months, was cancelled and replaced by a less powerful Inertial Upper Stage (IUS) booster. Furthermore, because of the need to use the Space Shuttle to launch Galileo to Jupiter during October 1989 when Magellan would otherwise have been launched, Magellan had to be sent on its way earlier. This meant that Magellan had to adopt a longer route to Venus, circling the Sun one and half times before arriving 15 months after launch. Magellan was successfully deployed from the Space Shuttle Atlantis on 4 May 1989 and after an uneventful flight arrived in Venus orbit on 10 August 1990.

Magellan was placed into a near polar, elliptical (250km × 8 000km) orbit around Venus. In a typical orbit, as Magellan swoops low over the planet, moving from North to South, its radar operates for 37 minutes, mapping a swath below the spacecraft and storing the data on a tape recorder. Each swath is about 20km wide and 17 000km long. Then, as Magellan climbs away from the planet it swings around, pointing its antenna back towards Earth so it can replay the recorded data back to the receivers of NASA's Deep Space Tracking Network. Once this is done, and checks of the spacecraft's pointing direction have been made, Magellan swings around again ready to map another swath of the planet. Each orbit lasts just over three hours and mapping is done on every orbit, except for periods when Venus moves behind the Sun and radio transmissions are restricted because of solar interference.

After a few early problems which led to the sudden loss of radio contact, the spacecraft performed well and in May 1991 completed its primary mission of mapping at least 80% of the planet's surface. The mission then entered an extended phase intended to complete coverage of the remainder of the planet, to provide stereoscopic coverage of selected regions and to re-image certain objects, such as what appeared to be volcanoes, to see if there was any evidence for changes. The extended mission is also intended to reveal subtle details of the planet's gravitation field by precision tracking of Magellan as it circles the planet. This will shed light on the nature of the planet's internal structure.

Venus revealed

The Magellan primary mission went very well, covering 84% of the planet with a resolution good enough to show details as small as 120m in places. Scientists reported finding evidence of extensive vulcanism

Synthetic Aperture Radar

This technique uses computer processing to produce radar images far better than apparently possible with Magellan's 3.5m-diameter dish. For each radar pulse the system measures the strength of the returned signal, the time each pulse takes to return to the spacecraft and any changes in frequency resulting from the movement of Magellan while the pulse went down to the surface and bounced back. In effect the computers treat the data as if it came from a single dish 3.5m wide, but as long as the swath swept out as Magellan moved along its orbit while the radar pulse was travelling. This produces a 'Synthetic Aperture' hundreds of metres long and provides very high-quality radar data. Once the radar data has been returned to Earth it can be processed to produce photograph-like images in which changes of shade and texture reflect not only physical relief but also differences in the nature of the planet's surface.

over much of the planet. Much of the magma released from these volcanoes is basaltic, like volcanic material on Earth, but other forms are also present. Venus also shows signs of explosive vulcanism, but this appears to be less significant than on Earth, probably because the dense atmosphere inhibits such activity. One of the volcanoes, Maat Mons which lies near the equator of Venus, appears to have a dull black appearance. Some scientists interpret this as meaning that the volcano is covered in fresh lava and if this is the case then Maat Mons is probably still volcanically active. Magellan images have also revealed evidence for tectonic deformations in the Venusian crust including rift zones several hundred kilometres wide and thousands of kilometres long.

Venus, like the other terrestrial planets, also has a number of impact craters and studies suggest that there are about a thousand over the entire planet. The smallest craters found during the early part of the mission were about 3km in diameter, probably because the thick atmosphere totally destroys small meteoroids during entry and only relatively large objects penetrate to ground level to form craters. At the other extreme, several basins with diameters in excess of 200km have been discovered. Comparing the number and size of the craters to the craters on the other planets suggests that that much of the surface of Venus is only about 500 million years old.

The images also revealed numerous types of lava flows, including lava rivers hundreds of kilometers long, and a few types of features that defied immediate explanation. One type of terrain which appears unique to Venus comprises a complex pattern of ridges and valleys called 'tessera', which reflects intense folding, faulting, shearing, compression and extension. An example of this type of terrain is found in the Alpha Regio area, 25° south of the planet's equator.

The spacecraft is named after Fernao de Magalhaes, captain and instigator of the first circumnavigation of the world. The actual spacecraft has outperformed its namesake; Magalhaes was killed en route but Magellan survived to complete its mission.

A radar image of the Lavinia region of Venus. The three impact craters are about 35–50km in diameter

Manned flights to Mars

Das Marsprojekt 1952, Ride Report 1987

To the red planet

Proposals for manned missions to Mars have been discussed for almost 50 years.

Although the first fictional voyages to Mars were written in the 1920s, it was not until 1952 that a serious study of a Mars mission was published. In his book *The Mars Project* Wernher von Braun proposed building 10 spaceships, carrying 70 men between them, which would fly to Mars in convoy. Fifty of the astronauts would land on Mars and remain there for over a year before returning to Earth. By 1956 von Braun, in collaboration with space writer Willy Ley, had reduced the scale of the Mars project to just 12 men, and two spaceships: a winged glider that would land on Mars and another which would remain in orbit, perhaps diverting to explore the Martian moons Phobos and Deimos. The first men on Mars would return to the mother ship, which would carry them back to Earth, using a small rocket brought down to the planet as part of their glider. Ley and von Braun described this mission, complete with details of possible orbits and the certain need for a washing machine, in their book *The Exploration of Mars* (1956) which was illustrated by artist Chesley Bonestell.

The idea of travel to Mars received a big boost when the USA decided to send men to the Moon. A variety of studies appeared in the 1960s, including a proposal by Ernst Stulinger, a former German rocket engineer employed by NASA, for a Mars mission using electric propulsion (ion drives) instead of chemical rockets. Unfortunately, lack of investment in the development of ion drives left this proposal stillborn and engineers concentrated on missions using chemical propellants. One example was a study of a Mars Excursion Module (MEM) performed in 1967 by the North American Rockwell Corporation. This vehicle, which was estimated to cost $4–5 billion dollars in 1967, would have supported four men on Mars for 30 days.

The heyday of these second generation Mars missions was probably 1969 when Apollo was landing on the Moon and US Vice President Spiro Agnew was urging a flight to Mars in the 1980s. US industry, seeing its involvement in the Apollo programme finishing, was gearing up for what it hoped would be the next big project, a manned flight to Mars. One scenario envisaged two identical spaceships, each with a crew of six, but able to carry 12 astronauts in an emergency, leaving Earth orbit on 12 November 1981. Each ship would be boosted towards Mars by two nuclear engines which would be separated and returned to Earth orbit once the ships were en route. The two Mars ships, which each had their own nuclear rocket, would go into orbit around Mars on 9 August 1982. Four men from each ship would use a MEM to descend to the surface for a 30-day stay, then use the upper portion of the MEM to return to Martian orbit and a rendezvous with their respective mother ships. On 28 October 1982 the ships would leave Mars orbit to start the journey back to Earth. Since by then the Earth would be almost on the opposite side of the Sun from Mars, the quickest return journey involved a flight that would use a close approach to Venus on 28 February 1983 to deflect the ships back to Earth, where they would arrive on 14 August 1983.

Unfortunately, the cost of the Vietnam war and the perception that, by beating the USSR to the Moon, America had won the space race, conspired against an expensive and dangerous flight to Mars. Development of the nuclear engine programme, essential to the planned Mars mission, was stopped around 1970 and the idea of sending Americans to the red planet faded from view. For the proponents of manned missions to the planets, a bleak decade lay ahead.

The Mars underground

Despite the setbacks of the early 1970s, interest in Mars still flickered in places and various proposed missions appeared from time to time in technical journals. Although there was an almost total lack of official interest, a meeting was organized in the spring of 1981 at Boulder, Colorado in order that interested parties could exchange ideas. As news of the meeting spread, more and more people came forward and an informal grouping calling itself 'The Mars Underground' came into existence. This group provided a means to study and lobby for a journey to Mars and gradually the idea began to gain momentum in official circles. Finally, on 20 July 1989, at a ceremony to celebrate the twentieth anniversary of the Apollo 11 moon landing, President Bush announced that the USA would begin planning for a mission to Mars that could take place in the next century.

Sprint or dawdle?

It is too soon to say what the US Mars mission will look like, or even when it will take place. Unlike President John F Kennedy, George Bush did not set a deadline for his vision, although a landing by about 2010 seems a reasonable estimate. At present, almost every aspect of the programme is open for debate and two opposing strategies have emerged.

The report of a commission chaired by astronaut Sally Ride favoured sprint missions in which a small crew would spend a relatively short time on Mars. This would be achieved by sending an unmanned craft carrying supplies and a MEM to Mars using a slow but fuel-efficient route to the planet. When the cargo ship arrived, a small personnel carrier with a crew of six would make a fast trip to Mars and dock with the cargo ship. The astronauts would transfer to the MEM and land, remaining on Mars for 10–20 days before returning to the orbiting personnel carrier for the journey home. The entire round trip would last about a year and Ride's report suggested that the first mission could take place as early as 2005 if the USA was prepared to triple NASA's budget and begin development soon.

The alternative scenario, described in a paper by Ivan Bekey of NASA headquarters, is more leisurely and allows a gradual build up to a permanent Martian base. The first mission would send a crew of three to explore the Martian moon Phobos and to deposit an empty landing craft on Mars before returning to Earth. The second mission would carry a crew of five who would land close to the module which had arrived earlier and would then spend about a year exploring Mars before returning. The third mission would land on Mars and also visit the moon Phobos, where it would emplace the first part of an automatic facility designed to manufacture rocket propellant, using material extracted from Phobos itself. The next flight, another year or so later, would complete the propellant factory, establishing a facility that could support regular visits to Mars and allow the setting up of a permanent base.

As yet no one knows which route will be taken, nor if various nations might combine to mount such a programme. The USSR often hinted that it had interests in sending astronauts to Mars, and the experience of long duration spaceflights it has gained aboard the Mir station would be invaluable for such a flight. Perhaps the two countries, and the developing space powers of Europe and Japan, will one day mount an international mission to Mars. That would indeed be 'a giant leap for mankind'.

> In 1982 science writer James E Oberg wrote 'If all the different types of Mars spaceships proposed in the last 30 years were actually built and then laid end to end, the resulting contraption might reach half way to the intended destination.'

Manned Manoeuvring Unit (MMU)

Prototype 1973, first flight 1984

A flying armchair

The 'Buck Rogers' flying backpack enabled Space Shuttle astronauts to repair and rescue crippled satellites.

The experiences of spacewalking astronauts during the US Gemini project showed that Extra Vehicular Activity (EVA) appeared easy, but that it was actually very difficult to move to a precise position in space and remain there to complete a particular task. Gemini 4 astronaut Ed White experimented with a hand-held gas gun, but that soon ran out of propellant even though his EVA lasted only 20 minutes.

The first attempt to provide a more sophisticated astronaut propulsion system was scheduled for the Gemini 9 mission. On this flight it was planned that astronaut Eugene Cernan would leave the Gemini cabin and, after carrying out some other tasks, move to the rear of the spacecraft and back himself into a small, rocket-powered, backpack called the Astronaut Manoeuvring Unit (AMU). The AMU, developed by Ling-Tempo-Vought at the direction of the US Air Force, weighed about 75kg and provided the astronaut with life-support systems, communications, telemetry, propulsion and automatic stabilization in all three axes. The propulsion system, which used hydrogen peroxide propellant, had 12 small thrusters at the corners of the backpack. The thrusters were operated by controls in the armrests. Once in the AMU, Cernan was supposed to pull down the armrests, connect himself to the AMU oxygen supply, detach the AMU from the Gemini and fly around using the AMU's propulsion system. Unfortunately, mission planners had underestimated the difficulty of the task and during his attempts to don the AMU Cernan became so hot that his space helmet fogged up, preventing him from seeing what he was doing and forcing him to abandon the experiment. Further tests of the AMU during later missions were cancelled while the problems of spacewalking were reconsidered.

There was no need for an AMU during the Apollo programme, but the concept reappeared during the Skylab space station mission of 1973-4. Skylab had a huge pressurized volume in which an AMU could be tested without the need for a spacesuit. To take advantage of this opportunity a new type of AMU, known as the Automatically Stabilized Manoeuvring Unit or experiment M509, was carried in the Skylab station. Similar in concept to the AMU, but using 14 thrusters powered by pressurized nitrogen instead of hydrogen peroxide (to avoid contaminating the atmosphere of Skylab), the M509 was tested inside Skylab by astronauts Alan Bean and Jack Lousma. M509 proved highly successful and a modified version was ordered for use during Space Shuttle missions.

The MMU

Like M509, the new Manned Manoeuvring Unit (MMU) was built by Martin-Marietta astronautics under a contract placed by the NASA Johnson Space Center in Houston. The MMU, which is about 120cm × 78cm × 114cm in size when the armrests are down, weighs about 150kg. Two tanks each contain 11kg of nitrogen propellant stored under pressure and these supply 24 thrusters (in two independent sets of 12) which each provide 0.75kg of thrust. Normal speed in use (relative to the Space Shuttle) is about 1km per hour. The MMU is carried in a special support station on one side of the Space Shuttle payload bay. The astronaut, wearing his or her own spacesuit, reverses into the MMU which latches automatically on to the astronaut's backpack. He or she then unfolds the armrests and releases two levers which allow the MMU to move away from the support station and fly freely in space.

Astronauts train for the MMU by flying a highly realistic simulator at Martin-Marietta's plant in Denver, Colorado. The simulator has controls that exactly reproduce those on a real MMU, but which pass commands not to thrusters, but to a crane that moves the astronaut about. This device allows astronauts to practice moving up to, and docking with, other spacecraft and was praised by the astronauts who actually flew the MMU in space.

MMU in action

The first flight of the MMU came during the tenth Space Shuttle mission (STS-41B). Launched on 3 February 1984 this mission deployed two satellites, both of which failed to go into the correct orbit, after which testing of the MMU commenced. Two separate MMUs were carried and astronaut Bruce McCandless was the first to try one. After checking the MMU in the payload bay, he moved out on excursions up to 100m from the Space Shuttle, becoming the first untethered human satellite. After flying the MMU for 1 hour 22 minutes McCandless handed it over to astronaut Robert Stewart who flew it for a little over an hour. After a day's rest both astronauts conducted a second series of tests, using the other MMU, and both declared themselves completely satisfied with its performance. This gave NASA confidence to use the MMU for a dramatic satellite rescue attempt on the next Space Shuttle mission.

Launched on 4 April 1984 the STS-41C mission deployed a single satellite and then moved to rendezvous with the Solar Maximum Mission, a scientific satellite crippled soon after launch by technical problems. It was intended that astronaut George Nelson would use an MMU to fly to the ailing satellite, dock with it using a special device and then tow the satellite back to the payload bay for repairs. In the event Nelson was unable to dock with the satellite due to a design error in the docking mechanism, but the MMU again performed flawlessly. The satellite was eventually grappled by the Space Shuttle's robot arm and was repaired successfully the next day.

The most spectacular MMU mission was that of STS-51A, launched on 8 November 1984, which rescued the two errant communications satellites deployed unsuccessfully on STS-41B earlier in the year. Astronaut Joe Allen used an MMU to fly to, and dock with, one of the satellites. He was then able to stabilize it and return it to the Space Shuttle, where it was secured in the payload bay. The next day his companion Dale Gardener repeated the feat with the second failed satellite. Both satellites were returned to Earth and repaired and, after being sold to a new operator, one of the satellites was relaunched by a Chinese rocket in 1990.

The STS-51A mission was the last use of the MMU. A more conservative approach to EVA introduced after the destruction of the Space Shuttle Challenger has curtailed any more MMU flights for the present. However, when construction of the space station Freedom begins, and NASA resumes regular EVAs, the MMU will almost certainly be brought out of its enforced retirement.

The USSR also developed an MMU, but this is known as the SPK or individual space transport vehicle. The Soviets also referred to it as a 'space motorcycle'. It was first flown on 1 February 1990 by cosmonaut Alexander Serebrov during an EVA from the Mir space station. A second flight took place on 5 February when Alexander Vitorenko flew about 45m from the station. During both flights the SPK remained attached to the Mir station by a winch cable in case the space transport vehicle should fail.

Mariner

Mariner 1 1962, Mariner 10 1973

Planetary explorers

The USA's Mariner series of unmanned probes explored Mercury, Venus and Mars.

Shortly after NASA was formed, it was decided that a series of unmanned planetary probes would be developed by the Jet Propulsion Laboratory (JPL) in Pasadena, California. The JPL, which was operated by the California Institute of Technology on behalf of NASA, called this programme Mariner. The Mariner spacecraft was designed to be stabilized in all three axes, that is, it would cruise through space as a stable platform using small thrusters to control its attitude in pitch, roll and yaw. Radio communications with Earth would be directed through a dish-shaped, high-gain antenna that would aim a narrow radio beam back to Earth. A low-gain antenna, which would broadcast a weaker signal in all directions at once, would be carried in case the narrow beam from the high-gain antenna should lose its lock with the receiving station on Earth. By early 1961 it had been decided to develop two versions of the craft: the Mariner A design, which would be sent to Venus in 1962, and Mariner B which would be sent to Mars in 1964. These spacecraft would weigh about 500kg each and would be launched by the then unproven Atlas-Centaur rocket.

First to Venus

It was soon apparent that development of the Centaur stage of the Atlas-Centaur rocket was falling behind schedule and that it would not be ready in time to send a Mariner-A-type mission to Venus in 1962. Since the space race was in full swing, NASA decided that it had to mount a Venus mission at the first opportunity and so authorized the rapid development of a smaller, lighter version of Mariner that could be launched by the less powerful Atlas-Agena rocket. To increase the chances of success, two identical Mariners were to be launched in quick succession and this meant a rapid programme to design, test and deliver spacecraft that could both explore interplanetary space and then fly close enough to Venus to investigate the planet's surface temperature, atmosphere and cloud structure.

The resulting 202kg spacecraft consisted of a hexagonal body 1.5m across to which a dish-shaped, high-gain antenna and two rectangular solar panels were attached. The solar panels, which were folded for launch, had a total span of 5m. Above the main body was a tubular tower-like structure topped by a low-gain antenna. The scientific instruments included an infrared spectrometer (to determine atmospheric composition) and a microwave radiometer (to measure the surface temperature of Venus). The total height of the spacecraft was 3m.

The project had an inauspicious start on 22 July 1962 as the Atlas-Agena B rocket carrying Mariner 1 went off course and was blown up by the range safety officer to prevent it falling on a populated area. After hasty modifications to the guidance equation in its autopilot, a second Atlas-Agena lifted Mariner 2 safely into an Earth parking orbit on 27 August 1962. Soon after, the Agena motor was restarted to send Mariner 2 on its four-month journey to Venus. Throughout its interplanetary cruise Mariner 2 returned data on magnetic fields, cosmic dust and the solar wind before flying past Venus on 14 December 1962. The flyby was a complete success, with Mariner's instruments reporting that Venus was much hotter than ever suspected, over 400°C at the surface, and that the temperature was much the same on both the day and night sides of the planet. Mariner 2 also reported that Venus had no significant magnetic field, a fact attributed to the planet's very slow rotation period (243 Earth days).

First to Mars

The next Mariner mission was in 1964 and this time the objective was the planet Mars. Once again two spacecraft were prepared, and the use of a more powerful Atlas-Agena D launcher meant that they could be 60kg heavier than their predecessors. Mariner 3 and 4 had an octagonal centre body 1.27m across to which four solar panels, which folded for launch, were attached. A high-gain antenna was mounted on the rear of the spacecraft and a single television camera projected from the other end. Other scientific instruments were attached to the spacecraft body. An innovation was the addition of small vanes to the ends of the solar panels. These were intended to perform the function of stabilizing the spacecraft by using the pressure of the solar wind.

The mission of Mariner 3 was short and unsuccessful. After a launch into Earth orbit, Mariner 3 did not achieve the correct velocity when the time came to re-ignite the Agena motor to send the craft to Mars. In addition, the solar panels did not unfold properly and the spacecraft was unable to find its reference star Canopus in order to stabilize itself. It was concluded that the nose shroud, which surrounded the spacecraft to protect it from the atmosphere during launch, had failed to jettison, preventing the solar panels from unfolding and obstructing the view of the star sensors. With no means of generating electricity from its solar cells, Mariner 3 fell silent when its batteries ran down.

After the failure of Mariner 3, technicians worked around the clock to modify the protective nose shroud for its sister craft and Mariner 4 was successfully sent on its way to Mars on 28 November 1964. After a mid-course correction on 5 December, Mariner 4 flew within 900km of Mars on 14 July 1965. During the flyby it was found that Mars lacked any significant magnetic field and that the atmosphere was much thinner than expected, and 21 historic television pictures of the surface were obtained. These images showed craters, but no sign of the fabled Martian canals and no evidence that life might exist on the planet.

Mariner 5

After the success of Mariner 4, NASA decided to modify the reserve Mariner 4 spacecraft and send it to Venus in 1967. To accommodate a flight towards, rather than away from, the Sun, Mariner's solar panels were turned around and shortened and a sunshade was added to protect the central body of the spacecraft. Since Venus was known to have a cloud-covered surface, the television camera was removed and other scientific instruments modified or added. The 245kg Mariner 5 was launched by an Atlas-Agena D on 14 June 1967 and flew past Venus on 19 October. The flyby was successful and generally confirmed the results from Mariner 2.

Return to Mars

Just as the race to the Moon was reaching its climax, Mariner 6 and 7 were launched towards Mars on 24 February and 27 March 1969. The two spacecraft were similar in general appearance to Mariner 4, but used the Atlas-Centaur launcher, which allowed them to be heavier (413kg) and to carry improved computers, better communications systems and more scientific instruments. Mariner 6 and 7 each carried two television cameras, one for wide-angle images and the other to take close-ups of smaller areas. The other scientific instruments included infrared and ultraviolet spectrometers to probe the composition of the Martian atmosphere and an infrared radiometer to measure the temperature of the surface.

The two craft approached Mars in late July and early August, only a few weeks after the first men had walked on the Moon, and returned an enormous amount of data, including a series of spectacular close-up images. These data appeared to confirm the Mariner 4 results that Mars was a dead world with a cold and cratered surface. The south polar cap of Mars was revealed as solid carbon dioxide, not water ice, and the thin atmosphere was found to be 98% carbon dioxide. Interesting geological features unique to Mars were found near the south pole, but the majority of the pictures showed dozens of craters. The overall impression created was that Mars, although having its own personality, was basically similar to the Moon. Amazingly, although the close-up pictures taken by Mariner 6 and 7 covered about 10% of the planet, they missed almost all of its interesting geological features. The true diversity of Mars was not revealed until a spacecraft was placed in orbit around the planet in 1971.

In Martian orbit

The next Mariner mission was planned to go into orbit around Mars so that the entire planet could be studied in detail. The spacecraft design retained the octagonal body and four solar panel design of its predecessors, but the need to add a rocket motor to slow the Mariner craft down so that it could enter Martian orbit increased the weight to over 1000kg. The scientific instruments, including a television camera and spectrometers, were mounted on a movable platform which could be pointed at various regions below the spacecraft without having to reorientate the spacecraft every time.

It was intended that the two spacecraft would enter rather different orbits around Mars and complement each other during a mission that would last 90 days after arrival at Mars. This was, however, not to be: on 9 May 1971, just as Mariner 8 appeared to be on its way to orbit, the upper stage of its Atlas-Centaur launcher failed and crashed into the Atlantic. After a detailed check of the second rocket, Mariner 9 was launched on 30 May without incident.

Mariner 9 arrived at Mars on 13 November 1971 and its motor fired for 15 minutes, capturing the spacecraft into an orbit ranging from 1400 to 18000km around Mars. The first television pictures revealed little detail because the entire planet was covered by a global dust storm that completely hid the surface. Gradually the dust began to settle and one by one a variety of interesting landforms emerged. The Mariner 9 mission was eventually extended until, 349 days after arrival, it ran out of gas for its attitude control system and began to tumble, breaking its radio link with Earth. It remains in orbit around Mars, but will eventually crash on the surface as its orbit decays.

During its mission, Mariner 9 revealed huge Martian volcanoes, a giant valley (subsequently named Vallis Marineris or Mariner Valley) which dwarfs the Earth's Grand Canyon and clear evidence that water had flowed across Mars at some time in the past. The absence of any water on the planet at present was explained in terms of complex global weather patterns lasting about 50000 years in which the water remains frozen below the soil for much of the time. After Mariner 9, Mars emerged as a fascinating world on which life might have developed after all. Interest in the Viking mission (see p248) to land on Mars in 1975 and search for life was revitalized.

Two planet mission

The last of the Mariner missions involved a single spacecraft and two planets. Taking advantage of a planetary alignment that occurred only every 10 years, Mariner 10 was designed to fly past Venus and then to use Venus's gravity to move closer to the Sun and rendezvous with the innermost planet, Mercury. By using this gravity assist manoeuvre Mariner 10 could be heavier than would have been possible if it has been launched directly to Mercury. Furthermore, the experience gained during the gravity assist would be useful in planning the more complicated missions to the outer planets that were then under development.

The 408kg spacecraft, built by the Boeing company, used the familiar layout of a hexagonal body with solar panels and a movable, dish-shaped, high-gain antenna. Television cameras and other instruments were mounted on a steerable platform, while additional instruments were either fixed to the main body or carried on deployable booms. A hydrazine-fuelled rocket motor was available for mid-course corrections. The main body was protected from the Sun's heat by a shade and the spacecraft's two rectangular solar panels were tiltable so that as Mariner 10 approached the orbit of Mercury the panels could be angled to reduce the amount of sunlight falling on them. Although designed to prevent overheating of the panels, this feature was to have an unexpected benefit later in the mission.

Mariner 10 was launched on 3 November 1973 by an Atlas-Centaur rocket and, despite a few technical problems, arrived at Venus on 5 February 1974. During its flyby 5800km above the planet, Mariner 10 returned over 3000 images, including some taken in ultraviolet light which revealed details of the circulation of Venus's clouds. It also returned data on the atmosphere and the planet's weak magnetic field, before settling down to the seven-week cruise to Mercury.

The Mercury flyby on 29 March 1974 was a complete success and Mariner 10 returned a series of images that covered about 40% of the planet's surface. The pictures showed that the surface of Mercury was very similar to that of the Moon, with many thousands of craters and a few, very large, impact basins. The largest of these was the 1 300km-diameter Caloris basin. After the flyby, Mariner 10 continued in its orbit around the Sun. However, this was not the end of the mission because orbital dynamics meant that Mariner 10 circled the Sun in 176 days, twice the length of Mercury's year. This meant that after 176 days the planet and the spacecraft were in the same relative positions again and so on 21 September 1974 Mariner 10 flew over Mercury again, this time passing 48000km above the planet's south pole. By now, attitude control gas was running short, but by tilting the solar panels to different angles, it was possible to improvise a solar sailing technique and keep Mariner stabilized long enough to allow a third visit to the planet on 16 March 1975. This passage took place at very low altitude, only 327km above the night-time surface of Mercury, so that scientists could study the planet's newly discovered magnetic field which the previous results had shown to be unexpectedly powerful. Eight days after the flyby the remaining control gas ran out and contact with Mariner 10 was lost. It was a successful end to a successful series.

The double success of Mariner 6 and 7 was almost thwarted at the last minute. A few hours before Mariner 6 was due to make its closest approach to Mars, contact with its sister craft was lost. Other NASA radio stations swung into action and a weak signal was heard from Mariner 7 just as Mariner 6 reached the climax of its mission. Control of Mariner 7 was soon regained and it was concluded that for some reason, possibly a hit from a small meteorite, the craft had lost its celestial reference and had been pointing its high-gain radio antenna in the wrong direction.

When contact with Mariner 9 was lost, its memory still contained about 50 images of Mars that it had not yet returned to Earth. It is interesting to speculate whether some future astronauts will ever recover the satellite and examine these 50 lost images.

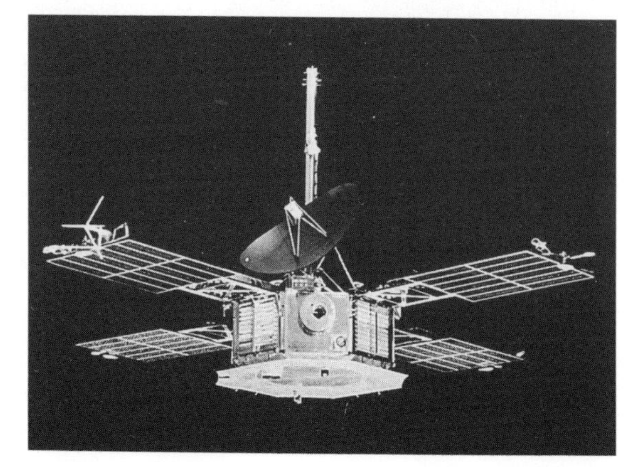

The Mariner 5 spacecraft

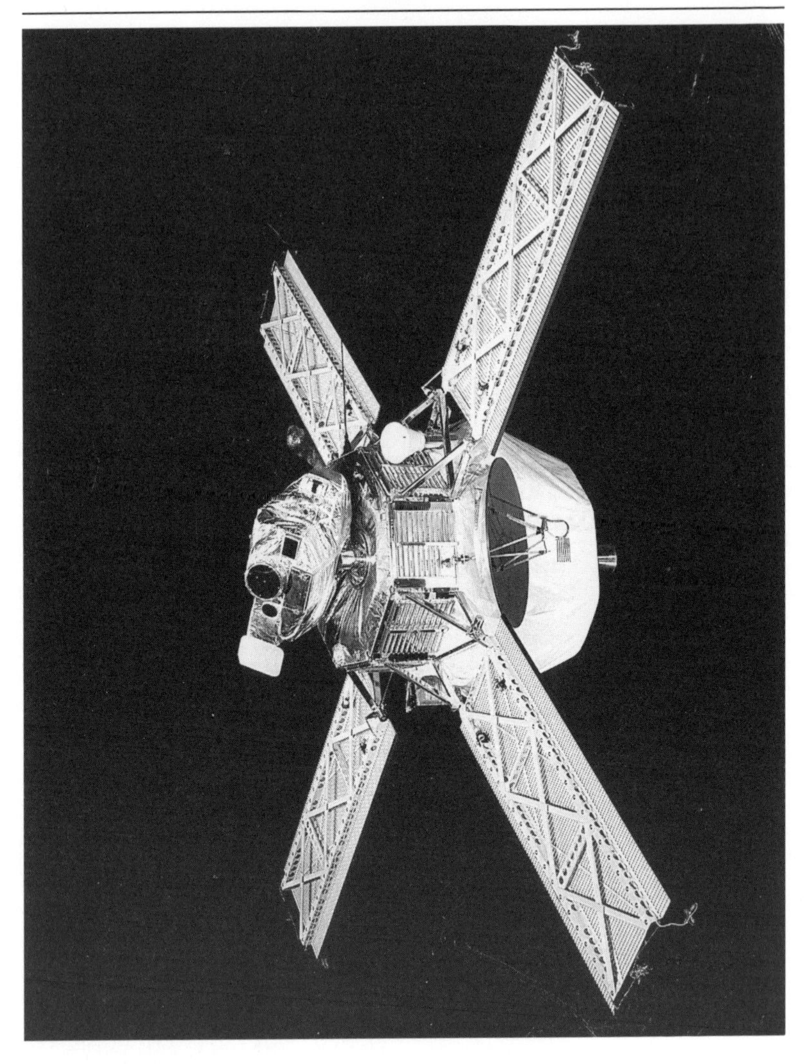

Mariner 9, the first spacecraft to orbit the planet Mars

Mars Observer

Launch September 1992

The USA returns to Mars

The US Mars Observer spacecraft will study the red planet from orbit and pave the way for more advanced missions.

NASA's Mars Observer will be the first US mission to Mars since the two Viking spacecraft were launched in 1975. However, unlike the Viking project, which was aimed specifically at searching for life on Mars, the new mission has a more general goal. Mars Observer will study the Martian atmosphere, surface and interior over a period of at least one Martian year (687 Earth days) and provide scientists with their first global view of the planet. This will enable them to determine how Mars has evolved over the last few billion years and to compare the planet's geology and climate with that of the Earth. Data from the Mars Observer will also be used to plan future projects, such as robotic sample-return missions, which are being planned as precursors of a manned flight to the planet in the next century.

The Mars Observer is the first of a new class of spacecraft called Planetary Observers. These are spacecraft designed for use in the inner solar system and which, in an attempt to reduce costs, use systems already developed for Earth orbiting satellites. The Mars Observer spacecraft, the body of which is box shaped, uses systems developed from those aboard the TIROS and DMSP weather satellites. Electricity is supplied by six solar panels which form a rectangular array attached to the spacecraft body on a stalk. Scientific equipment, such as a camera which can resolve details as small as 1.5m, a laser altimeter to monitor the height of the terrain underneath the spacecraft and instruments to probe the composition and temperature of the atmosphere and surface, are attached to the spacecraft either directly or via long deployable booms. The instruments will be controlled directly from the home institutions of the scientists involved, by means of computer links with the NASA mission control room at the JPL in Pasadena, California.

The Mars Observer was originally expected to be launched in 1990 from the Space Shuttle, but the plans changed after the explosion of the Space Shuttle Challenger in 1986. The Mars Observer will now be launched in September 1992 by a Titan III rocket and boosted towards Mars by a new upper stage called the Transfer Orbit Stage. The cruise to Mars will take about 11 months, and when the Mars Observer arrives it will use its main rocket engine to place itself into an elliptical orbit around the planet. Over the next few months the orbit will be adjusted until the spacecraft is in a 400km, circular orbit which passes over the Martian poles. From this orbit Mars Observer will survey the planet every day from December 1993 to October 1995.

The Mars Observer spacecraft will also be fitted with a radio system which can be used to relay data from future Soviet Mars missions back to Earth. Some of these missions are expected to release balloons into the Martian atmosphere and the data from these will be transmitted to the Mars Observer and stored there until it can be beamed back to Earth.

Mars programme

Mars 1 1962, Mars 7 1973

A Red Star on Mars, but barely

The USSR's Mars missions were far less successful than its series of Venus probes.

The 1960s were the most frantic era of the space race as the USA and USSR tried to leapfrog each other with more and more impressive space achievements. Having been the first nation to send spacecraft to the Moon, the USSR also wanted to be the first to reach the planets and so, in the autumn of 1960, missions to both Mars and Venus were prepared for launch. The Mars probes weighed about 500kg and carried simple cameras and other instruments. Two spacecraft were launched in September 1960 during the visit of Soviet premier Nikita Khruschev to the USA, but in both cases the third stage of the launcher failed and the probes fell back to Earth. Khruschev was furious and it has long been rumoured that a third probe was hurriedly assembled for a final attempt before the launch window closed. On 23 October 1960 the rocket failed to lift off on time and the Soviet commander, Marshall Nedelin, insisted that technicians examine the rocket at once. Ignoring normal safety precautions the launch team swarmed around the fully fuelled rocket which suddenly exploded, killing Nedelin and almost the entire launch team. Although denied for years, this story leaked out to the West and became known as 'The Nedelin Catastrophe'.

Mars 1

The USSR tried for Mars again on 24 October 1962 but this launch failed when the upper stage of the rocket broke up. A measure of success was achieved the following week when Mars 1 was placed en route to the red planet, but a third attempt, on 4 November, failed to escape from Earth orbit.

Mars 1 weighed 894kg and consisted of a cylindrical body 3.3m long with a large, umbrella-like radio antenna, an engine for mid-course corrections and a set of solar panels 4m across. The scientific payload included cameras, meteoroid detectors and radiation meters. For a time all went well; regular radio communications were established and a mid-course correction adjusted the probe's trajectory so that it would pass within 6 000km of Mars. Then, on 21 March 1961, when Mars 1 was only three months away from its target, contact was lost for ever.

More failures

A further Mars attempt, called Zond 2, was launched in 1965 but once more contact was lost before the probe reached its target. The USSR tried again in 1969, but the first spacecraft was destroyed in a launch failure and a second attempt was cancelled while the launcher design was reviewed. In contrast, as the Soviet Mars programme hobbled from disaster to disaster, the US Mariner 4, 6 and 7 missions returned hundreds of photographs of the planet during their flybys in 1964 and 1967.

Mars 2 and 3

Three Soviet Mars craft were prepared for launch in 1971. The first of these, launched on 10 May 1971, failed to leave Earth orbit and was given the cover name Cosmos 419. The jinx was broken with the launch of Mars 2, on 19 May, and Mars 3 on 28 May. Talk of a space race revived since, although the US Mariner 8 had crashed on 9 May, its twin, Mariner 9, was already on its way to Mars. Mariner 9 was intended to go into orbit around Mars, but the Soviet Mars 2 and 3 had more ambitious aims and were designed to achieve a soft landing. The new Mars craft were large and heavy and were surmounted at one end by a landing probe inside a shallow conical heatshield. A retro-rocket, designed to place the mother craft into orbit around Mars, projected from the other end. The plan was for each probe to be released directly into the atmosphere of Mars during the approach to the planet; the rest of the craft would then go into orbit to study Mars for an extended period.

It was a cruel misfortune that when the three probes reached Mars the planet was in the grip of a tremendous dust storm. This did not affect Mariner 9, which could wait until the dust settled, but the Soviet Mars craft were obliged to release their landers and hope for the best. The Mars 2 lander was released on 27 November 1971, cut into the atmosphere at about 6km per second and vanished. Although no data were received, the capsule undoubtedly lies somewhere on the Martian deserts, and one day it may be found and placed on display as the first man-made object to land on Mars.

Mars 3 released its lander on 2 December 1971 sending it into the Martian dust storms. The heatshield slowed the craft until its parachutes could open, and the 450kg capsule began to descend towards the surface. Just above the ground the parachute was jettisoned so that it could not fall and smother the capsule and then, at 16:50:35 Moscow time, the capsule landed. One and a half minutes later the lander began to transmit the first television picture from the surface of Mars. The triumph was incredibly short lived, just 20 seconds later, and before any useful data could be received, Mars 3 fell silent and was never heard from again. Either the probe was damaged by the Martian dust storm, or the radio relay from the orbiter failed. It was unbelievably bad luck.

Despite the failure of the two landers, both orbiters were successful and continued to transmit data until September 1972. They returned photographs, provided data on the temperature of Mars and probed the composition of the Martian atmosphere. Despite these achievements, they were totally eclipsed by the US Mariner 9 which returned thousands of high-quality images of the entire planet.

The 1973 Mars fleet

With the US Viking Mars lander delayed until 1975 by budget cuts, 1973 offered a second chance for the USSR to beat the USA to a successful soft landing on Mars. A simple repeat of the Mars 2 and 3 mission was not possible because, due to the eccentricity of Mars's orbit, the distance between the Earth and Mars was greater than in 1971 and the mass of each spacecraft had to be reduced to avoid overloading the launch vehicle. This meant that each probe could either carry a lander, or go into orbit around Mars, but not both. Mars 4 and 5 were the orbiters, Mars 6 and 7 carried the landers. Compared with the earlier missions the 1973 fleet got off to a tremendous start. All four craft were launched without incident on 21 and 25 July and 5 and 9 August. They arrived at Mars the following spring, but the result was a fiasco.

Mars 4, an orbiter, failed to fire its retro-rocket and sailed past the planet into deep space. Mars 5 went into orbit and prepared to act as a relay station for the oncoming landers, but it was hardly worth it; the lander carried by Mars 6 missed the planet completely and although the Mars 7 probe did enter the atmosphere and began to return data, contact was lost before touchdown. Three out of the four missions were total failures and despite Mars 5 returning some useful data from orbit, the Soviet Mars programme was scrapped.

Mars 94 and Mars 96

After the failures of the original Mars series it was not until the late 1980s that Soviet interest in Mars revived, with the launch of the Phobos mission (see p149) and publication of an ambitious series of projects intended to lead to a manned landing on the planet by the year 2015. The proposed series began with a two-spacecraft mission in 1992 which would orbit the planet, drop penetrators (spear-like probes which embed themselves in the ground to study conditions a few metres below the surface), land a small Mars rover and deploy balloons into the Martian atmosphere. This was to be repeated in 1994 and then, in 1996 there was a chance of launching an unmanned mission to scoop up samples of Martian soil and return them to Earth. All of these new projects were to involve considerable international collaboration and were called Mars 92, Mars 94 etc. As is common with space projects, before long these schedules had been adjusted and the 1992 mission was delayed to 1994 and so on.

The economic and political turmoil which led to the dissolution of the USSR in 1991 caused further changes to be made to these expensive projects. In November 1991 it was reported that the Mars 94 mission would be restricted to a single spacecraft with two penetrators and two small surface stations. The Mars 96 mission was also reported to be cut back to a single spacecraft which would deploy two penetrators, one balloon and a single unmanned Mars rover.

It was suggested in 1991 that the USSR's failed Mars probe Cosmos 419 was different from Mars 2 and 3. According to this source Cosmos 419 did not have a landing capsule and so was lighter, enabling it to fly a faster trajectory to Mars and arrive in Martian orbit ahead of the US Mariner craft. This would have allowed the USSR to claim the record for the first spacecraft to orbit Mars.

Mars balloons

The balloon system for the new Soviet Mars missions will be provided by the French space agency CNES (Centre National d'Etudies Spatiales). The helium-filled balloon will sample the thin Martian air and return television pictures of the ground below as it drifts across the surface. At night, when the temperatures drop, the balloon will lose its buoyancy and descend until its instrument package, suspended below the balloon envelope, lies on the ground. In the course of the night the instrument package will determine the composition of the Martian soil. In the morning, the sun will heat the gas in the balloon causing it to expand and generate extra lift, lifting the experiment package back into the air. The balloon system is expected to operate for about 10 days.

The Mars orbiters

Mission	Apogee(km)	Perigee(km)
Mars 2	24 938	1380
Mars 3	190 333	1500
Mars 5	32 586	1300

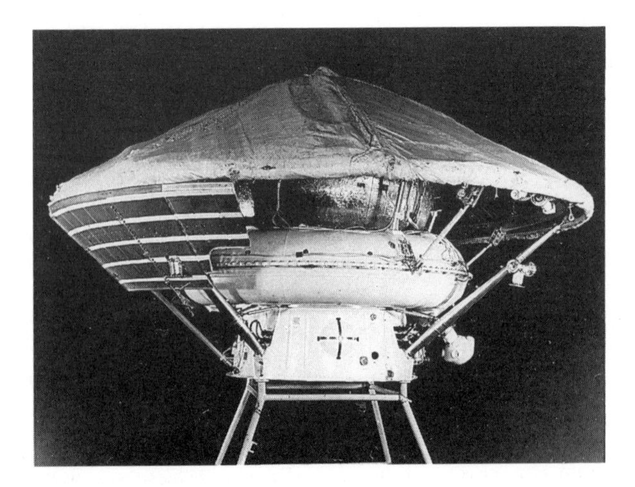

Descent capsule of the automatic station Mars-3

Medical aspects of spaceflight

Doctor in Space 1964

Shrinking bones, wasting muscles

Before humans went into space there was concern that they could never survive the experience.

Amongst the hazards facing the first astronauts were the unknown effects of acceleration, vibration, weightlessness, radiation, disruption of the normal biological rhythm, the psychological stress of isolation and the rigours of re-entry and landing. Today most of these fears have been proved groundless, or less severe than first thought, but the biomedical effects of really long spaceflights remain unclear.

Acceleration

As a rocket accelerates, its passengers appear to be pushed back into their seats. These forces are usually measured in units of 'g', where 1g is the acceleration due to gravity at the Earth's surface. Studies carried out in centrifuges, which mimic acceleration by swinging a test subject on a high-speed roundabout, have shown that resistance to g forces is greatly influenced by the position of the subject. The body is most resistant to g forces applied from the chest to the back (know technically as +gx or more graphically as 'eyeballs in') and so astronauts lie in couches for launch. In this position it is possible to remain conscious at up to 10g or even, for short periods, 25g. Acceleration in the direction from the head to the feet (+gz or 'eyeballs down') makes it difficult for the heart to pump blood upwards towards the brain and at about +5gz leads to blackouts. Experience has shown that the forces of launch are not intolerable, although at high g there is some discomfort from what feels like a steady pressure on the chest.

There are other effects of acceleration which spacecraft design affect. The most obvious is that it is harder for the astronaut to move his or her arms making it difficult to manipulate switches and controls. Peripheral vision is also affected and there may be temporary tunnel vision. The Mercury astronauts were subjected to up to 10g during a launch, but the Space Shuttle accelerates at 3g, providing a more comfortable ride.

Vibration and noise

A rocket launch is noisy and subjects astronauts to considerable vibration. Fortunately, this must only be endured for a short period, about 10 minutes, and so the effects are not normally serious. Although US astronaut Joe Allen described the Space Shuttle cabin as tremendously noisy during launch, Michael Collins was surprised by how quiet it was during the Apollo 11 launch towards the Moon. Vibration can be uncomfortable, because a human body consists of many organs connected together and each has its own natural frequency. If a spacecraft vibrates at this particular frequency then parts of the body will vibrate in sympathy and this could damage muscles or tissue. However, the effects seem to be limited to discomfort rather than serious injury. Vibration can also affect vision, causing the astronaut's view of his or her instruments to go slightly out of focus.

Radiation

In space astronauts are no longer shielded by the Earth's atmosphere and are subjected to ionizing radiation such as cosmic rays, the solar wind (which may increase to dangerous levels during solar flares), and electrons and protons trapped in the Earth's Van Allen belts. There is also a constant flux of gamma rays, X-rays and ultraviolet radiation from the Sun, but most of this does not penetrate the spacecraft walls.

Ionizing radiation is so called because as it passes through the body it ionizes some of the atoms that make up human tissue and can disrupt the biological processes involved in cell division. High levels of radiation, such as might be experienced by a space-walking astronaut exposed to a powerful solar flare, could be fatal. Apollo moonwalkers were vulnerable to this and had contingency plans to make a rapid return to the relatively well-shielded command module should a flare occur. The effects of prolonged exposure to low-level radiation are more difficult to determine. Although research has indicated 'safe' levels of exposure, radiation effects are cumulative and long exposure to low levels of radiation could be dangerous. Particular hazards are damage to the blood-forming organs, which may lead to leukaemia, and to eye tissue, which may cause cataracts. Since radiation exposure can also cause sterility or increased risk of birth defects there have been concerns about astronauts reproductive capability after their return to Earth, but these have proved groundless so far.

Drugs have been developed to help protect against radiation and recommendations have been made on the maximum levels to which astronauts on space stations should be exposed. Exposure during a three-year mission to Mars need not exceed the safe limits, but some sort of radiation shelter will be necessary in case of a large solar flare.

Weightlessness

The major hazard of long spaceflights seems to be prolonged weightlessness. The crew of the USA's Gemini 7 survived a 14-day mission in 1965, but the 18-day flight of the USSR's Soyuz 9 left its crew unable to stand. Fortunately the Soyuz 9 experience was not typical and the space endurance record was extended to 84 days by the final Skylab crew in 1973–4. Since then the record has passed to the USSR who, after progressively longer missions, kept a crew aboard the Mir space station for an entire year between December 1988 and 1989. From these missions various conclusions have been drawn.

On arrival in space, about two-thirds of astronauts experience some space adaptation syndrome (space-sickness) which may produce nausea, headaches and vomiting. The symptoms usually pass within a day or so and drugs have been found that relieve the problem.

In zero gravity blood that normally pools in the legs rises to the upper body. The heart, thinking that it is over-supplied with blood, secretes hormones which reduce the level of body fluids. The blood plasma that remains is rich in red blood cells and so no new ones are produced until the proportion is correct again. On landing, blood flows back into the legs, reducing the supply to the brain which can cause dizziness or fainting.

Next, with less work to do, muscles begin to waste away. The solution is regular exercise for one to two hours a day. This can be boring, and lacking a shower, some astronauts found that washing afterwards was difficult. Soviet cosmonauts sometimes use spring-loaded suits that try to fold up the body unless the muscles work against the springs. These are called 'penguin suits' because of their effect on the cosmonaut's posture.

The most serious problem may be calcium loss from the bones. This varies from person to person, but is worse on longer flights. Calcium is lost from the astronaut's bones and passed out of the body in the urine. The reasons for this are still not understood. Although not yet a problem, it could become so if astronauts returned from Mars with bones so weakened that they could not live on Earth.

In space astronauts may grow several centimetres as their spine relaxes. It is said that astronaut Charles Conrad was delighted that for the first time in his life he was taller than his wife.

The first doctor in space was Boris Yegorov who flew on the Voskhod 2 mission in 1964. US physician Dr Joe Kerwin flew with the Skylab 2 crew in 1973 and Dr Valeri Poliakov visited the Mir space station in 1988 to monitor the health of two cosmonauts undertaking a one-year stay in space.

Mercury

Alan Shepard 1961, Gordon Cooper 1963

The first Americans in space

The tiny Mercury capsule laid the foundations for the US manned space programme.

In 1959, before President John F Kennedy had committed the USA to sending men to the Moon, the newly formed NASA had already ordered its first manned spacecraft. Drawing on work done by its predecessor, the National Advisory Committee for Aeronautics, NASA asked US companies to submit designs for a one-man space capsule to be named 'Mercury'. The objective of the project was to develop the technology to place a man into space, test his reactions and then return him, unharmed, to Earth. The winning design came from the McDonnell Aircraft Corporation (now part of McDonnell-Douglas).

The Mercury capsule

McDonnell produced a bell-shaped capsule 2.9m tall and 1.88m in diameter at the base. A square hatch, cut in the side, allowed the spacesuited astronaut to squeeze into the capsule and on to a couch that had been specially tailored to his body. Once strapped into his seat the astronaut faced a control panel containing various switches, dials and instruments. Early Mercury capsules had two small round portholes, but following complaints from the astronauts, a large rectangular window was installed in later versions. A retractable periscope was also fitted. The capsule's attitude could be controlled in all three axes by 18 small thrusters which were linked to a controller operated by the astronaut's right hand. Since it was vital that the capsule was orientated correctly for retro-fire and re-entry, an extra margin of safety was assured by arranging that the thrusters could be operated in various combinations by different control systems. The capsule was filled with pure oxygen at about one-third of atmospheric pressure and the astronaut usually kept his helmet visor open. Only if the cabin pressure fell would he lower his visor and use his spacesuit's independent oxygen supply.

The pressurized cabin was made from titanium and the outer shell of the capsule from a nickel alloy called Rene 41. Around the base was the heatshield, a fibreglass reinforced laminated plastic, that would ablate, or char, during re-entry. After re-entry, the heatshield was detached and dropped about a metre to form the base of a pneumatic cushion that softened the impact of splashdown. A pack of three, solid-fuelled, 5.16kNewton retro-rockets was held at the centre of the heatshield by three metal straps. When it was time to return to Earth the three rockets were fired in rapid succession and the retro-rocket pack was jettisoned. Above the cabin was a cylindrical section containing the main and reserve parachutes. Topping the whole capsule, at least during launch, was a lattice-work tower supporting a solid rocket escape motor with three canted nozzles. In an emergency the escape motor would fire and drag the capsule away from disaster and high enough for the parachutes to open in time for landing.

The Mercury capsule was designed to be launched by either Redstone or Atlas rockets. The Redstone rocket was more reliable than the new Atlas vehicle (which was still experiencing development problems) but it could only be used for sub-orbital missions since it was less powerful than the Atlas.

The test programme

The development of the Mercury capsule proceeded through a series of unmanned test flights involving different launch vehicles. Early, so-called boiler plate, versions were lifted by solid-fuelled Little Joe rockets to test the structure of the capsule and the rocket escape system. Boilerplate capsules were also tested using Redstone and Atlas rockets. Some of these tests were successful, but others ended in disaster as the rocket veered off course or exploded. As more experience was gained, attention switched to unmanned flights of real Mercury capsules, but even these did not all go as planned. One particularly embarrassing failure occurred during an attempt to launch the MR-1 flight when the Redstone rocket lifted a few centimetres off the ground then suddenly shutdown its motor and settled back on to the launch pad. The Mercury capsule electronics detected the shutdown of the main engine and, acting as if it were already in space, jettisoned the escape tower in a flash of smoke. Next, sensing that it was at low altitude, the capsule deployed its main parachute and set off its flashing recovery beacon. Fortunately the capsule was undamaged and, after being fitted to a new Redstone rocket, the mission took place as planned.

Eventually things began to improve and Mercury began to pass important milestones on the route to a manned mission. On 31 January 1961 the chimpanzee Ham survived a sub-orbital flight to an altitude of about 250km and, although an Atlas rocket carrying an astronaut simulator blew up on 25 April 1961, NASA grew increasingly confident that the Redstone rocket was safe enough to be used to launch a manned sub-orbital mission.

The sub-orbital flights

The first manned Mercury mission took place on 5 May 1961 when Alan B Shepard was launched by a Redstone rocket on a sub-orbital flight. After separation from the booster, Shepard, who had named his capsule Freedom 7, took over control to test his ability to point the capsule in any desired direction. This proved simple and he was able to place the capsule within 5° of the correct angle before test firing the retro-rockets. During his descent Shepard experienced acceleration forces of 5g before his parachute opened to lower Freedom 7 into the Atlantic Ocean. His flight lasted 15 minutes 22 seconds and he reached a peak altitude of 185km. Although beaten in the race to be the first man in space by the Soviet Yuri Gagarin, Shepard did confirm that the Mercury capsule design was sound and did something to restore US prestige.

On 21 July 1962 Virgil Grissom made the second sub-orbital Mercury flight in his capsule Liberty Bell 7. The flight was broadly similar to Shepard's mission except that it had a near disastrous ending. After splashdown, and while Grissom was floating in the water waiting for the recovery team to arrive, the explosive bolts which secured the capsule's hatch were triggered for some, still unexplained, reason. Water flooded in and Grissom was forced to abandon the sinking capsule. A helicopter attached a line to the capsule, but the weight of water was too much and the helicopter was forced to release its grip. Liberty Bell 7 sank in 6000 metres of water. Grissom nearly drowned as his spacesuit became waterlogged, but he was eventually rescued by another helicopter.

Into orbit

Despite the success of the sub-orbital missions, NASA did not get a Mercury capsule into orbit until 13 September 1961 when a capsule carrying an astronaut simulator was successfully launched by an Atlas rocket. After one orbit the retro-rockets were fired and the capsule landed safely, east of Bermuda. This was followed by the two-orbit mission of the chimpanzee Enos on 29 November after which NASA felt that the Mercury system was ready for a manned orbital mission.

After a number of false starts, Marine Colonel John Glenn was launched aboard his capsule Friendship 7 on 20 February 1962. Glenn went into a 160km × 261km orbit and during his five-hour flight he tested the attitude control system, observed the Earth and sampled a tube of food paste. He also observed that, as his craft came into sunlight from the night side of the Earth, it was surrounded by thousands of tiny luminous particles. These were later found to be particles released from the capsule and reflecting sunlight. During the latter part of the flight ground controllers began to suspect that the heatshield was loose and so instructed Glenn not to release the retro-rocket after retro-fire. It was hoped that the straps of the retro-rocket pack would keep the heatshield in place. As a result Glenn had a dramatic re-entry as pieces of retro-rocket pack burned away and flew past his window. Fortunately, the heatshield remained fixed and the landing was normal.

There were three further missions after Glenn's flight. On 24 May 1962 Malcolm Scott Carpenter conducted a three-orbit flight in a capsule named Aurora 7. Carpenter tried to carry out a number of scientific observations during his flight and, as a result of the amount of manoeuvring he did, his capsule was dangerously short of fuel when he began his re-entry. Although the re-entry was successful, the capsule landed 350km off target and there was an anxious wait at mission control before Aurora 7 was located. On 3 October 1962 Walter M Schirra made a successful six-orbit flight in Sigma 7 and the programme drew to a close with the 22-orbit flight of L Gordon Cooper's capsule Faith 7 on 15 May 1963. Cooper suffered a number of equipment problems during his mission but overcame them all and made a manual re-entry to land within sight of the recovery aircraft carrier Kearsarge.

A three-day Mercury flight with Shepard as the astronaut, was considered, but was cancelled to enable work to go ahead with the next programme, the two-man Gemini spacecraft.

> Seven Mercury astronauts were chosen, but only six flew missions. The other, Donald K Slayton, would have flown the second orbital mission, but was grounded because of an irregular heart beat. Slayton did not leave NASA but instead became head of the Astronaut Office and played a major part in selecting new astronauts and assigning them to specific missions. Throughout the 1960s he continued to argue that his heart irregularity was not dangerous and eventually persuaded NASA to return him to flying status. He finally flew in space on the international Apollo–Soyuz mission in 1975.

The seven Mercury astronauts. Front row; (left to right) Schirra, Slayton, Glenn, Carpenter. Back row; Shepard, Grissom, Cooper

Military satellites

UN Outer Space Treaty 27 January 1967

IDSCS, Fltsatcom, DMSP, Transit, GPS

Although there has never been a war in orbit, there are a number of military uses of space.

At the beginning of the space age there were a number of attempts to demilitarize space which culminated in the UN Outer Space Treaty, signed on 27 January 1967. The treaty declared that no nation could appropriate a celestial body and banned the placing into orbit of weapons of mass destruction. There are, however, ways in which space can be used for military purposes without violating these treaties, and these have been exploited by both the USA and the USSR. In this section satellites which could serve civilian as well as military users are described; other more overtly military projects are described elsewhere (see p219).

Communications

The possible value of satellites for communications was not lost on the Pentagon and projects such as SCORE and Courier (see p35) were the forerunners of many US military communications satellites. The first of these were the 26 Initial Defense Satellite Communications System (IDSCS or DSCS-I) which were 45kg, 26-sided polygons launched in batches of up to eight by Titan 3C rockets. The IDSCS satellites were placed below geostationary orbit so that, although they remained in sight of a single ground station for several days, they appeared to drift slowly across the sky. This was so that should one satellite fail, another would soon drift into range.

NATO-1 and -2, which employed an improved and heavier (129kg) version of the IDSCS design, were launched into geostationary orbit on 20 March 1970 and 2 February 1971. The Western alliance subsequently ordered four of a more advanced series known as NATO-3 and these were launched between 1976 and 1984. A new NATO-4 series was developed by British Aerospace and the first example was launched in 1991.

The IDSCS system was replaced by the Defense Satellite Communications System (DSCS) which involved satellites in geostationary orbit. These satellites were 'hardened' to resist both jamming and the effects of nearby nuclear explosions. An operating DSCS system requires four working satellites in orbit, and to ensure this, spares are placed into orbit ready to be brought into action should one of the primary satellites fail. Fourteen of the spin-stabilized DSCS-II series were launched in pairs between 1971 and 1979, although several of them failed to reach their intended orbits. The larger, and still more powerful DSCS-III series are three-axis stabilized and have a design life of 10 years. DSCS-III satellites use encryption techniques and advanced anti-jamming systems. The first of these was orbited in 1982 along with a fifteenth DCSC-II satellite. Further satellites in both DSCS series were launched throughout the 1980s.

The US Navy also relies heavily on communications satellites. The Fltsatcom series, which became operational in 1981, provides global communication for US nuclear forces. The satellites, which weighed about 1 000kg, were launched by Atlas-Centaur rockets. The Fltsatcoms are supplemented by the Leasat system under an arrangement in which a subsidiary of the Hughes Aircraft Company orbits the satellites and recoups its investment by charging the US Navy for the use of the system. So far the project has not been very successful since a number of the Leasat craft (which are also known as Syncom-IVs) have failed quite soon after launch.

Other US military communications

satellite programmes include Tacsat, designed for tactical communications using small dish aerials carried by individual units; the Satellite Data System, a series of satellites in eccentric, high-inclination orbits, used to communicate with US forces in the polar regions; and Milstar, a new system that may come into service around the end of the century. SKYNET is a series of satellites bought for the British services. Early examples suffered a number of failures of both satellite and launch vehicle, but the later SKYNET-4 series has proved more reliable.

Meteorological

Meteorological satellites have also been developed for military use. The TIROS weather satellites probably had military uses since they were developed jointly by NASA and the US Department of Defense. The TIROS programme eventually assumed a civilian guise, but since then a number of dedicated meteorological satellites have been launched under the umbrella name of the Defense Meteorological Satellite Programme (DMSP). These have included small satellites called Calspheres and others known as DMSP Block 4A, 4B, 5A, 5B, 5C. The Block 5D series, also known as AMS (Advanced Meteorological Satellites), are built by the RCA corporation and probably resemble the Advanced TIROS and NOAA civilian weather satellites.

Navigation

The ability to navigate precisely is vital to both civilian and military enterprises. Cargo ships wish to reach port as quickly as possible and the captain of a nuclear submarine needs to know his position so that he can aim his missiles. Satellite navigation systems are often used by both groups, with the civilian users allowed to operate with equipment that provides lower precision than that used by the military.

The first navigation satellites were the US Navy Transit series, intended to enable Polaris submarines to fix their position to within 150m. Each Transit carried a very stable radio oscillator and a suitable receiving station could estimate its position by combining the doppler shift of the radio signal from the satellite with detailed knowledge of the satellite's orbit. The first Transit was launched in 1963 and transmitted for three months. By 1968 a total of 23 satellites, comprising three different varieties of Transits, had been launched. A further 15 satellites were built in the 1970s; the first few proved very reliable and the remainder were kept in storage on the ground to act as spares. Some of these were launched during the 1980s. Nova was an improved version of Transit, the first of which was launched in 1981.

Another space navigational system is the Global Positioning System (GPS) which uses satellites called NAVSTAR. The GPS series began with several experimental satellites called Timation and the Navigational Technology Satellite. The original plan was based on a constellation of 24 satellites which would have provided positions accurate to 1m. Budget cuts reduced this to a system of 18 which has degraded the accuracy of the system to about 6m.

The NAVSTAR satellites, which weigh about 462kg, are placed in polar orbits 17700km high. Each NAVSTAR transmits information about its position in orbit, together with very precise time checks. Since the speed at which radio signals travel is constant, it is possible for a user equipped with a suitable terminal to listen to the signals from a number of satellites (usually four) and compute the range to each of them. This allows him, or rather the computer in his terminal, to determine his own position by triangulation.

Probably the most notorious military communications experiment was project West Ford in which 350 million copper needles were released into orbit to see if they would reflect military radio signals. The project was widely attacked by radio astronomers, but went ahead in 1963. It was never repeated, so was presumably not considered a viable system.

The satellites referred to here are all from the Western powers, mostly from the USA. The USSR has also operated satellites for military purposes, but has disguised this by giving them all the cover name Cosmos. It is estimated that in 1982, which was a fairly typical year, the USSR launched 89 military satellites of all types.

The UK military communications satellite SKYNET

Military space stations

MOL 1965–9, Salyut 2 1973, Salyut 3 1974, Salyut 5 1976–7

Soldiers in space

Both the USA and the USSR considered military space stations, but only the Soviets ever built any.

Before Sputnik, most prophets of spaceflight envisaged manned space stations being developed for both civilian and military purposes. However, the advent of microelectronics, computers and other miniaturization techniques meant that much could be achieved by unmanned satellites. Nonetheless, both the USA and the USSR investigated putting men into space for military purposes. Some ideas, such as using the Moon as a base for nuclear missiles, soon fell by the wayside, but the idea of putting men into orbit for reconnaissance purposes had a longer life.

Manned Orbiting Laboratory

The US Air Force Manned Orbiting Laboratory (MOL) (approved by President Johnson in August 1965), was intended to develop reconnaissance techniques using the combination of advanced sensors and astronaut observers. The MOL design was for a cylindrical space station about 12.5m long and 3m in diameter which would have comprised living quarters for a crew of two and an experimental laboratory. It was to have been launched by a Titan 3M rocket with the two astronauts carried in a modified Gemini capsule attached to the top. Once in orbit, the astronauts would transfer to the MOL via a hatch cut into the Gemini heatshield and perform a flight lasting about 30 days. At the conclusion of their mission, the astronauts would use their Gemini capsule to return to Earth. The MOL would be abandoned, and probably instructed to re-enter so that its secret equipment could not fall into the hands of another nation. Additional missions would require the launch of another MOL since it was not possible to use the same MOL more than once.

Considerable work was undertaken, including the launch of a dummy MOL to check the performance of the Titan 3M rocket, but the project was cancelled in 1969 before any manned flights occurred. Much of the technology developed for MOL is believed to have been transferred to the unmanned spy satellite programmes such as Big Bird and KH11.

The military Salyut programme

In parallel with the development of civilian space stations the USSR developed a military version of Salyut whose objectives apparently mirrored that of the US MOL. However, unlike MOL, the military Salyut station programme did progress to a series of manned flights.

Although sharing the same name, the military and civilian Salyut stations were rather different in design and mode of operation. For example, the three military Salyuts were placed into lower orbits, probably because such orbits are better suited to reconnaissance, and little was ever disclosed about their objectives. The crews which flew to these stations were composed exclusively of military pilots and the stations featured unmanned re-entry capsules, probably used to return exposed film of strategic targets, that were not carried on their civilian counterparts.

The first launch in the military series was Salyut 2 which was placed into orbit on 3 April 1973. Over the next few days the station made a series of manoeuvres that left Western observers expecting the launch of a manned mission to the station but on

14 April the station's propulsion system exploded, wrecking Salyut 2 and sending it tumbling out of control. No attempt was made to launch a crew to Salyut 2 and it re-entered on 28 May. At the time, this disastrous failure was covered up by Soviet space authorities who indicated that the station was merely conducting unspecified tests of equipment.

Salyut 3

Salyut 3 was launched into a low (219km × 270km) orbit on 25 June 1974. This time there were no problems and Pavel Popovitch and Yuri Artyukin were launched in Soyuz 14 on 3 July and docked successfully with the station 32 hours after launch. During the mission little was said about what the crew were doing and only a few, blurred television pictures were released. Popovitch and Artyukin remained aboard Salyut 3 for two weeks before returning to Earth in Soyuz 14, leaving Salyut 3 ready to receive further visitors.

Unfortunately, those visitors never came; although Gannadi Sarafanov and Lev Demin were launched successfully in Soyuz 15, they were unable to rendezvous with the station and were forced to make an early return to Earth as fuel and power for their Soyuz spacecraft began to run low. Soviet officials tried to cover up the failure by claiming that the purpose of the mission was to test the compatibility of cosmonauts of different ages (Sarafanov was 32, Demin 48) and to practice emergency landings at night, but few Westerners were fooled. Salyut 3 continued in orbit, operating under ground control, and on 23 September the re-entry capsule was undocked and returned to Earth. The station eventually re-entered and burned up over the Pacific Ocean in January 1975.

Salyut 5

The final military Salyut was launched on 22 June 1976 and followed the successful flight of the civilian Salyut 4. The first crew were Boris Volynov and Vitally Zholobov who were launched aboard Soyuz 21 on 6 July and docked with Salyut 5 the following day. As with the flight of Salyut 3, little was said about what the cosmonauts were doing, although it was announced that they were carrying out experiments in materials processing and biology. The mission may have been planned to last up to 66 days, but the crew left hurriedly after only 48 days in orbit, landing at night and far from the normal recovery zone. Problems with the air supply were cited as the reason for the sudden evacuation.

Despite the problems which had affected Volynov and Zholobov, a second crew, Victor Gorbatko and Yuri Glazhkov, were launched aboard Soyuz 24 on 7 February 1977. The spacecraft docked with Salyut 5, but the crew did not attempt to board the station immediately. Instead, they remained in the Soyuz capsule for 24 hours, during which time they valved off the air from the station and replaced it with fresh supplies, confirming suspicions about the reason for the sudden termination of the previous flight. This unusual procedure seems to have solved the problems with Salyut 5, for the two men then boarded the station and remained aboard until 25 February.

During their visit the cosmonauts continued with the materials processing experiments and observed the Earth. The precise purpose of the mission is unclear, but the day after Soyuz 24 landed, Salyut 5 released its return capsule, which was recovered almost immediately. It is possible that a key objective of the Soyuz 24 flight was to recover material left behind during the emergency departure of the Soyuz 21 crew.

Salyut 5 burned up on re-entry on 28 August 1977 and no further military Salyuts were launched. Perhaps, like the Americans, the Soviets abandoned manned military space stations in favour of unmanned satellites or perhaps the military work was blended into the programmes of the ostensibly civilian Salyut 6 and Salyut 7 stations.

The unmanned Gemini capsule used during the test of the US MOL/Titan 3M launcher had been previously used for the unmanned Gemini 2 mission. It was the first spacecraft to complete two missions.

Mir

Launched 20 February 1986

Third generation space station

The Soviet Mir programme has demonstrated the in-orbit assembly of a modular space station.

When the USSR realized that it could not win the race to the Moon it decided to develop a permanently manned space station instead. Experiments began with the Soyuz programme in 1970 (see p184) and the series of Salyut space stations launched from 1971 onwards (see p163). The first five Salyuts could be occupied by only one crew at a time, but the last two, Salyut 6 and 7, featured a docking port at each end. This allowed the crew to receive supplies delivered by unmanned Progress tankers and permitted brief visits by additional cosmonauts. The success of the Salyut programme led to the Mir space station, which was launched on 20 February 1986.

The core station

The Salyut stations crammed both living space and scientific experiments into a single spacecraft and as the missions grew longer, the cosmonauts demanded a more convenient layout which would separate living and working space. As a result the Mir space station consists of a central core, used as far as possible for living space, to which specialized modules containing scientific and technological experiments can be added. Externally this core module is quite similar to the design of the later Salyut stations, but it has extra docking ports and larger solar panels.

At the front of the station is a docking unit with five docking ports. One of these is along the axis of the station and is used to receive the Soyuz spacecraft which bring the crews to the station. The other four, at right angles both to each other and to the axis of the station, are reserved for the experimental modules. The docking unit also serves as an airlock from which the cosmonauts can perform spacewalks. Various antennas, used for communications and for guiding other craft towards manual or automatic dockings, are attached around the outside.

Behind the docking module is the work compartment, which consists of two cylinders joined end to end. The forward section, 2.9m in diameter, contains the main control console from which the station is operated. The console features two 'seats' which are actually little more than padded metal frames around which the cosmonauts can hook their legs to stay in position whilst working. Various systems for attitude determination, navigation and communications are mounted around the walls of this section. Two large rectangular solar panels (total area 76sq m), which provide Mir's electricity, are attached to opposite sides of the work compartment.

Next comes the living space, where the diameter increases to 4.2m. There are two small cabins where each crewmember can retire to sleep, work or relax in privacy. An individual sleeping bag is fixed to the 'wall' of each cabin. The living section also features an exercise bicycle, treadmill, work table, washbasin and toilet. A colour coding, intended to help the cosmonauts orientate themselves, is applied to the walls, ceiling and floors.

At the rear of the Mir core module is a sixth docking port which is surrounded by a donut-shaped propulsion compartment. This compartment houses two (300-Newton) rocket motors used for orbital adjustments and 32 smaller (14-Newton) attitude control jets. Provision is made for unmanned Progress cargo ships to dock with the rear of the station and transfer fuel to the station's tanks. Bulky cargo can be unloaded by the cosmonauts via the circular docking port hatch.

Since its launch Mir has received three experimental modules, and two more are planned.

The Kvant module

The first module to be docked with Mir was the Kvant (Quantum) astrophysics module. Launched on 31 March 1987, the Kvant module featured a pressurized section, an unpressurized section containing a number of experiments for ultraviolet, X-ray and gamma ray astronomy and a propulsion module or 'space tug'. Kvant was flown to Mir by remote control and, after a failed docking attempt a day or so after launch, it was finally connected to the rear of Mir on 6 April. Unfortunately, the module was unable to make an airtight seal and so, on 12 April, cosmonauts Yuri Romanenko and Alexander Laveikin, who were already aboard Mir, made a spacewalk to examine the situation. They found a piece of cloth fouling the docking mechanism and removed it, allowing the Kvant module to be joined permanently to Mir. Kvant's propulsion module was then jettisoned, revealing another docking port suitable for Progress and Soyuz spacecraft. Kvant also carried a set of 'gyrodynes', or reaction wheels, which can be used for attitude control of the Mir station without consuming precious rocket fuel.

Kvant did not become fully operational at once because of a shortage of electrical power. This situation was rectified in the summer when additional solar panels, ferried to Mir by Progress 29, were installed by the cosmonauts during two successive spacewalks.

Kvant-2

The 20-tonne Kvant-2 module was launched on 26 November 1989. Initially one of its solar panels failed to deploy properly, but this was freed on 30 November and the module continued towards Mir under ground control. A docking attempt on 2 December was aborted due to a computer problem, but a second attempt, on 6 December, brought Kvant-2 smoothly into the forward docking port of Mir. Two days later a small crane (known as Ljappa) mounted on the docking module grappled Kvant-2 and moved it from the forward docking port to one of the side ports, freeing the forward port for future visiting spacecraft.

Kvant-2, which is also known as the 're-equipment module', is 13.7m long, 4.35m in diameter and has an internal volume of 59cu m. It comprises a research section and an airlock. The airlock hatch is 1m in diameter, somewhat larger than the hatches in the docking module, and is intended as the main airlock for spacewalking cosmonauts. The hatch also serves as a means of exit for cosmonauts using the Soviet 'space motorcycle', the Mir equivalent of the NASA Manned Manoeuvring Unit. The space motorcycle was launched inside Kvant-2. The module is equipped with attitude control jets and a set of six extra gyrodynes. The interior of the module features a shower for the cosmonauts and a variety of technological and biological experiments. Other experiments, including equipment for Earth observations, are mounted externally.

The Kristall module

The third Mir module is similar in size and shape to Kvant-2 and is known as Kristall. It comprises two sections, an instrument compartment and a docking compartment with hatches that can accommodate the Soviet space shuttle Buran. Kristall was launched on 31 May 1990 and, after an abortive attempt on 6 June, docked with Mir on 10 June. The next day the module was relocated to a side port opposite Kvant. This gave the complex a T-shaped layout.

The instrument compartment houses equipment for materials research, such as semi-conductor manufacture, and for medical experiments. Its total payload is 7 000kg. The docking compartment has two hatches of a new design. A third opening in this section is equipped with Earth observation experiments.

Future modules

The next module expected to be sent to Mir is called Spektr and will carry equipment for observing the Earth. The following, and possibly final, module will be the Priroda module with advanced remote sensing equipment.

Mir has been translated as 'peace' or 'community living in harmony'.

An impression of the Mir space station as it will appear with all four specialized modules attached

Mir programme

1986–1991

A permanent presence in space

The Soviet Mir programme achieved, for the first time, the goal of a permanently manned space station in Earth orbit.

The first visit to Mir began on 13 March 1986 when Soyuz T15, flown by Leonid Kzim and Vladimir Solovyev, lifted off from the Baikanour cosmodrome. The two men docked with Mir two days later. Once aboard, Kzim and Solovyov activated Mir's systems and received and unloaded supplies sent up in two unmanned supply ships, Progress 25 and Progress 26. On 5 May 1986, after almost two months on Mir, the two men boarded Soyuz T15 and flew to the Salyut 7 space station, which had been orbiting empty since 21 November 1985. The cosmonauts spent 50 days on Salyut 7 during which time they made two spacewalks to conduct tests of space assembly techniques. On 25 June they returned to Mir and stayed for another three weeks before landing on 16 July. Their mission lasted 125 days.

The next crew, Yuri Romanenko and Alexander Laveikin, were launched on 5 February 1987 in an improved version of the Soyuz called Soyuz TM2 (Soyuz TM1 had been an unmanned test flight). As well as solving the docking problem that arose with the Kvant module (see p132) the two men carried out spacewalks to add extra solar panels, delivered by Progress 30, to Kvant.

Crew exchange

On 24 July Romanenko and Laveikin received a visit by the Soyuz TM3 crew, which included Syrian guest cosmonaut Mohammed Faris. It was planned that, following the practice developed during the Salyut programme, the visiting crew would return to Earth in Soyuz TM2 leaving Soyuz TM3 to provide a 'fresh' Soyuz for the crew on board Mir. However, in this case, doctors had noticed some irregularities in Laveikin's heartbeat and ordered him to return with the visiting crew. Alexander Alexandrov replaced him aboard Mir.

Romanenko and Alexandrov remained in orbit until December allowing Romanenko to establish a new space endurance record of 326 days. The replacement crew were launched on 21 December in Soyuz TM4 which carried Vladimir Titov, Musa Manarov and Anatoli Levchenko to Mir. Levchenko, who was assigned to the Buran space shuttle programme and was on the mission to gain space experience, returned to Earth with Romanenko and Alexandrov on 29 December. Titov and Manarov remained on Mir.

A year in orbit

The new crew were to attempt the first year-long space mission. During this marathon there were numerous supply flights by Progress tankers and three visits by other cosmonaut teams. The first was Soyuz TM5, which brought Alexander Solovyev, Alexander Alexandrov and Bulgarian Victor Savinyth for a week-long stay in June. Next came the flight of the Afghan cosmonaut A Mohmand who accompanied Vladimir Liakhov and Varery Poliakov to Mir in August. Poliakov was a doctor; he examined the long-stay cosmonauts and remained aboard when his fellow crewmen returned to Earth. The final visit began when Soyuz TM7 was launched on 26 November 1988. This crew comprised Alexander Volkov, Sergei Krikalev and Frenchman Jean-Loup Chretien. Chretien had previously visited Salyut 7, and on this mission would become the first European to make a spacewalk.

The Soyuz TM7 visit was longer than average to accommodate the French cosmonaut's six-hour spacewalk and lasted a total of 26 days. Chretien returned to Earth with Manarov and Titov, who had survived their year-long flight quite well. The doctor, Poliakov, remained onboard until all three cosmonauts left Mir and returned to Earth on 27 April 1989.

After almost three years of continuous occupancy Mir was then left unmanned for the first time. This was because of delays in preparing the new technological modules; Soviet space officials decided that the costs involved in keeping Mir occupied could not be justified until the new modules were ready.

The second phase

The second phase of missions to Mir has involved two-man crews each living on the station for about six months. The crew replacement process may involve a two-man flight which culminates in a simple crew exchange or the use of a three-man Soyuz which brings the two cosmonauts who will form the next crew together with a third crew member making a short visit. All five cosmonauts work aboard the station for about a week, then the two existing Mir crewmembers, and the short-stay cosmonaut, return to Earth. In this way Mir is continuously manned and foreign passengers can be given brief rides in space in return for an injection of cash into the space programme.

The first crew to return to Mir were Alexsandr Vitorenko and Alexsandr Serebrov, who were launched aboard Soyuz TM8 on 5 September 1989. The objectives of the mission were stated to be projects involving a new experimental module, but the launch of the module was delayed for technical reason, disrupting the planned programme. The module, Kvant-2, eventually arrived on 6 December 1989. The highlight of the six-month flight was probably the first use of the Soviet space motorcycle (similar to NASA's Manned Manoeuvring Unit) on 1 February 1990 by Alexsandr Serebrov. During this test Serebrov flew up to 33m from the Mir station while Viktorenko remained near the hatch observing and filming the exercise on video. Viktorenko himself tried out the device during a spacewalk on 5 February during which he withdrew about 45m from the station.

Emergency in space

The next crew sent to Mir were Anatoli Solovyov and Alexsandr Baladin who were launched aboard Soyuz TM9 on 11 February 1991. Their arrival allowed Viktorenko and Serebrov to return to Earth on 19 February, after a delay caused by poor weather in the recovery zone.

However, after Soyuz TM8 had returned to Earth it was revealed that, as Soyuz TM9 was docking with Mir, television pictures broadcast to ground control had shown that some of Soyuz TM9's external insulation had come loose and was flapping about. The damage, which had probably occurred when the launch shroud had been released from Soyuz about 160 seconds after lift-off, was not immediately dangerous but, since the loose material could obstruct the view of vital sensors required to orientate the craft during re-entry, it was decided that the two men would need to conduct an emergency repair. Special equipment for this repair was loaded into the next experimental module, Kristall. The module was intended to be launched in March, so the Soyuz TM9 crew could use it to conduct materials processing research, but the launch was delayed several times and Kristall eventually arrived at Mir on 31 May 1990.

The emergency spacewalk was made on 17 July, with the cosmonauts leaving Mir via the Kvant-2 hatch and using ladders to reach the affected regions of their Soyuz spacecraft. Two damaged insulation blankets were rolled up and secured, a third was cut away and, rather later than planned, the cosmonauts returned to the Kvant-2 airlock. Here they discovered another problem, they could not close the airlock hatch and their spacesuit supplies were running dangerously low. Eventually, mission controllers depressurized the next section of the Kvant-2 module allowing the exhausted crew to use that section as an airlock and leave the faulty hatch open. Despite the near disaster caused by the problems with the hatch, the two men had to make another spacewalk to remove the tools and ladders left behind during the repairs to the Soyuz. This they did on 26 July during a three and a half hour spacewalk that included a temporary repair to the damaged hatch. The repairs to the Soyuz were successful and the two men returned to Earth on 10 August 1990, having been replaced by Gennadi Manakov and Gennadi Strekalov, who were launched in Soyuz TM10 on 1 August 1990 and docked with Mir on 3 August.

Routine operations

After the problems with Soyuz TM9, the Mir programme settled into a routine. Manakov and Strekalov were replaced in December 1990 by Victor Afanasyev and Musa Manarov on the Soyuz TM11 flight that carried Japanese journalist Toyohiro Akiyama on a week-long passenger flight. Manakov and Strekalov had a busy but relatively uneventful mission during which they carried out several spacewalks. On 7 January 1991 the two men went into space and repaired the damaged hatch on Kvant-2 by replacing a hinge which had been bent out of shape. On a subsequent spacewalk they assembled a small crane which would later be used to move the solar panels from one module to another to improve the supply of electricity to the complex. The crew then made a final, unscheduled spacewalk to examine the automatic docking system's radio antenna which had apparently been damaged during earlier spacewalks. The two cosmonauts were replaced in May 1991 by Anotoli Artsebarsky and Sergei Krikalyev during the Soyuz TM12 mission which carried passenger Helen Sharman, the first Briton to go into space.

Artsebarsky and Krikalyev had an arduous mission which featured a series of major spacewalks. The first of these was on 24 June 1991 when they replaced the damaged antenna of the docking system. This was followed by a series of spacewalks to test space construction techniques. The six-month flight ended for Artsebarsky with the arrival of Soyuz TM13, carrying Austrian Hans Vietiboeck as a passenger, in October 1991. Artsebarsky returned to Earth with Vietiboeck and Kazakh cosmonaut Toktar Aubakirov, who was making a short flight, leaving Krikalyev aboard Mir with Soyuz TM13 commander Alexsandr Volkov. Krikalev finally returned to Earth in March 1992 during a crew rotation flight which carried German astronaut Klaus-Dietrich Flade as a passenger.

> On 14 April 1991 Musa Manarov became the first person to spend a total of 500 days in space.

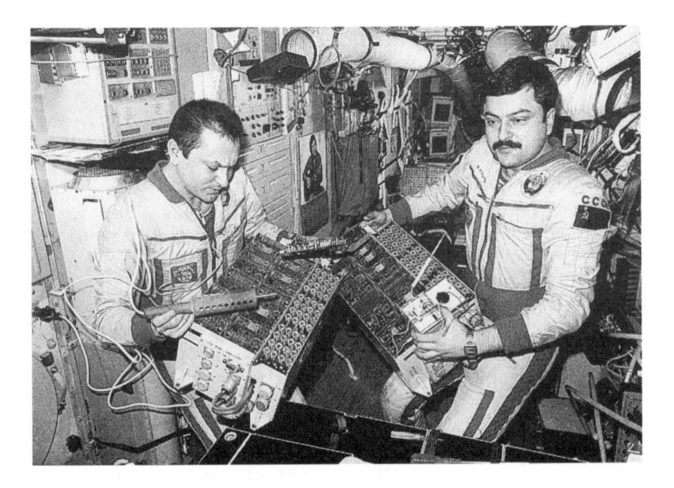

Soviet cosmonauts Vladimir Titov (left) and Musa Manarov (right) mend radio communication units aboard the orbital station Mir

Moon bases

Space Exploration Initiative announced 1989

At home on the Moon

Permanent bases on the Moon are a staple of the science fiction writer, but when will fact catch up with fiction?

Perhaps the first reference to establishing a base on the Moon was made by Bishop John Wilkins in 1638. In his book *A Discourse Concerning the New World and Another Planet* the bishop stated his belief that men would one day learn to fly and would plant a colony on the Moon. Despite this early start, it was not until after World War II that technically-sound descriptions of lunar bases began to appear, some as thinly disguised fiction and others as technical books and articles by visionaries like Arthur C Clarke, Willy Ley and Wernher von Braun.

Post-Apollo Moonbases

In the USA detailed planning for a lunar base began in the early 1960s when the Apollo programme seemed to herald the beginning of a systematic campaign of lunar exploration. Most of these concepts assumed the use of Saturn 5 rockets to send advanced lunar spacecraft and prefabricated habitats to the lunar surface. Typical of these was the Lunar Exploration Systems for Apollo (LESA), which proposed developing modules weighing about 11 tonnes and able to support astronauts (perhaps six) during long stays (eg six months) on the lunar surface. The LESA modules would also have served as the nucleus for a permanent lunar base. In 1971 it was suggested that a lunar base could be constructed using modified versions of modules developed for a space station and delivered to the Moon by the NASA Space Transportation System (which was then envisaged as comprising a space shuttle plus space tugs to send payloads further into space). However, the massive cutbacks which the US space programme suffered in the early 1970s, and which led to the cancellation of the last three Apollo flights, effectively ended any hope of establishing a lunar base before the turn of the century. Only once the Space Shuttle was in regular service, and the USA had committed itself to the development of a permanently manned space station in Earth orbit, did attention begin to swing back to the Moon.

Return to the Moon

In 1989, US President George Bush gave notice that it was time for the USA to begin planning new missions to the Moon with the aim of establishing a lunar base. At present it is too early to say what such a lunar base will look like, or even where on the Moon it is likely to be placed, for much depends on why the base is established. It is quite clear that the USA will not return to the Moon simply to repeat Apollo-type landings. The USA's return to the Moon, whether or not it is done in partnership with other nations, is expected to be a part of a much wider programme, which may involve using the Moon as a training ground for a manned flight to Mars, or as a scientific base or as a source of resources. Several possible scenarios were described by a group headed by veteran Apollo astronaut Tom Stafford in 1991 and what, if anything, happens next depends on which of the scenarios is eventually adopted.

If the Moon is simply to be used as a testing ground for Mars exploration then only a small number of missions need be performed and these may be aimed at landing sites already explored by Apollo astronauts. A few flights might take place between 2005 and 2011 before the Moon is abandoned again as attention switches to Mars.

If the Moon is to be explored for wider scientific reasons then a different approach will be required. The first priority will be to survey the Moon thoroughly in order to identify a small number of potential landing sites which are of special scientific interest. This would be done by an unmanned spacecraft which would be placed in a polar orbit around the Moon in about 1999. Such a mission, which has been proposed several times for purely scientific reasons, would be an essential precursor of any new lunar exploration programme. After about 2003, when suitable landing sites have been identified, a series of missions will take place annually. Each mission will carry five astronauts to the lunar surface and leave one in lunar orbit with a 'mother ship'. At first the astronauts will spend only one lunar day (14 Earth days) on the Moon, but once a site has been selected for a permanent scientific base then the missions will last longer, with supplies being delivered by unmanned cargo craft. This will enable the astronauts to remain on the Moon for periods of first 90 then 180 days, allowing them time to conduct major scientific programmes. As the lunar base builds up, geological expeditions many hundreds of kilometres from the base will be undertaken in mobile laboratories. At the same time the astronauts at the base will erect astronomical instruments such as infrared and radio telescopes and detectors for X-rays, gamma rays and cosmic rays. By about 2015 a full scientific base on the Moon might be in operation, although the astronauts will remain dependent on the steady arrival of supplies from Earth.

If the intention is to colonize the Moon, developing the capability to use lunar materials to produce food, building materials and other necessities, then a more dramatic strategy will follow the initial unmanned missions to survey the Moon. Once a site has been found, astronauts will land but, instead of conducting detailed scientific programmes, they will concentrate on establishing a permanent base, building a prefabricated habitat which can support later visits. After one or two more visits the habitat should be able to support a crew throughout a complete lunar night, using nuclear power to provide electricity, and the first experiments in food production and waste recycling should be underway. By about 2007 the base should have grown sufficiently to support 18 astronauts, and three missions, each carrying six astronauts, will be sent to the Moon every year. As each ship arrives, six of the existing team will return to Earth, having spent a tour of duty of one year in the base. By about 2010 the lunar base may be well on the way to self sufficiency in food and may also be extracting at least some of its own oxygen and water from the lunar soil. The colonization of the Moon will have begun.

As an alternative to establishing a permanent colony on the Moon, it may be that the lunar explorers will concentrate on exploiting lunar resources. This will involve setting up automatic processing plants to mine the lunar soil and extract useful materials which can be used in other space projects. Examples of the materials which might be extracted from the lunar surface are hydrogen and oxygen (used for rocket fuels), a rare isotope of Helium called Helium 3 which is used in nuclear fusion reactors, or materials required for building space habitats such as those described on p198.

No one really knows what a lunar base will look like, but it is unlikely to comprise the clear domes envisaged by the science fiction writers of the 1950s. The base will probably be built from cylindrical sections, perhaps partly buried in the lunar surface. To protect the astronauts from harmful radiation, and to provide insulation to keep in heat during the long cold lunar night, the modules will probably be covered in a thick layer of lunar dust. It has even been suggested that lunar dust mixed with epoxy resin could be formed into a kind of concrete that could be shovelled into position by spacesuited construction workers.

Observing the Earth

ERTS-1 23 July 1972

Getting the whole picture

Satellites observing the Earth can provide warning of such problems as crop disease, flooding and icebergs and can help to monitor pollution.

In 1963, when US astronaut L Gordon Cooper claimed to have seen roads and buildings from his orbiting Mercury capsule many people thought that he had been hallucinating. However, his reports were soon confirmed both by other astronauts and by photographs which showed considerable detail of the Earth below. To investigate whether this discovery could be put to practical use the Apollo 9 mission, which was an Earth orbiting test flight, carried a special suite of cameras which took pictures of the ground below in several different wavelengths simultaneously. These photographs showed that this technique, which became known as multispectral imaging, could be used to distinguish between healthy and diseased vegetation and could provide information for map making, studies of land use, pollution monitoring and mineral prospecting. The success of the Apollo 9 experiments confirmed the value of using satellites to monitor the Earth and so NASA's Earth Resources Technology Satellites, or ERTS, were launched to develop new techniques and instruments to exploit the full potential of Earth observation from space.

The principle behind the ERTS project, and the missions that followed it, is that of taking multispectral images of the same area at regular intervals. Interpretation of the images is easier if they are all taken under similar lighting conditions and this requires that the satellite is placed in an orbit which remains constant relative to the Sun and to the Earth below; such an orbit is said to be Sun synchronous. It is usual to use an orbit which passes over the poles and which crosses the equator almost at right angles so that the orbit of the satellite remains fixed in space and the Earth rotates underneath it. Since the same areas are covered again and again, scientists can see changes that have occurred between images taken on different days and can monitor both short- and long-term changes.

The first ERTS, weighing 891kg, was launched on 23 July 1972 by a Delta rocket. It was placed in a 900km-high polar orbit from which it could observe the same area once every 18 days. The spacecraft was based on the Nimbus weather satellite and consisted of a central core containing the experiments and other systems, with a 3m-high solar panel on either side. This gave ERTS-1 a butterfly-like appearance. ERTS-1 proved highly successful and showed the potential for this type of satellite. Soon the programme was renamed Landsat and the basically similar Landsat-2 was launched on 22 July 1975. Landsat-3 followed on 5 March 1978.

Landsat-4 and Landsat-5 were of a new design based on a special multi-mission spacecraft equipped with improved sensors. They were each about 6m long, 2m high and weighed about 2 000kg. Unfortunately, Landsat-4 developed communications problems soon after its launch on 16 July 1982 and later suffered a partial failure of its solar panels, making it unable to operate all its instruments due to power restrictions. By 1991 Landsat-4 was still operating, but its capability was severely degraded. Landsat-5 was launched on 1 March 1984 and should remain in service until Landsat-6 is launched in 1992. Planning for Landsat-7 is now underway.

Since 1985, Landsat data (including data

taken before then and archived) have been marketed by EOSAT (Earth Observation Satellite corporation) which is a joint venture of the aerospace giants General Electric and the Hughes Aircraft Company. EOSAT can provide users with photographs or electronic images suitable for display and image processing on desktop computers. These sorts of data are used throughout the world for monitoring irrigation, flood damage, oil spills and land erosion, for observing waterfowl habitats and for applications such as mineral prospecting, land-use control and real estate development.

SPOT

The success of Landsat led the French space agency CNES (Centre National d'Etudies Spatiales) to authorize, in 1978, the development of a French remote sensing satellite to be known as Satellite Pour l'Observation de la Terre (SPOT). The satellite design was modular with a platform carrying all of the systems, such as attitude control, power and communication, required to operate the satellite and a separate payload module which carried two high-resolution cameras. The cameras were able to show details as small as 10m across, making them superior to the instruments on the Landsat satellites.

SPOT-1 was launched by an Ariane rocket on 22 February 1986 into an 825km-high, circular, polar orbit. The satellite's ground track was designed to repeat every 26 days, but an ingenious arrangement of tilting mirrors means that the instruments could look out to either side of the path directly under the satellite to obtain more frequent images if required. This ability to look off to the side also means that SPOT can produce stereoscopic images for map-making purposes. A second craft, SPOT-2, was launched on 22 January 1990 and placed in an orbit diametrically opposite to SPOT-1. This orbital location means that while both satellites are operating, only 13 days elapse between SPOT satellites passing over a given point on the ground. SPOT-3 is scheduled for launch in 1992 and a fourth satellite has been ordered for a 1994 launch. This will ensure that the system remains in operation until at least the year 2000.

SPOT images are marketed by a special company called SPOT Image which was set up to exploit the commercial potential of the satellite. By 1990 the company had achieved sales of $32 million and with the market for remote sensing increasing all the time, SPOT Image hopes to become financially self sufficient by the mid 1990s.

SEASAT and ERS-1

SEASAT was a US satellite launched in 1978 which was designed specifically to observe the oceans. The spacecraft, which was based on an Agena rocket stage, was equipped with five instruments, the largest of which was a Synthetic Aperture Radar which could provide radar images of the water surface in any weather conditions. The other four instruments returned data on sea temperatures, wave heights and wind speeds. Although first results from SEASAT were very impressive, the satellite failed after only four months in orbit.

ERS-1 is an Earth Resources Satellite developed by the European Space Agency (ESA) and is similar in concept to SEASAT. The satellite's main objective is observations of the oceans and coastal areas. Like SEASAT it carries five instruments, the largest of which is an active microwave instrument which can operate as a Synthetic Aperture Radar able to see details as small as 30m on the ocean below. The other ERS-1 instruments provide information on sea temperatures, wave height and on the satellite's precise altitude.

ERS-1 was launched by an Ariane 40 rocket on 16 July 1991 and placed in an orbit 777km high which repeats every 15.33 days. Results from the satellite were excellent and detailed images of both the sea and the land were soon being released by ESA. ERS-1 is expected to remain in operation for three years and a follow-up satellite, ERS-2, is scheduled for launch in 1994 or 1995.

Two other missions to observe the Earth with radar are planned. The joint US/France TOPEX/Poseidon mission, which is intended to study the oceans by means of a radar altimeter, is due for launch in 1992 and the Canadian Radarsat project (see p23) is expected to be in orbit by about 1994.

Almaz, Okean and Resurs

The USSR also had a programme of Earth observation satellites, details of which are now being revealed as the veil of secrecy is removed from the Soviet space programme.

Almaz is an unmanned satellite based on military versions of the Salyut space station. Its main payload is a large radar system able to produce images that show objects about 15–30m across. After a long delay the first Almaz-type satellite, called Cosmos 1870, was launched into a 250km-high orbit on 22 July 1987 and de-orbited in 1989. Almaz-1 was launched on 31 March 1991. The USSR then offered to carry instruments from a variety of Western nations on later craft in the series.

Another Soviet radar satellite is the Okean series, first launched in 1983 under the covername Cosmos 1500. These carry a radar with 2km resolution to monitor the state of the polar icecap and other Earth observation instruments. A typical orbit for an Okean satellite is 670km high and inclined at 82.5° to the equator. An improved version is expected to come into service in 1993.

Earth observation is also the mission of the RESURS-F1 and -F2 satellites, the first of which was launched in 1977 under the covername Cosmos 1099. These satellites are based on the Vostok capsule used 30 years ago by Yuri Gagarin and carry cameras in what would have been the manned section of the Vostok. After a flight of two to four weeks the capsule, still containing the cameras and exposed film, is returned to Earth for examination. Newer versions, called RESURS-O, are based on a different satellite design and return data electronically without the use of a re-entry capsule. They operate from a Sun synchronous orbit and have a planned lifetime of two years. An improved version will be introduced in about 1992.

Satellite images can be useful for the World's press. Images of the 1986 nuclear disaster at Chernobyl and the 1991 Gulf War were in demand by television and other news organizations. Sometimes political factors prevent such images being released and this has led to the idea of MEDIASAT, an observation satellite which would provide images of world troublespots without risk of the censorship which can be exercised by the governments which operate national satellites.

The USA's Upper Atmosphere Research Satellite (UARS) was launched on 12 September 1991 from the Space Shuttle Discovery. The objectives of the satellite include monitoring of the Earth's ozone layer from its 600km-high, circular orbit. The UARS has 10 scientific instruments, some of which were provided by Canada, France and the UK.

The first Earth resources technology satellite

Observing the Sun

OSO 1 7 March 1962

Studying the nearest star

Our local star has been monitored from space for three decades.

Although it is natural to think of it as something special, the Sun is a fairly ordinary star which just happens to lie very close to the Earth. This makes the Sun relatively easy to study and so astronomers know much more about it than about any other star. Much can be learned using ordinary telescopes but, by taking experiments into space, it is possible to observe the Sun at wavelengths which are normally absorbed in the atmosphere. For example, ultraviolet, X-ray and gamma ray observations can reveal details of the Sun's tenuous outer layers which are impossible to study from the ground. Satellites can also be used to study the solar wind, the stream of charged particles which flow steadily outwards from the Sun. The solar wind is usually relatively quiescent but, from time to time, solar flares send a flood of extra particles into space. These blasts of particles may be channelled by the Earth's magnetic field into the upper atmosphere where they cause aurorae and radio blackouts. Solar physics can thus be divided into two disciplines: the study of the Sun itself and the study of solar–terrestrial relationships such as the interaction of the solar wind with the Earth's magnetic field.

OSO, ATM and SMM

The first spacecraft designed specifically for observing the Sun were the NASA Orbiting Solar Observatories (OSO). These were intended to observe the Sun in the ultraviolet, X-ray and gamma ray regions over an entire 11-year solar cycle. With the technology available in the early 1960s it was unreasonable to expect one craft to remain active for the entire 11 years so a series of OSOs were built, with the first launch taking place in 1962 and the last in 1975.

The OSO had an unusual, two-part design. The body of the spacecraft was a relatively flat, six-sided box which rotated several times a minute. On the top face of the box was a semicircular sail which rotated in the opposite direction to the box below. The result was that the sail, which was covered in solar cells, faced the Sun permanently, while the body rotated below it. This design meant that instruments in the sail remained pointed towards the Sun while those in the box scanned the Sun every rotation. The first six OSOs had a spherical fuel tank located at the end of each of the three arms which extended radially outwards from the box. The final two spacecraft, OSO 7 and 8, were larger and dispensed with the arms.

The Apollo Telescope Mount (ATM) was a collection of solar telescopes attached to the Skylab space station and was operated regularly during 1973–4. An interesting feature of the ATM was that some of its instruments recorded their results directly on to photographic film and so it was necessary for the astronauts to replace the film canisters by means of regular spacewalks. The ATM also marked the first use of grazing incidence telescopes (see p263) for X-ray astronomy. These telescopes, originally intended for an advanced OSO satellite, produced stunning X-ray images of the Sun.

The Solar Maximum Mission (SMM) was a project which followed the OSO series and which was, as its name implies, intended to study the Sun during the period of maximum activity in the Sun's cycle. Unlike the OSOs, the SMM was stabilized in all three axes so that it could point its seven main instruments at the Sun with great precision. The 2 300kg satellite was launched by a NASA Delta rocket on 14 February 1980.

In December 1980 the SMM suffered a serious problem in its attitude control system which made it impossible to point the satellite with the required accuracy. Although it was still possible to make some observations, the scientific productivity of the mission was drastically reduced and so, in 1984, astronauts from the Space Shuttle Challenger rendezvoused with the SMM and replaced several of the failed components. This restored the satellite's control systems and the revitalized SMM remained in service until 1989 when atmospheric drag caused it to re-enter and be destroyed.

Other projects

NASA has also carried solar physics experiments aboard the Space Shuttle, for example in 1985 Challenger carried a set of solar telescopes as part of the Spacelab-2 mission. Solar instruments were also carried on the French D2A and D2B satellites (launched in 1971 and 1975) and the USSR fitted their Salyut 4 space station with a large solar telescope. Japan has also launched satellites for solar observations. On 21 February 1981 they orbited the scientific satellite Hinotori and in 1991 the Solar-A (Yoko) mission was launched. In the 1990s the European Space Agency (ESA) expects to launch SOHO, the Solar and Heliospheric Observatory, which will observe the Sun directly while the Cluster mission, to be launched at about the same time, makes observations of the solar wind.

Solar–terrestrial relationships

The study of the physics of the solar wind is one which gets little attention since it concerns an interaction between two essentially invisible things. Except in the case of aurorae, either natural or artificial, it is difficult to produce dramatic images which relate to these phenomena. Nonetheless, the subject is important and involves much interesting physics, so a number of spacecraft have been launched to help scientists develop a better understanding of solar–terrestrial relationships.

Six box-shaped Orbiting Geophysical Observatories (OGO) were launched by NASA between 1964 and 1969 to complement the observations made by the Interplanetary Monitoring Platforms launched during the Explorer programme. The Soviet Electron satellites of the 1960s carried out similar studies and members of the Soviet Prognoz (Forecast) series are still being launched. Several small Japanese satellites were also devoted to studies of the solar wind, as were early satellites developed by the European Space Research Organisation (ESRO), for example ESRO-2B, and HEOS-1 and -2.

In the 1970s an international study of these phenomena was mounted using data from satellites such as the International Sun–Earth Explorers (three spacecraft), the European GEOS 1 and 2 satellites and Soviet Prognoz spacecraft.

Despite the progress which was made by these missions, much remains to be done and a series of new missions will take place in the 1990s. The European Cluster mission will involve four spacecraft flying in formation and sampling the solar wind at different locations, while the Japanese GEOTAIL will send a probe into the Earth's geomagnetic tail. The results from these missions will be combined with data from US and Soviet spacecraft which will be in orbit at the same time.

Some important solar physics missions

Mission	Date	Notes
OSO 1	7 Mar 1962	
OSO 2	3 Feb 1965	
OSO C	25 Aug 1965 (launch failure)	
OSO 3	8 Mar 1967	
OSO 4	18 Oct 1967	
OSO 5	22 Jan 1969	
OSO 6	9 Aug 1969	
OSO 7	29 Sep 1971 (improved design)	
OSO 8	21 Jun 1975	
SMM	14 Feb 1980 (in-orbit repair)	
ISEE 1/2/3	See p48	
IMP 1–10	See p47	
Prognoz 1	14 Apr 1972	USSR
Prognoz 2	29 Jun 1972	USSR
Prognoz 3	15 Feb 1972	USSR
Prognoz 4	22 Dec 1975	USSR
Prognoz 5	25 Nov 1976	USSR
Prognoz 6	22 Sep 1977	USSR
Prognoz 7	30 Oct 1978	USSR
Prognoz 8	25 Dec 1980	USSR
Prognoz 9	1 Jul 1983	USSR

Mission	Date	Notes
OGO 1	4 Sep 1964	
OGO 2	14 Oct 1965	
OGO 3	7 Jun 1966	
OGO 4	28 Jul 1967	
OGO 5	4 Mar 1968	
OGO 6	5 Jun 1969	
ESRO 2A	29 May 1967	(launch failure)
ESRO 2B	17 May 1968	European
ESRO 1A	3 Oct 1968	
HEOS 1	5 Dec 1968	Highly Eccentric
HEOS 2	31 Jan 1972	Orbit Satellites
Denpa	19 Aug 1972	Japan
Ume	29 Feb 1976	Japan
Kyokko	4 Feb 1978	Japan
Jikiken	16 Sep 1978	Japan
Hinotori	21 Feb 1981	Japan
Ohzora	14 Feb 1984	Japan
Akebono	21 Feb 1989	Japan
Yoko	30 Aug 1991	Japan

The Solar Maximum Mission satellite

Passengers in orbit

Vladimir Remek March 1978

Political junkets, paying passengers

As space travel becomes routine, there are more opportunities for non-astronauts to fly in space.

In the 1960s and 70s spaceflight was restricted to career astronauts, usually experienced jet pilots, but by the late 1970s NASA realized that the Space Shuttle demanded the selection of new types of astronauts. Pilots would still be needed to control the Space Shuttle, but different skills would be required to perform spacewalks, operate the Space Shuttle's robot arm and deploy satellites. These tasks were given to a class of astronauts called mission specialists, who were specifically trained to operate the Shuttle's complex systems. Another class of astronaut, called payload specialists, were selected to carry out specific tasks requiring skills not found within the regular astronaut corps. The payload specialists were expected to be scientists who would fly a mission and then return to their laboratories to analyse their observations.

The first Space Shuttle missions carried just two pilots, but gradually the size of the crews increased. The first mission specialist flew on the fifth flight and henceforth all missions carried two or three. The first payload specialists flew on the ninth Space Shuttle mission when Columbia carried the European Spacelab into orbit. Spacelab was a pressurized laboratory carrying a wide range of scientific equipment and Ulf Merbold and Byron Lichenberg were the payload specialists carried to work in the Spacelab module. The next payload specialist, McDonnell-Douglas employee Charles Walker, was carried on the twelfth Space Shuttle mission to operate an experiment provided by his company. Canadian Marc Garneau and US Navy employee Paul Scully-Power accompanied five astronauts on mission 13, but the boundary between astronaut and passenger was already beginning to blur because the roles of the two payload specialists on this flight were not entirely clear. Canada had built the Space Shuttle's robot arm, and planned to develop a similar system for the international space station, and the US Navy had an interest in observing the ocean from space, but both men were banished 'below decks' when the rest of the crew were engaged in complex operations on the flight deck.

Shuttle passengers

By the mid 1980s the Space Shuttle was flying regularly and NASA found itself under increasing criticism that the system was uneconomical. To deflect this, NASA tried to use the Space Shuttle as a commercial launch service, but it found itself in competition with the European Ariane launcher. Ariane had technical advantages over the Space Shuttle, but NASA had an important card to play; it could offer customers a ride into space to see the Space Shuttle in action.

First to take advantage of this was US Senator Jake Garn whose flight, in April 1985, was probably the world's most expensive political joyride. A second politician, Bill Nelson, flew aboard Columbia in January 1986. The lure of a national astronaut also appealed to customers trying to choose between Ariane and the Space Shuttle. The eighteenth mission deployed the multinational Arabsat-1B communications satellite and Prince Sultan Al-Saud, a member of the Saudi Royal family, was carried as an observer. Similarly, Mexican Rudolfo Neri Vela watched the deployment of the US-built, but Mexican-owned, Morelos satellite from Atlantis in November 1985. Britain selected the Space Shuttle to launch several of its

SKYNET military satellites and was promised that a UK astronaut could make a flight in 1986.

US companies were not immune to the lure of having an astronaut on the staff; Robert Cenker (from RCA) flew alongside Bill Nelson in January 1986 and Hughes Aircraft company employee Gregory Jarvis was twice promised a flight, only to be pushed down the queue by Congressmen Garn and Nelson. Jarvis was assigned to another mission, alongside teacher Christa McAuliffe who was selected as part of a 'citizen in space' programme that was expected to orbit journalists, artists and poets. The journalists had expected the first flight in this programme, but during the 1984 presidential election campaign Ronald Reagan announced that the first citizen in space would be a teacher. Over 11 000 teachers applied, but only McAuliffe was aboard the Challenger when it exploded on 28 January 1986, killing her, the luckless Gregory Jarvis, and five career astronauts.

The explosion of Challenger killed more than its crew; support for passengers on the Space Shuttle died too. The citizen in space programme was postponed indefinitely, and the astronaut observers lost their chance to fly as NASA transferred satellites to unmanned launches.

Guest cosmonauts

The Soviet guest cosmonaut programme began during the Salyut space station programme. As the missions grew longer it was felt unsafe to leave a Soyuz spacecraft docked to the space station for many months in case it suffered a failure which might leave the cosmonauts stranded. So, from time to time, a 'fresh' Soyuz spaceship was sent to replace the one already docked to the space station. This could be done automatically, the unmanned Soyuz 20 was an example, but it could also be done by sending a visiting crew to the station who would travel up in one Soyuz and return in the old one. Each such mission provided a chance to give a ride to a guest cosmonaut from a friendly nation who could carry out a few experiments then return to Earth.

The first such cosmonaut was Vladimir Remek of Czechoslovakia who flew to Salyut 6 aboard Soyuz 28 in March 1978. He was followed by representatives from the rest of the Warsaw Pact countries including Miroslaw Hermaszewski (Poland), Sigmund Jahn (East Germany), Georgi Ivanov (Bulgaria) and Bertalan Farkas (Hungary). The series soon expanded beyond Eastern Europe with flights by cosmonauts Pham Tuan (Vietnam), Arnaldo Mendez (Cuba), Jugderdemidyin Gurragcha (Mongolia), Dimitru Prunariu (Roumania), Jean-Loup Chretien (France), Rakesh Sharma (India), Muhammad Faris (Syria), Abdul Ahad Mohamand (Afghanistan), Alexander Alexandrov (Bulgaria) and a second flight for Jean-Loup Chretien. In every case the guest cosmonauts were assigned experimental tasks, but in many cases it is believed that these were nominal and the main objective of the flights was to strengthen political ties between the USSR and other nations.

Japanese journalist

All pretence of a scientific role for guest cosmonauts was abandoned when it was announced that the USSR would fly a Japanese journalist to the Mir space station for a fee of about $12 million. The two candidates for the flight were reporters from the TBS television station, Toyohiro Akiyama (aged 48) and 26-year-old camera woman Ryoko Kikuchi, whose hopes of a flight were dashed when she was rushed to hospital for an emergency appendectomy. Akiyama was launched aboard Soyuz TM11 in December 1990 and made a number of live television and radio broadcasts from the orbiting station. There were also some scientific experiments, including one reportedly chosen on the basis that the tree-frogs which formed the subject of the tests had cute faces which would appeal to children.

British cosmonaut

Akiyama's flight was followed by the Juno mission of British cosmonaut Helen Sharman. The Juno project began as a private venture which was to be sponsored by industry and by the British media, but the project was seen by many as a stunt and the necessary sponsorship was not forthcoming. Although Sharman flew to Mir in

May 1991, the British experiments originally planned for the flight were abandoned and little of consequence was achieved. An Austrian cosmonaut visited Mir in October 1991 and a German in March 1992 under similar commercial schemes.

> The rocket which lifted Toyohiro Akiyama into space was painted with advertisements, including one for a company which marketed babies nappies.

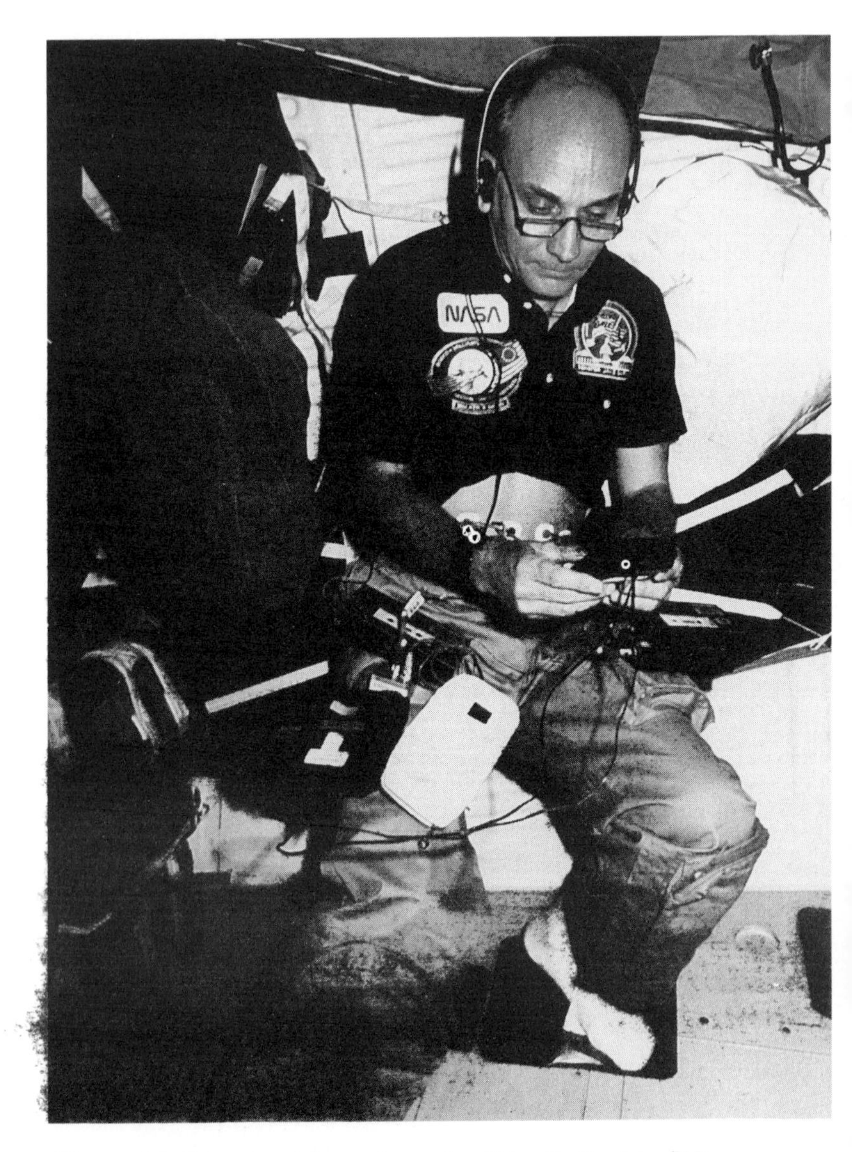

US senator Jake Garn was carried as an observer aboard Space Shuttle mission 51-D

Phobos

Phobos 1 launch 7 July 1988

So near, but so far

The Soviet Phobos project to explore Mars and its largest satellite came tantalizingly close to success.

The Phobos project was an attempt to restore interest in the red planet after the failure of the Soviet Mars programme during the 1970s (see p118). The objectives were to study the surface and atmosphere of Mars, to examine the Martian moon Phobos at close quarters and to study both the Sun and the interplanetary medium. The project was an international one, involving two Soviet spacecraft carrying scientific experiments from more than a dozen other countries.

The Phobos spacecraft was of a totally new design. It comprised a two-stage propulsion system attached to toroidal and cylindrical equipment compartments. The first stage of the propulsion system was used for mid-course corrections and to insert the spacecraft into orbit around Mars. The second stage was intended for orbital adjustments and attitude control once the spacecraft was closing in on Phobos. Above the propulsion module was the donut-shaped instrument compartment and above this was a cylindrical section topped by a steerable, dish antenna. Two rectangular solar panels attached to the instrument section were used to provide electricity.

Scientific instruments for observing the Sun were mounted on the exterior of the cylindrical section, and most of the instruments for studying Phobos were clustered around the toroidal section. Each spacecraft carried two small landers designed to be dropped on to Phobos as the main spacecraft cruised past only 50m above the surface. One of these, the Long Term Automated Lander, was intended to anchor itself to Phobos using a harpoon-type penetrator and then act as a semi-permanent geophysical station. The other, a spherical capsule known as the Hopper, was designed to investigate the chemical composition of Phobos's surface at a number of positions by landing, making measurements and then hopping to another location.

Phobos 1 was launched on 7 July 1988 by a Proton booster. Phobos 2 followed five days later. Contact with Phobos 1 was lost in September after an incorrect command sent by mission control caused it to switch off its orientation system and begin to tumble. With its solar panels no longer facing the Sun, Phobos 1 ran out of electrical power and fell silent.

Phobos 2 reached Mars on 29 January 1989 and went into a highly elliptical orbit around the planet. Over the next few weeks the orbit was changed to a circular one just outside the orbit of Phobos itself. From this orbit the probe transmitted a series of pictures which were used to refine the relative position of the spacecraft and its tiny target so that the final rendezvous sequence could begin. In late March preparations were made to send the probe towards Phobos so that it could complete its flyby and release the two lander capsules. Then, on 27 March 1989, contact with Phobos 2 was lost and could not be re-established.

The precise cause of the failure is not known, but it came when the USSR was about to make the first ever landing on Phobos. Although interesting results had already been obtained on the cruise to Mars and during the two months in orbit around the planet, the failure of Phobos 2 so close to its final destination was a cruel blow to the mission's scientists.

Phobos, Mars's innermost satellite, is an irregular object about 27km along its longest axis.

Pioneer

Moon 1958, Jupiter 1973, Venus 1978, Saturn 1979

Into deep space

The US Pioneer programme had an inauspicious start, but reached a grand finale.

The Pioneer programme originated in the Advanced Research Projects Agency of the US Department of Defense which, in 1958, authorized the launch of five spacecraft towards the Moon. Three of these spacecraft were the responsibility of the US Air Force and two were the responsibility of the US Army.

The first to be launched were the drum-shaped US Air Force probes which each weighed 38kg and contained a variety of simple scientific instruments. The probes were intended to go into lunar orbit and so were fitted with small, solid-fuelled rockets which could be used to adjust their course en route to the Moon and then to insert the craft into lunar orbit. It was an ambitious plan which had a bad start on 17 August 1958 when the first launch attempt failed after just 77 seconds.

The next attempt, designated Pioneer 1, was launched by a Thor-Able rocket on 11 October 1958 and was the first mission launched under the auspices of the newly formed NASA. It did not achieve sufficient velocity to escape from the Earth and, instead of flying to the Moon, the spacecraft reached an altitude of 113854km before falling back to destruction in the South Pacific. Pioneer 2 even was less successful: launched on 8 November 1958 the second stage of its launcher failed and the probe only reached an altitude of 1500km before falling back to Earth. On 6 December 1958 Pioneer 3 suffered the same fate as Pioneer 1 and fell back to destruction after climbing to 101000km.

A measure of success was achieved with the first Army probe, the 5.9kg, cone-shaped, Pioneer 4. This tiny probe, which was not intended to go into lunar orbit, was launched by a Juno 2 rocket on 6 December 1958 and passed about 60000km from the Moon. Since the flyby distance was much larger than planned, the probe did not return any lunar data. A further launch, on 26 November 1959, was unsuccessful.

A new design was then tried, a spherical capsule with solar panels mounted on paddle-like appendages. This weighed 175kg and was intended to be launched by the more powerful Atlas-Able rocket. No less than three attempts to send one of these probes to the Moon failed before the programme was abandoned.

Pioneer 5–9, into interplanetary space

After the dismal performance of the Pioneer Moon probes, the next Pioneers were given objectives that were less visible on the world stage. Each craft was sent into orbit around the Sun to report on conditions in interplanetary space. Pioneer 5, a 43kg, 66cm-diameter sphere with four solar panels, was launched on 11 March 1960 and sent into an orbit which ranged between the Earth and Venus. This highly successful flight provided the first experience in communicating with a spacecraft at what was then the very great distance of 37 million km.

Pioneer 5 was followed by a new series of drum-shaped, spin-stabilized spacecraft. Each spacecraft, which weighed about 63kg, had solar cells wrapped around the sides of the drum and was equipped with sensors to monitor cosmic rays, magnetic fields and the solar wind. The intention was that each Pioneer would operate for at least 180 days, sufficient time for it to make a journey half way around the Sun. A total of five of these were built and the first four, Pioneer 6–9, were launched on 16 December 1965, 17 August 1966, 13 December 1967 and 8 November 1968 respectively. The last one, called Pioneer E, was destroyed during a failure of its Delta launch vehicle on 7 August 1969. Pioneer 6 and 9 were placed in paths that took them inside the orbit of the Earth and to within 118 million km of the Sun. Pioneer 7 and 8 were sent outwards into orbits that took them 15 million km beyond the Earth. All four missions exceeded their design life many times: Pioneer 9 remained in operation until 1983 and the others were still operating in the 1990s.

The longevity of the Pioneer spacecraft enabled them to make important contributions to our understanding of interplanetary space. In particular, physicists could draw on near simultaneous data from all four craft to produce a comprehensive picture of conditions in the inner solar system.

Pioneer 10 and 11, first to Jupiter

The next Pioneers were sent further into uncharted space. Their objectives were to traverse the asteroid belt, to report on conditions in deep space and to investigate conditions close to the giant planet Jupiter. This would provide important scientific data and pave the way for the more complicated missions, like the Voyager project, which were by then being planned.

Pioneer 10 (like its almost identical twin Pioneer 11) was very different from the previous Pioneers, although the design drew heavily on previous experience. The dominant feature was a 2.75m-diameter dish antenna required to beam radio signals back to Earth from Jupiter. Scientific instruments and the spacecraft control systems were grouped together on the back of the dish, as were the structures which joined Pioneer to its Atlas-Centaur launch vehicle. A single TE-M-364-4 solid-fuel motor was used to give Pioneer 10 an extra boost once it had separated from the Centaur stage. Electrical power was provided by four Radioisotope Thermoelectric Generators or RTGs (see p153) which use the heat produced during the decay of radioactive material to generate electricity. RTGs were fitted because solar panels could not generate enough electricity to operate the spacecraft when it had reached as far as Jupiter. The RTGs were mounted in pairs on long booms so that they were removed as far as possible from the delicate scientific instruments. The entire spacecraft was stabilized by spinning it around the axis of the radio dish. This kept the dish pointed accurately back to Earth and allowed Pioneer's scientific instruments to scan across the sky as it travelled through space.

Pioneer 10 was launched on 2 March 1972 and flew through the asteroid belt without incident, showing that the risk from interplanetary dust and other debris in this region was far less than originally thought. The spacecraft flew past Jupiter on 4 December 1973 and revealed new details about the magnetic field and radiation belts surrounding the planet. Although the spacecraft did not carry a television camera, it was possible to return some images of Jupiter by joining together data from a photometer which scanned the planet in strips every time the spacecraft rotated. These images, although relatively crude compared with pictures from the later Voyager spacecraft, were better than any ever taken from the Earth.

Pioneer 11, launched on 5 April 1973, flew past Jupiter on 5 December 1974 and used Jupiter's gravitational field to send it on to a rendezvous with Saturn on 1 September 1979. As well as returning the first spacecraft images of Saturn, the Pioneer 11 encounter of the planet provided important information required to plan the trajectories of the two Voyager spacecraft through the complex system of moons and rings that surround Saturn.

After their encounters with the giant planets, both Pioneer 10 and 11 continued to operate, returning data on the outermost regions of the solar system. Pioneer 10 crossed the orbit of Neptune on 13 June 1983, becoming the first spacecraft to leave the solar system. On 22 September 1990 it was 50 times further from the Sun than the Earth, at which time radio signals from the probe took nearly 14 hours to reach ground control. It is hoped that contact with Pioneer 10 will be maintained until about the year 2000. Pioneer 11 crossed Neptune's orbit on 23 February 1990 and, if technical problems encountered in late 1990 can be solved, may remain in service until about 1995.

Pioneer Venus

The final flights in the Pioneer series involved two spacecraft sent to Venus. The first, known as the Pioneer Venus Orbiter was launched on 20 May 1978 and went into orbit around Venus on 4 December. As well as observing the details of Venus's atmosphere and the interaction of the solar wind with the planet, the orbiter carried a radar altimeter that was used to map the planet's topography. The orbiter far exceeded its original mission and was still in service in 1991, although by that time the more sophisticated Magellan radar mapper (see p105) was also in orbit around the planet.

The other spacecraft, the Pioneer Venus Probe, consisted of a carrier spacecraft and four entry probes. It was launched on 8 August 1978 and arrived at Venus on 9 December. The entry probes, three small and one large, were released from the carrier and plunged into the atmosphere at widely separated points on both day and night sides of the planet. As it fell towards the surface, each probe, which comprised a roughly spherical shell tucked behind a shallow conical heatshield, returned data on the temperature, pressure density and chemical composition of the atmosphere. Unlike the Soviet Venera missions (see p245) the Pioneer probes were not expected to survive the impact with the surface, although one did transmit for a few minutes after landing. The carrier spacecraft burned up as planned in the upper Venusian atmosphere.

As the first probes to leave the solar system, Pioneer 10 and 11 carry plaques which that will allow anyone who finds them to trace them back to the Earth using pulsars as astronomical signposts. The plaques, 152mm × 229mm in size, are of gold-anodized aluminium plate into which is etched a diagram of the solar system, a pulsar map and pictures of a man and a woman. The plaques are mounted in a position on the spacecraft's antenna mount which is expected to protect them from erosion by interstellar dust for at least 100 million years.

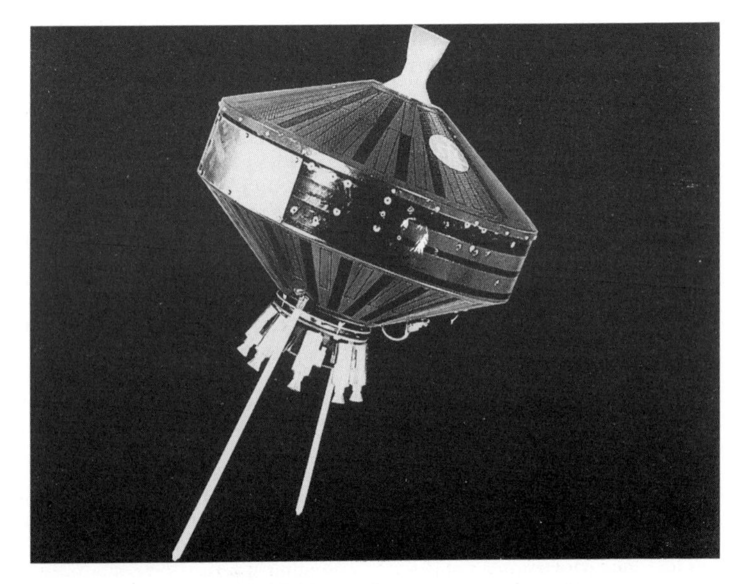

A model of Pioneer 1, the first US Moon probe

Power for spacecraft

Batteries, cells and generators

What a space mission can accomplish may be decided by the amount of electrical power that is available to operate instruments and to return data to Earth. The means of producing this power is up to the spacecraft designers, but several basic rules apply to all space power systems: they must be small, light and reliable yet must consume little fuel and must operate in the vacuum of space.

Batteries

Batteries are simple and reliable so are a sensible choice for short space missions. The first satellite, the Soviet Sputnik, used batteries as did the first US satellite, Explorer 1. Similarly the first manned spaceships, the USSR's Vostok and the USA's Mercury, relied on batteries because the energy required to operate these one-man craft was quite low, and the missions were short.

Today, batteries seldom form the main power supply for satellites, but they are still required for a number of secondary purposes. For example, although many spacecraft use solar cells (see below), batteries may be necessary to provide power before the craft has extended its solar panels or if, due to some emergency, the solar panels lose their lock on the Sun. In the latter case the satellite will normally operate from battery power while its computer, or ground control, tries to rectify the situation.

Rechargeable batteries are also needed for satellites which rely on solar panels, because sunlight is cut off on the night side of the Earth and satellites in low orbits can be in darkness for about half of the time. The batteries are used to supply the power needed when the satellite is in darkness and are recharged whenever the satellite is in sunlight. Various types of rechargeable battery are suitable for space use including nickel-cadmium, silver-cadmium and silver-zinc types. Some deliver low power, but can be recharged many times, others provide more power, but will survive only a few hundred charging cycles.

Fuel cells

A fuel cell is a sort of continuously operating battery. However, unlike a conventional battery, which carries its chemical reactants within a sealed case, the fuel cell draws its reactants in from outside. Although a fuel cell is more complicated and heavier than a single battery, it can go on producing power until its fuel supply runs out. Thus, when the total amount of energy required during a mission is large, it is better to carry one or two fuel cells than many heavy batteries. For missions requiring considerable power over a moderately long period, such as the USA's manned Apollo and Space Shuttle missions, fuel cells are a natural choice.

The most commonly used fuel cell combines hydrogen and oxygen to produce electricity. The hydrogen and oxygen are stored as cryogenic liquids, but are turned into gas before they enter the fuel cell, which consists of two porous electrodes surrounded by an electrolytic solution. The hydrogen and oxygen diffuse into the electrolyte via the porous electrodes, where they react to produce electricity. Water is produced as a by-product of this reaction and on the Apollo missions this was supplied for the astronauts to drink. Each Apollo service module carried three fuel cells, any one of which was able to provide enough power to keep the spacecraft operating in an emergency, although two were required during the most active portions of a normal mission.

Solar cells

Most Earth satellites use solar cells which convert sunlight directly into electricity. They are usually made from sheets of silicon to which small quantities of boron have been added. One side of the sheet is also treated with small quantities of phosphorus. Boron and phosphorus have a different atomic make-up to silicon, and when sunlight falls on the phosphorus-rich side of the cell the phosphorus tends to give up electrons (subatomic particles which carry a negative charge) and these try to flow towards the boron-rich region. This flow constitutes an electric current which can be used to provide power. Each cell produces only a tiny amount of power and many hundreds, or even thousands, must be joined together to produce enough electricity to operate a large satellite. None the less, solar cells are reliable and use no fuel other than sunlight, so are suited to missions lasting many years.

Unfortunately, solar cells are not very effective beyond the orbit of Mars. This is because the amount of sunlight falling on a solar panel falls rapidly as its distance from the Sun increases. In mathematical terms, the power falls as the square of the distance from the Sun: double the distance and the power falls to one-quarter, triple it and the power falls to one-ninth and so on. Spacecraft that travel far from the Sun would need impossibly large solar panels, and so use nuclear power instead.

Radioisotope Thermoelectric Generators (RTGs)

The RTG is a nuclear power generator which uses the heat generated when unstable radioactive isotopes decay. In its simplest form it consists of a cylindrical fuel rod of radioactive material (such as plutonium 238) which is surrounded by semiconductor devices called thermoelectric generators which convert some of the heat from the radioactive decay directly into electricity. The power produced depends on the heat being released from the radioactive material and, since the radioactivity decays with time, the total power produced falls steadily throughout the life of the RTG.

RTGs were used on spacecraft like the USA's Voyager and Pioneer, which ventured far from the Sun, and for its Viking spacecraft which landed on Mars. RTGs were also used to power the automatic monitoring stations which the Apollo astronauts left behind on the Moon. Only a few have been used in Earth orbit because of the risk that when the satellite re-enters it will spread radioactive fuel over a wide area. Just such an accident befell the two RTGs on a US Nimbus satellite which crashed in May 1968 due to a launch failure.

Nuclear reactors

Nuclear fission reactors, in which the decay of one atom triggers the release of several more in a chain reaction, have also been used as space powerplants. The fission reactor can produce more heat, and hence more electricity, than the decay of an unstable isotope. Also, unlike the RTG, the rate of the fission reaction can be controlled by inserting moderators into the nuclear material as is done for terrestrial nuclear powerplants. The heat produced by the fission reaction can be converted into electricity directly, as in an RTG, or it can be used to turn a working fluid (such as mercury) into a vapour which is then driven through a turbine to produce electricity, condensed back into a liquid and then returned to the reactor to repeat the cycle. Fission reactors can produce large quantities of power over long periods, but they are heavy and require considerable cooling to dispose of waste heat.

The first US nuclear reactor was called SNAP-10A and was launched on a test mission on 3 April 1965. Since then, US interest in space reactors has dwindled, although the Strategic Defence Initiative (Star Wars) has rekindled work in this field. The USSR uses a nuclear reactor design called Topaz to power some of its military satellites and it was one of these, Cosmos 954, which crashed and scattered radioactive debris across Canada in 1978.

> The nuclear fuel for the Apollo lunar stations was launched inside an armoured container to stop the radioactive material being spread across Florida in the event of a launch failure. The astronauts had to insert the fuel core into the RTG once on the Moon.

This 4m-wide, 32m-long solar panel was deployed from the Space Shutle in 1984 as a test for future applications

Radio astronomy satellites

Karl Jansky 1932, RAE-1 1967

Beyond the ionosphere

Radio astronomy from space demands very large antennas, or ingenious observing techniques.

Radio waves from beyond the Earth were first recognized in 1932 by Karl Jansky, a physicist at the Bell Telephone Laboratory in Homdel, New Jersey. Surprisingly, Janksy's discovery aroused little interest at the time and it was an amateur astronomer, Grote Reber, who was first to build a radio telescope and explore the sky at radio wavelengths. After World War II, physicists who had been developing military radars turned their newly found expertise back to scientific research and radio astronomy was born.

Radio astronomy is possible because the Earth's atmosphere is transparent to radio wavelengths between a few millimetres and a few tens of metres. Ground-based radio telescopes, like Jodrell Bank in Britain, all operate in this wavelength range. Radio waves with longer wavelengths are reflected by the ionosphere, a region of charged particles (ions) extending from about 90km to about 500km above the Earth. The reflecting properties of the ionosphere make it possible to bounce radio messages from continent to continent, but also prevent long-wavelength radio waves reaching the Earth from space. To explore the sky at these wavelengths, astronomers must place their radio receivers beyond the ionosphere.

The first detections of long-wavelength extra-terrestrial radio emissions were made with satellites in the Alouette (Canada/UK), Ariel (UK/US) and Electron (USSR) series, even though these projects were primarily interested in the properties of the ionosphere, not of astronomical objects. A few simple astronomical experiments were also carried out on unmanned lunar and interplanetary probes which, of necessity, travelled far beyond the ionosphere. These early experiments revealed the spectrum, that is the variation of intensity with wavelength, of the cosmic radio emissions, but could tell little or nothing about the directions from which the emissions came.

Radio Astronomy Explorers

One of the difficulties with early radio experiments was the lack of angular resolution (the ability to separate sources close together) possible with the relatively small antennas on the early satellites. Since the angular resolution of a radio telescope is determined by the ratio of the size of the radio antenna used to the wavelength of the radio waves under study, the only solution was to develop satellites with very long antennas. The first of these were NASA's two Radio Astronomy Explorers (RAE).

Each RAE was a 1m-tall, cylindrical body with solar cells on four extending paddles. The first, known as RAE-1 and also as Explorer 38, weighed about 200kg. It was launched on 4 July 1968 by a Delta rocket and was placed into a circular orbit, 5 800km high and inclined at 121° to the equator. Once in its final orbit RAE-1 began a slow, but spectacular metamorphosis as long thin radio antennas began to sprout from the satellite's body. Each antenna was comprised of thin metal tape which unrolled from a drum in the satellite and, once clear of the satellite's body, curled up to form a tube 1.3cm in diameter. Four such tubes formed an X-shaped array with arms over 450m long. RAE-1 was orientated so that half of the X was towards the Earth and the other half pointed into space. Every time the satellite went around the Earth, the upper part of the X swept out a strip of sky and recorded the radio signals detected.

During its four-year lifetime RAE-1 detected radio emission from our galaxy, probably emitted by electrons travelling through space and spiralling around under the influence of magnetic fields, from the Sun and from the planet Jupiter. A similar radio astronomy experiment was carried on Explorer 43. This satellite, whose main purpose was to study interplanetary space, was launched into a highly eccentric Earth orbit on 13 March 1971. The results from its radio experiment were generally similar to those from RAE-1.

Radio interference from the Earth, particularly from thunderstorms, had been a major problem for RAE-1, so it was decided to place the second RAE in orbit around the Moon. The 328kg spacecraft (200kg plus the motors to place it into lunar orbit) was launched by a Delta rocket on 10 June 1973 and entered lunar orbit five days later. It was the last US spacecraft to go into orbit around the Moon and, apart from improved instruments, was similar to RAE-1. RAE-2 (also known as Explorer 49) took advantage of periods when the Moon shielded it from terrestrial radio signals to produce maps of the radio emission from our galaxy.

The Salyut KRT-10

The Soviet Cosmic Radio Telescope (KRT-10) was attached to the Salyut 6 space station in 1979. This large, deployable, radio telescope was carried to the station by the Progress 7 supply craft. The folded antenna was then fitted to the rear docking port of Salyut 6 in such a way that it projected into the cabin of the still attached Progress 7 craft. When Progress 7 had undocked and pulled away from Salyut 6, the dish-shaped antenna was unfurled by remote control. The radio telescope was used for three weeks before the cosmonauts attempted to jettison it so that the docking port could be used by another supply ship. Unfortunately, the dish became snagged on the docking port and the cosmonauts were forced to make an emergency spacewalk to cut it free.

Space VLBI

As already noted, the long wavelengths of radio waves means that even large radio telescopes cannot distinguish between radio sources that are close together. To surmount this obstacle, radio astronomers have learned how to use pairs of telescopes to produce better angular resolution than possible with a single dish. By combining signals from two dishes that were both looking at the same object at the same time, astronomers can produce radio pictures as clear as they would have obtained from a single dish as large as the distance between the two dishes they used. This technique is called Very Long Baseline Interferometry (VLBI) and can produce results that mimic those from a telescope as large as the diameter of the Earth. To get better resolution requires a baseline (the distance between the two telescopes) larger than our planet and to do this it is necessary to put one of the telescopes into space.

No space VLBI mission has yet taken place, but several are being considered. The European Space Agency (ESA) studied a mission called QUASAT (short for QUAsar SATellite) which would have involved a dish 15m in diameter placed in an orbit which might have extended as far as 36 000km into space. QUASAT was not actually built, and a modified proposal called IVS, the International VLBI Satellite, which was proposed to ESA a few years later, was not selected for development. The USSR was planning a similar project called Radioastron which would employ a 10m dish in an orbit extending 200 000km from the Earth. Canada has agreed to participate in the Radioastron mission by supplying various equipment to the USSR. Japanese astronomers are also considering a Space VLBI satellite with a dish 10m in diameter. It is not clear which of these missions will actually occur and it is possible that some of them may be merged to save money. This would be entirely logical since VLBI is already an international activity which regularly uses telescopes spread across different continents. An international space VLBI satellite would be a natural extension of the worldwide cooperation already existing amongst radio astronomers.

> When fully extended the antenna of the RAE-1 and RAE-2 satellites were taller than the Empire State Building in New York.

Ranger

Ranger 1 1961, Ranger 9 1965

Crashlanding a camera

The US Ranger spacecraft provided the first close-up photographs of the Moon as they plunged to destruction.

After the unsuccessful Pioneer moonprobes of the late 1950s, NASA's Jet Propulsion Laboratory continued to develop new lunar missions. The first of these was a relatively large spacecraft, stabilized in three axes, which would fly to the Moon and return pictures and other data via a steerable, high-gain radio antenna. It was called Ranger and comprised a tubular body attached to a hexagonal base containing the equipment for inflight control and communications. Two rectangular solar panels, used to produce electricity, were attached on opposite sides of the hexagonal base. The dish-shaped, high-gain antenna was also mounted, via a hinge, to the hexagonal section. The precise design of the spacecraft evolved during the nine-mission programme, but each weighed about 350kg, had a solar panel span of about 5m and stood 3m tall.

In a typical mission each Ranger would be placed into a low parking orbit by an Atlas-Agena rocket. After a brief (about 15-minute) coasting flight while tracking data was checked and the trajectory to the Moon recalculated, the Agena rocket ignited to boost the spacecraft out of its parking orbit. The Ranger then separated from the Agena, unfolded its solar panels, and used gas jets to manoeuvre around until a light-sensitive sensor located the position of the Sun. Ranger then stabilized itself so that the Sun was shining on its solar panels and pointed its antenna back to Earth. It was then ready to return any data which it collected as it approached the Moon. A small rocket motor was carried to make any mid-course corrections required during the flight.

Ranger 1 and 2, engineering tests

Ranger 1 and 2 were not intended to fly to the Moon, but rather to test out the design of the spacecraft and the operating procedures required for the planned lunar missions. Ranger 1 was launched by an Atlas Agena rocket on 23 August 1961 and successfully placed into a 174km × 280km parking orbit. Unfortunately, the Agena stage failed to re-ignite properly, leaving Ranger 1 trapped in low orbit. Although the spacecraft did separate from the Agena and attempted to orientate itself towards the Sun, its attitude control system was confused when the craft went behind the Earth and the Sun sensor could no longer detect sunlight. Ranger 1 re-entered the atmosphere about a week later after completing 111 orbits. Ranger 2, launched on 18 November 1961, suffered a similar fate when its Agena failed to restart. It re-entered two days later.

Ranger 3–5, rough landing attempts

Despite the problems with the Agena, it was decided to continue with the Ranger programme and to attempt to obtain scientific data from the Moon. To do this, the next Rangers were fitted with a small television camera developed by the Radio Corporation of America (RCA). The camera was expected to return pictures showing objects as small as 3.5m across, as the Ranger headed directly towards the Moon. The camera would cease transmitting when the Ranger was about 25km above the lunar surface and only 8 seconds from impact so that a capsule, intended to make a rough, but survivable landing on the Moon, could be separated from the parent craft. Once clear of the spacecraft, a solid-fuelled retro-rocket attached to the capsule was to fire, slowing it down so that it came to a stop about 330m above the Moon. The rocket would then be jettisoned, allowing the capsule to fall on to the lunar surface. The speed at which the capsule would impact was estimated at about 175km per hour. The capsule, which contained a seismometer (a device for measuring moonquakes) floating in a bath of oil to cushion the shock of landing, was then expected to roll to a stop, release the shock-absorbing oil, and begin to transmit data back to Earth. It was an ambitious and daring plan.

Ranger 3 was launched on 26 January 1962, but was thwarted by a problem with the Agena rocket. This time, instead of failing to ignite, the Agena sent the Ranger away from Earth at too great a velocity. The mid-course engine was unable to put Ranger back on to the correct track and it missed the Moon by 36 785km. Attempts were made to take photographs, but the high-gain antenna failed to point towards the Earth and no usable pictures were received. Ranger 4 was launched on 23 April 1962 but, due to failure of its onboard timer, it did not deploy its solar panels. The timer problem also meant that Ranger 4 could neither receive nor transmit information and it was silent when it crashed on the far side of the Moon 64 hours later. Ranger 5, launched on 18 October 1962, was also a failure. This time a problem developed with the solar cells themselves and contact was lost when the spacecraft's chemical batteries ran down. Ranger 5 missed the Moon by 725km.

Ranger 6–9, flying cameras

The objective of the final Ranger missions was changed to concentrate on obtaining the high-quality photographs of the lunar surface needed by the designers of the Apollo spacecraft. To do this, the original television camera and landing capsule were removed and replaced by a battery of six cameras, four with telephoto and two with wide-angle lenses. The camera package, which included the electronics required to operate the system, weighed 170kg and was installed in a conical aluminium shroud which sat atop the usual hexagonal central body. The six cameras looked out of a cut-out near the top of the shroud.

The flight of Ranger 6, launched on 30 January 1964 by an Atlas-Agena B, looked as if it would vindicate the programme at last. All went well as the spacecraft headed towards the Moon. Twenty minutes before impact Ranger 6 was pointed in the correct direction for picture taking and the television cameras were switched on so that they could warm up. The first pictures should have been transmitted a few minutes later, but none ever came and the Ranger smashed into the Moon with its camera inoperative. It was eventually concluded that the cameras had been accidently switched on for 67 seconds during the launch and that electrical arcing had destroyed crucial circuits. Ranger 6 had been blind almost from birth.

The final three missions, however, laid the ghost of the earlier failures. Ranger 7, launched on 28 July 1964, was completely successful. A total of 4 316 pictures were returned by the camera pack during the last 13 minutes of the 68.5-hour mission before Ranger 7 struck the Sea of Clouds at about 9 300km per hour. Ranger 8, launched on 17 February 1965, repeated the success, returning 7 137 pictures before it smashed into the Sea of Tranquillity. Ranger 9, which was launched on 21 March 1965, plunged into the Moon near the crater Alphonsus after returning 5 814 pictures.

The results from the last three Ranger missions were very well received. The images returned a few moments before destruction showed objects as small as 1m across, providing information for the designers of the Apollo lunar module. Astronomers were also very interested in the pictures, which provided views a thousandfold better than possible from a ground-based telescope. Today the Ranger missions are almost forgotten, but they played an important role in the eventual success of the Apollo programme.

The crash sites of the Ranger probes were:

Ranger 6	9° 24′ North	21° 30′ East
Ranger 7	10° 38′ South	20° 36′ West
Ranger 8	2° 43′ North	24° 38′ East
Ranger 9	12° 58′ South	2° 22′ West

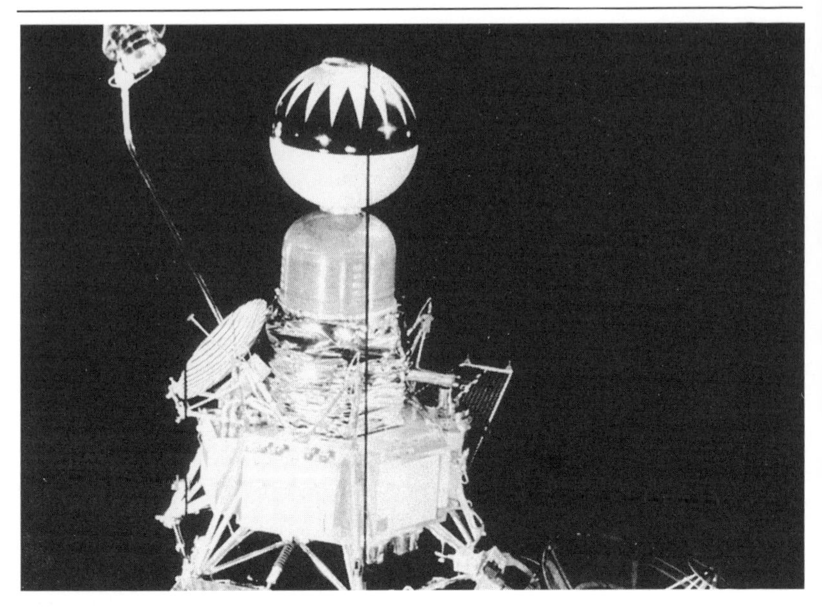

The Ranger 3 spacecraft with its spherical landing capsule on top

An artist's impression of one of the later Ranger spacecraft about to crash on the Moon

Routes to the Moon

The great debate 1961–1962

Four ways to go

Although the techniques used by the USA's Apollo to send men to the Moon now seem obvious, various alternatives are possible.

The method known as direct ascent is at first sight the most obvious way of sending men to the Moon. In its simplest form a large moonship would take off from Earth and fly directly towards the Moon. As the spaceship approached the lunar surface it would turn and lower itself on to the Moon using its main engine as a brake. After a period of exploration, the astronauts would use the same engine to escape from the Moon and return to Earth. This was the method featured in George Pal's famous film *Destination Moon* but in reality direct ascent presents major engineering challenges. The single-stage moonship portrayed by Pal could not be built with any conceivable chemical rockets because it would be too heavy to lift itself into space. However, this problem could be overcome by using a multi-stage rocket and so the technique is worth considering.

One disadvantage would be the precise navigation that would be required for the moonship to arrive directly over its designated landing site during its rapid approach to the Moon. The moonship would then have to turn around and fire its braking rocket with great precision so that it could be brought to a stop just before touchdown in a single smooth rocket burn that would begin far above the Moon and finish at an altitude of a few metres. This was done by the unmanned US Surveyor and Soviet Luna craft and, although one Surveyor and several Lunas crashed during this critical rocket firing, others landed safely.

The moonship would, however, be rather larger than a Surveyor. It would have to carry a braking rocket and a rocket to lift the ship off from the Moon with the astronauts, their supplies and all the equipment needed for landing back on Earth. Since the spaceship would have to be streamlined for its initial climb through the Earth's atmosphere, it would almost certainly be tall and thin and so might be in great danger of toppling over if it landed on a sloping surface. A further disadvantage of its shape would be the distance that the astronauts would have to descend to the surface to carry out their exploration.

However, if the size of the launching rocket required, and the difficulties of reversing a large rocket smoothly on to the Moon could be overcome, direct ascent does have advantages. For example, no complicated rendezvous or docking operations would be required and the lift-off from the Moon and return to Earth would be relatively straightforward since the rocket would simply blast off directly into the correct trajectory.

Direct ascent was considered by NASA in the early 1960s for its Moon landing programme. Use of this method would have required a launch vehicle much larger than the Saturn 5 rocket, and studies of such a vehicle were conducted under the name Nova. Although never designed in detail, because the direct ascent technique was not adopted, Nova would have been a five-stage rocket with a take-off weight of about 3000 tonnes.

Lunar Surface Rendezvous

One method of reducing the size of the moonship, which makes it easier to launch and easier to land on the Moon, would be to send it with only enough supplies for a one-way trip. In this scheme the manned moonship would have to land close to a cache of supplies deposited on the lunar surface by an automatic supply ship. The astronauts' first task once on the Moon would be to locate this supply ship and refuel their own spacecraft ready for the return to Earth. Needless to say, such a mission would be fraught with risks. For example, the astronauts might fail to find the supplies, or might be unable to use them for some reason, and would be stranded on the Moon. In this regard it is chilling to consider that during the first Moon landing the Apollo 11 lunar module 'Eagle' missed its planned landing point by about 5km.

Earth Orbit Rendezvous

The method originally favoured by many rocket engineers, and which frequently appears in the books and paintings of the 1950s, is the technique of Earth orbit rendezvous. In this method several sections of the moonship would be launched separately into Earth orbit and assembled there. The moonship would set off for the Moon, land and then take off again in the same manner as that used for the direct ascent. A variant of this method would be to launch the moonship whole, but empty of fuel, and top it up before it set off for the Moon.

The main advantage of Earth orbit rendezvous is that the size of the rockets required to lift the necessary equipment from the Earth is dramatically reduced. NASA estimated that if Earth orbit rendezvous was adopted for Apollo then two Saturn 5 rockets could do the job, avoiding the need to develop the massive Nova booster. Another advantage is that the relatively dangerous (and in the early 1960s untried) rendezvous and docking manoeuvres would all be done in Earth orbit where there would be a chance of rescue if something went wrong.

The disadvantages of Earth orbit rendezvous include the need to launch two or more large rockets in quick succession, and the complexity of space assembly and refueling in the weightless conditions of Earth orbit. It would also have been expensive, since it required two Saturn 5 rockets per mission. Although initially favoured by NASA's Wernher von Braun, this method was dropped in favour of the lunar orbit rendezvous technique.

Lunar Orbit Rendezvous

The lunar orbit rendezvous technique requires the use of two different spacecraft, but each of these can be much smaller than a single craft required to carry out an entire mission by itself. One of the two vehicles would be responsible for carrying the astronauts from the Earth to lunar orbit and back, the other would be used to descend to the Moon and return to the orbiting mother ship. The advantage of this technique is that the lunar landing vehicle can be designed specifically for its part of the mission and need not carry any unnecessary weight. In particular, all the equipment and supplies for the return to Earth could be left in lunar orbit rather than being carried down to the surface and then back up again. Once the landing was over, and the astronauts had returned to the orbiting mother ship, the lander could be discarded, thus reducing the size of the rocket needed to boost the mother ship out of lunar orbit for the journey home.

The main disadvantage of lunar orbit rendezvous is that the complicated rendezvous and docking manoeuvres required would all take place in orbit around the Moon. This would make accurate tracking by Earth-based radars difficult and offers no possibility of rescue if, for any reason, the astronauts in the lunar lander were unable to rendezvous with the mother ship.

Despite the risks, this was the method finally adopted for the Apollo programme, since calculations indicated that the entire mission could be launched by a single Saturn 5 rocket. Von Braun, after initial resistance to the idea, eventually realized that lunar orbit rendezvous offered the best chance of completing a Moon landing mission within the deadline set by President John F Kennedy and threw his considerable influence behind the plan. The correctness of the decision to adopt lunar orbit rendezvous was shown by the eventual success of the Apollo programme.

Salyut 1–5

Salyut 1 1971, Salyut 5 1976–1977

Early Soviet space station programme

The early Salyuts laid the foundations for permanently manned space stations.

As it became obvious that the USA was going to win the Moon race, the emphasis of the Soviet space programme shifted away from lunar exploration and towards Earth orbiting space stations. The first of these, known as Salyut (Salute) 1 was launched by a Proton rocket on 19 April 1971. Salyut was a small space station designed to support a crew of three men for a mission lasting about three to four weeks. It was probably intended to receive several sets of visitors, who would each fly to and return from the station using a Soyuz spacecraft as a ferry. Unfortunately, it suffered a tragic setback (see p193) and only one crew ever occupied the station.

Salyut 1

Salyut 1 is best described as a series of four cylinders of different diameters and lengths. At one end was a transfer compartment about 2m in diameter and 3m long. This compartment had a docking port at which the Soyuz ferry craft would arrive and through which the crew could enter the station. It may also have had an airlock to allow the cosmonauts to perform spacewalks. Beyond the transfer compartment was a work compartment 2.9m in diameter and 3.9m long. This area, which had a table and drinking water tank, was used by the cosmonaut crew during their off-duty periods. It also held the main control console from which Salyut's systems were operated. The work compartment widened out into a cylinder 4.15m in diameter and 4.1m long which included a treadmill and sanitation facilities as well as the bulk of the scientific experiments carried on the station. Beyond the rear wall of the work compartment, and inaccessible to the cosmonauts, was a final cylinder 2.2m across and 2.17m long. This housed the propulsion system used to modify Salyut's orbit and to provide attitude control. Electrical power was supplied by four solar panels, two on either side of the transfer compartment and two on either side of the propulsion module. The total area of the solar panels was about 28sq m.

The first mission to Salyut was that of Soyuz 10, launched on 22 April 1971. The crew of three, led by veteran cosmonaut Vladimir Shatalov, docked with Salyut the next day. However, for some reason that has never been fully explained, the Soyuz 10 crew did not enter Salyut 1, but instead undocked again and returned to Earth. Some sort of problem with the docking system seems to have thwarted their attempts to enter the station. A second crew was launched aboard Soyuz 11 on 6 June 1971 and this time the docking and crew transfer were successful and the three men entered Salyut 1 for a mission lasting just over three weeks. Tragically, the three cosmonauts perished during their return to Earth (see p193) and this catastrophe ended the Salyut 1 programme. The station was eventually commanded to re-enter and burn up on 11 October 1971. It was the start of a black period for the Soviet space station programme.

Salyut 2, Salyut 3 and Salyut 5

Although it was not clear at the time, it is now known that the Salyut programme was split into two branches, one military and one civilian. Salyut 2 (which failed in orbit and was never occupied), Salyut 3 and Salyut 5 were military missions and are described on page 129. Little has been released about the design of these stations, but they seem to have been smaller than the civilian Salyuts, comprising only two cylindrical compartments and a single set of solar panels. They also featured an unmanned capsule which was used to return material, possibly camera film, to Earth. Salyut 3 received one visiting crew (Soyuz 14) and Salyut 5 two crews (Soyuz 21 and 24). Soyuz 23 attempted to visit Salyut 5, but failed to dock and returned to Earth after about 24 hours only.

Cosmos 557 and Salyut 4

What was probably intended to be the next civilian Salyut station was launched on 29 July 1972, but this spacecraft failed to reach orbit following a failure of the second stage of its Proton booster. Another station was launched on 11 May 1973, but it appears that something went very wrong almost immediately and instead of being acknowledged as a new Salyut, this mission was given the name Cosmos 557. The satellite re-entered after 11 days.

Salyut 4 was launched into a 215km × 252km orbit on 26 December 1974. The new space station was basically the same as Salyut 1 except that the solar panel arrangement had been changed. Instead of two pairs of panels, one at each end of the station, Salyut 4 had three, larger, panels mounted on the smaller section of the work compartment. The station also featured a large solar telescope mounted so that it looked through the floor of the wider section of the work compartment. The first crew to visit Salyut 4 was Alexi Gubarev and Georgi Grechko, who were launched aboard Soyuz 17 on 10 January 1975. The Soyuz docked with Salyut without any problems and the two men entered the station and began a month-long mission of scientific experiments. The cosmonauts returned to Earth on 11 February after completing what was then the longest ever Soviet spaceflight.

On 5 April 1974 the next crew intended for Salyut 4 were launched, but something went seriously wrong. The second stage failed to separate from the third stage properly and the rocket went out of control. The Soyuz spacecraft was blasted clear of the doomed rocket and the cosmonauts quickly separated the re-entry module from the rest of the Soyuz and made an emergency landing in Siberia just short of the border with China. Soyuz 18A, as it became known, reached a peak altitude of 192km and travelled 1574km during its brief sub-orbital flight.

The final manned mission to Salyut 4 was launched on 24 May 1975. The crew was Pytor Klimuk and Vitaly Stevastyanov and their spacecraft was designated Soyuz 18. The docking took place during darkness, with the crew taking over manual control when Soyuz approached to within about 100m of the station. The Soyuz 18 mission lasted 63 days and the two, almost forgotten, cosmonauts remained in orbit during the international Apollo–Soyuz mission which took place in mid July. Although the flight seems to have been successful, and extended the Soviet space endurance record again, there have been rumours that conditions aboard Salyut 4 became damp and uncomfortable during the flight. In particular, reference has been made to the growth of green mould on the walls of the station.

Soyuz 20

There was one final mission to Salyut 4, but it was unmanned. Soyuz 20 was launched on 17 November 1975 and docked automatically with Salyut 4 two days later. It remained in space until 16 February 1976 and then returned to Earth under remote control. Soyuz 20 carried a number of biological experiments, but the main purpose of the flight seems to have been to evaluate new automatic docking systems and to prove that a Soyuz could be kept in orbit for up to 90 days without failing. The significance of this was to emerge later in the Salyut programme.

Salyut 4 burned up during its re-entry on 2 February 1977.

The name Salyut is said to have been chosen as a 'salute' to Yuri Gagarin who was launched almost exactly 10 years before Salyut 1.

Salyut 6 and 7

Salyut 6 1977–1981, Salyut 7 1982–1986

Operational space stations

Building on the experience of the earlier Salyut space stations, Salyut 6 and 7 featured add-on modules, automatic re-supply missions and visiting cosmonauts.

The Salyut 6 and Salyut 7 space stations were similar and represented a gradual evolution from the earlier civilian Salyuts described on page 163. Although differing in detail, for example in the scientific equipment carried, the transfer and work compartments of the new stations were basically the same as those of Salyut 4. The main improvement was at the rear of the station where, unlike Salyut 4, Salyuts 6 and 7 had a second docking port. To allow cosmonauts to reach this extra docking port the propulsion module was redesigned to fit around a cylindrical tunnel instead of occupying the whole of the rear compartment.

The extra entrance gave the later Salyuts greatly increased flexibility. For example, it was possible to host visiting cosmonauts while retaining one crew aboard the station for a long mission. The advent of the second docking port also made it possible to send automatic supply ships to restock Salyut with fuel, food and water. This was a major breakthrough because it meant that Salyut missions were no longer limited by the amount of supplies that could be packed into the station itself when it was launched.

Progress tanker

A second element of the new generation Salyut programme was an automatic supply ship called Progress, similar to the Soyuz spacecraft from which it was derived. The propulsion section was almost identical to that of a Soyuz, save for the inclusion of systems to replace those normally carried in the manned, or descent, module of a Soyuz. The Soyuz descent module was replaced by a sealed unit which could carry about 1 tonne of propellants for the Salyut's propulsion system. The Soyuz orbital module was transformed into a compartment for cargo. This section, which also carried the docking system and connecting hatch, was the only portion of the Progress tanker which could be entered by the Salyut crew. A Progress tanker had a mass of about 7 000kg, including 1 000kg of propellant for Salyut and 1 300kg of cargo.

In a typical mission a Progress tanker took about two days to rendezvous and dock with the rear docking port of Salyut. It usually remained docked to the station for several weeks while the cosmonauts unloaded it and then filled it with rubbish and other unwanted items. Sometimes the engines of the Progress would be fired to raise the orbital altitude of the Salyut, making use of any spare fuel in the tanks of the Progress's own propulsion system. The Progress would then be undocked and commanded to re-enter and burn up in the atmosphere, thus disposing of the cosmonauts' rubbish.

Star modules

Star modules were unmanned satellites which were docked to a Salyut to provide temporary extensions to the basic station. After the module had made an automatic docking to one of the Salyut docking ports, the cosmonauts could work in it until it had outlived its usefulness or until it was necessary to clear the docking port to receive another spacecraft. When this was necessary the star module was undocked from the station. The precise details of the modules have never been fully revealed, but they were about 13m long, weighed about 20 tonnes and may have been based on the military Salyut space station design. Each carried a pair of solar panels and usually had a re-entry capsule able to return about 500kg of samples to Earth. In 1991 it was revealed that the re-entry capsule was a hitherto secret type of spacecraft very similar to a US Gemini capsule re-entry module. Although never flown with a crew, the Soviet capsule could have accomodated two cosmonauts.

The first star module was Cosmos 929, which carried out an independent test flight between 17 July 1977 and 2 February 1978. The next was Cosmos 1267, which docked with Salyut 6 after the final cosmonaut crew had abandoned the station. Cosmos 1443 was launched on 2 March 1983, docked with the front port of Salyut 7, and was used by the crew launched in Soyuz T9. The Cosmos module was undocked from the station on 14 August and released its return module on 23 August. The main Cosmos module remained in orbit until it was commanded to re-enter on 19 September. Cosmos 1686 was the last star module; it was launched on 27 September 1985 and docked with Salyut 7 five days later. This module did not carry a recoverable capsule but had a battery of astronomical telescopes fitted instead. It was used by the crews of Soyuz T14 and T15.

The successful use of these star modules during the Salyut programme proved that it was possible to extend the capabilities of a space station by using add-on modules. This concept was later exploited by the Mir space station programme.

The Salyut 6 mission

Salyut 6 was launched on 29 September 1977, but the first manned visit failed when the crew of Soyuz 25, which was launched on 9 October 1977, were unable to dock with the station and were forced to return to Earth. The Soyuz 26 crew, launched on 10 December 1977, docked to the rear port of Salyut 6 and made a spacewalk to check for any damage to the forward port which might have explained the problems encountered by Soyuz 25. No damage was found and the Soyuz 26 crew settled down for a 96-day flight that set a new space endurance record. This mission saw the arrival of the first Progress tanker and also some of the first in a regular series of visits by crews sent up for a short (usually eight-day) mission. Each visiting crew, which sometimes included a foreign guest cosmonaut (see p146), brought fresh supplies for the Salyut 6 crew but, more importantly, they also brought a new Soyuz spacecraft to the station. The visitors usually returned to Earth in the Soyuz that was already in orbit, leaving their original one behind. This meant that the permanent crew did not have to rely on a Soyuz that had been parked, unused, in space for many months, but had a 'fresh' one available.

Salyut 6 enjoyed a long period of routine operations during which manned visits alternated with periods of automatic flight. In all, Salyut 6 saw the arrival of numerous Soyuz craft (the last was Soyuz 40, although Soyuz 33 failed to dock and Soyuz 34 was unmanned), 12 Progress tankers and several of the improved Soyuz T spacecraft (see p185). The space endurance record was extended first to 140 days by the crew of Soyuz 29, then to 175 days by Soyuz 32, then to 185 days by the Soyuz 35 cosmonauts. The last crew left Salyut 6 on 25 April 1977 and, after experiments with the Cosmos 1267 star module, the station re-entered the atmosphere and was destroyed in July 1982.

Salyut 7

After its launch on 19 April 1982 the mission of Salyut 7 followed the same pattern as Salyut 6, that is, progressively longer flights by a resident crew, with regular short visits to replace the Soyuz ferry craft. Salyut 7 was visited by Soyuz T5–T12 and received supplies from Progress 13–24. The crew of Soyuz T5 stayed aboard for 211 days during 1982; their record was broken by the 237-day flight of the Soyuz T10 cosmonauts in 1984.

The long series of missions included several noteworthy events. The world's second female cosmonaut, Svetlana Savitskaya, made two visits to Salyut 7. The first was in 1982 aboard Soyuz T7 and the second in 1984 aboard Soyuz T12. During this second visit she became the first woman to make a spacewalk and she conducted a series of welding experiments in open space. The Soyuz T9 crew undertook the first in-orbit construction in 1983 when, during two spacewalks, they added new solar panels to the station.

A dramatic interlude occurred in early 1985 when, during a period of unmanned flight, Salyut 7 suffered major technical problems and ground controllers lost contact with the station. On 6 June 1985 a specially formed crew, Vladimir Dzhanibekov and Victor Savinykh, was launched in Soyuz T13 and, in a courageous but little publicized mission, docked with the derelict space station and repaired it so that it could be used for one further long duration mission. The final Salyut 7 crew left the space station on 11 November 1985 when the commander, Vladimir Vasyutin, became depressed and spent hours staring out of the window.

However, this was not the last time Salyut 7 was used. In March 1986 the station was visited by the crew of Soyuz T15, who had begun their mission at the new Mir space station, but flew to Salyut 7 for a six-week visit before returning to Mir. Salyut 7 was then placed in a parking orbit in case it should be needed again, but its orbit decayed faster than expected and it was destroyed during re-entry on 7 February 1991.

Female cosmonaut Svetlana Savitskaya made her 1982 flight just before the first US woman astronaut, Sally Ride, was launched on the seventh Space Shuttle mission and her 1984 spacewalk occurred shortly before Kathy Sullivan became the first US woman to walk in space.

When the Mir space station (see p131) went into operation a new version of the Progress tanker, called the Progress-M, was brought into service. The Progress-M craft had a recoverable capsule which could be used to return material, such a samples of crystals grown in the weightless conditions of space, to the Earth for analysis.

Cosmonauts Vladimir Remek, Alexei Gubarev, Georgi Grechko and Yuri Romanenko (left to right) preparing to open the first post office in space

Saturn rockets

Saturn 1 1961, Saturn 1B 1966, Saturn 5 1967

The moonrockets

For years the world's most powerful rockets, the USA's Saturn series never failed.

Even before the USA orbited its first satellite, her rocket designers were already thinking of a launch vehicle with a take-off thrust of 6700kNewtons which would be developed by clustering together a number of smaller rockets. The idea was formally proposed in 1957 by Wernher von Braun's group who, at the time, were working for the US Army. The project, known originally as Juno-5, was approved in 1958 and used components from the already existing Jupiter and Redstone rockets. In 1959 the US Department of Defense decided that it had no need of such a rocket and the project, by then renamed Saturn, was handed over to NASA. NASA decided that the project should continue and that upper stages should be developed using liquid hydrogen and liquid oxygen propellants. President John F Kennedy's decision to send men to the Moon gave a new impetus to the Saturn programme and, after a number of design studies, the production of a family of three different Saturn vehicles was authorized.

Saturn 1

First in the series was Saturn 1, which was intended to launch unmanned versions of the Apollo command and service modules into Earth orbit for test flights. The first stage, called the S-I, was 6.5m in diameter and used eight engines similar to those used in the Thor and Jupiter programmes. The engine was designated the H-1 and had an approximate thrust of 836kNewtons at sea level. The second stage, known as the S-IV, was equipped with six RL-10 engines which burned liquid oxygen and liquid hydrogen and each delivered about 67kNewtons of thrust. With the spacecraft attached, a fully assembled Saturn 1 was about 58m tall and weighed 505 tonnes at lift-off.

The first launch of a Saturn 1, comprising a first stage and dummy upper stage, took place on 27 October 1961. The flight, designated SA-1, was successful and was followed by three similar missions. The first flight of a complete Saturn 1 (SA-5) on 29 January 1964 placed a 17000kg payload into Earth orbit. The next two missions, in May and September 1964, placed development versions (the so-called boiler-plate models) of the Apollo capsule into orbit. The final flights, SA-8, 9 and 10, all launched Pegasus meteoroid detection satellites. Each Pegasus comprised a set of folding wings (total span 30m) which extended from either side of the S-IV stage once in orbit. Sensors on the panels counted the number of times each panel was hit by cosmic dust.

Saturn 1B

Development of the Saturn 1B was authorized in 1962 as a means of bridging the gap between Saturn 1 and Saturn 5 rockets. By combining the existing S-I stage with a new, more powerful version of the S-IV stage already under development for the Saturn 5, NASA realized that it could speed up the Apollo programme by using the intermediate Saturn 1B to launch Apollo spacecraft on Earth orbiting test flights. The S-I stage was modified by providing new, smaller fins and by making other changes that saved about 9 tonnes. In parallel the H-1 engine was improved to provide 7% more thrust. The new second stage, called the S-IVB, was taller (23.7m) and used a single J-2 engine developing 1000kNewtons of thrust in place of the six RL-10 engines in the S-IV stage.

The S-IVB stage was topped by an instrument unit (an improved version of one used on the last six Saturn 1s) that carried computers and guidance equipment to control the rocket during its flight. With an Apollo spacecraft attached, a Saturn 1B stood 68.2m high and weighed about 590 tonnes.

The first Saturn 1B flight (designated AS-201) took place on 26 February 1966. Subsequent unmanned missions were used for a variety of purposes. AS-203 (5 July 1966) was a test to study the behaviour of the liquid fuels within the S-IVB stage after it had reached orbit; AS-202 (25 August 1966) was a test of the Apollo command and service modules; and AS-204 (22 January 1968) orbited an unmanned Apollo lunar module. Nine months later, on 11 October 1968, a Saturn 1B launched the first manned Apollo mission (Apollo 7) into Earth orbit.

The Saturn 1B played no further role in the Apollo project, but its useful life was by no means over. Three times in 1973 Saturn 1B rockets were used to launch crews to the Skylab space station and once again the rocket performed flawlessly. The final Saturn 1B flight was on 15 July 1975 when the last Apollo capsule was launched as part of the Apollo–Soyuz international mission.

Saturn 5

Since studies indicated that a lunar mission using Saturn 1 rockets was barely feasible, NASA decided to develop the massive, three-stage, Saturn 5 launcher. The first stage, which was 10.1m in diameter and 42m high, was designated S-IC although it bore no resemblance to the S-I stage used in Saturn 1. It was powered by five F-1 engines which burned liquid oxygen and kerosene and each developed 6819kNewtons of thrust. The outer four engines swivelled to keep the rocket on course during its flight. In a typical launch the S-IC operated for just over two minutes. The next stage, known as the S-II, had the same diameter and was 24.8m tall. The S-II was powered by five liquid oxygen/liquid hydrogen J-2 engines (like the one in the S-IVB) which each developed 1026kNewtons of thrust. The S-II stage operated for about six minutes. The final stage was an S-IVB topped by an instrument unit and was essentially the same as that used by the Saturn 1B. The single J-2 engine in the S-IVB was required to fire twice during each mission. The first time it was used to insert the S-IVB plus the entire Apollo spacecraft into Earth orbit immediately after separation from the S-II stage. About 90 minutes later the engine was restarted to push Apollo out of Earth orbit and towards the Moon.

Above the instrument unit, four panels formed a truncated cone that joined the Saturn 5 to the Apollo spacecraft. Known as the Spacecraft LM (Lunar Module) Adaptor this section also formed a garage in which the fragile lunar module could be protected from aerodynamic forces during launch. Once en route to the Moon the lunar module was removed from its protective housing and the S-IVB was either sent into orbit around the Sun or directed to crash itself on to the Moon.

Fifteen Saturn 5 rockets were built. The first, known as AS-501, was launched on 7 November 1967 and sent the unmanned Apollo 4 spacecraft on a test flight. The second unmanned launch, which was on 4 April 1968, suffered a number of potentially serious problems. These problems were identified and cured before the third flight which, on 21 December 1968, launched Apollo 8 on the first mission into lunar orbit. Nine more Saturn 5s were used during the Apollo programme, one to place Apollo 9 into Earth orbit, the remainder to send spacecraft to the Moon. The thirteenth, and final, launch was in 1973 when the Skylab space station (itself a converted S-IVB stage) was placed in Earth orbit. Cuts in the NASA budget led to the cancellation of the final Apollo missions and the Saturn 5 rockets originally assigned to these flights are now just rusting museum pieces at Cape Canaveral and at the mission control centre in Houston.

> Since it marked the largest rocket that could be reasonably assembled by joining together components designed for other rockets, the Saturn 1 design was sometimes referred to as 'Cluster's last stand'.

A Saturn 1 rocket on launch complex 37 at the Kennedy Space Center, Florida

Skylab

1973–1974

The USA's first space station

Built inside the fuel tank of a Saturn rocket, Skylab weathered a major crisis at launch to become a productive scientific mission.

The idea of a manned orbiting workshop arose in the mid 1960s when NASA was considering missions intended to build on the technology developed during the Moon landing project. The initial concept was to use the upper (S-IVB) stage of a Saturn 1B rocket modified so that, after it had injected itself into orbit, it could be converted into a manned laboratory. The conversion would be done by astronauts who would fly up to the empty stage in an Apollo capsule, drain the S-IVB of any remaining fuel and then clamber inside. McDonnell-Douglas, builders of the S-IVB stage, set about modifying a normal S-IVB by adding a docking adaptor, airlock and special fittings that would convert the hydrogen tank of the S-IVB into a two-storey space laboratory. The plan also called for a package of astronomical instruments, installed in a modified lunar module known as the Apollo Telescope Mount, to be launched by another Saturn 1B and docked with the orbiting workshop.

However, in July 1969, perhaps worried by the risk of astronauts entering a live rocket stage, NASA announced a new plan. The workshop, to be renamed Skylab, would be fully prepared before launch, inside an S-IVB stage that had been stripped of its rocket engine. Skylab would then be placed into orbit by the first two stages of a Saturn 5 rocket. Three astronaut crews would fly up to Skylab for a series of missions lasting 28, 56 and 56 days, using modified Apollo capsules launched by Saturn 1B rockets. To provide the Saturn 5, the Apollo 20 mission was cancelled.

The Skylab design

The main feature of Skylab remained an orbiting workshop within the hydrogen fuel tank of the S-IVB stage, a domed cylinder about 15m long and 6.6m in diameter. The smaller liquid oxygen tank of the S-IVB was not occupied, but served as a dustbin into which the crew threw all their rubbish via a small airlock. The workshop was divided into two floors by a wire mesh grid. The lower section contained the living quarters, including a wardroom, toilet, kitchen area, three individual bedrooms and an area for exercising. The upper section was set aside for scientific experiments. Two small airlocks were set into opposite sides of the workshop so that experiments could be poked through to look either down at Earth or out into space. Connected to the top of the workshop was an airlock module 5.5m long and 3.5m in diameter. This featured a hatch, based on the one used in the Gemini spacecraft, which allowed the astronauts to leave Skylab to conduct spacewalks. At the other end of the airlock module was a cylindrical Multiple Docking Adaptor (MDA) 5.3m long and 3m across. This was equipped with two docking ports, one along the centreline of the MDA, and a second, for emergency use, on the side. The Apollo Telescope Mount (ATM), mostly comprising instruments for observing the Sun, was mounted above the MDA for launch, but swung 90° to the side once in orbit. Many of the experiments in the ATM recorded their results on film and, from time to time, the astronauts were required to make spacewalks to retrieve and replace the film. The ATM control console was within the MDA. Electrical power for Skylab was to have been supplied by two large solar panels, one on each side of the main workshop, and a windmill-like arrangement of four panels on the ATM. All of the solar panels were folded away for launch and were intended to be deployed once Skylab reached orbit.

In its orbital configuration Skylab was 36.1m long and weighed 75 048kg. The internal volume, including that of the Apollo ferry, was about 330cu m, about the same as a two-bedroom house.

Problems on launch

The unmanned space station, designated Skylab 1, was launched on 14 May 1973. Within seconds it was in serious trouble when a thin meteoroid shield, intended to swing a few centimetres away from the wall of Skylab once in orbit, deployed too early. Aerodynamic forces tore off the shield, carrying away one of the two main solar panels. The other panel was jammed almost completely closed by a metal strip which snagged across the panel. Stripped of its outer blanket (which also served as a sunshade) and with its main power generation system useless, Skylab was in very serious trouble. Fortunately, the ATM deployed normally and its solar panels provided enough power to operate Skylab, although temperatures in the workshop began to rise as sunlight fell directly on to the outer walls of the station. The manned launch, originally scheduled to take place 24 hours later, was postponed as mission controllers tried desperately to salvage the crippled mission.

Repairs in orbit

With a shout of 'we fix anything' the first Skylab crew was launched on 25 May 1973 and rendezvoused with Skylab seven and a half hours later. Before docking, the astronauts tried unsuccessfully to unjam the remaining solar panel by opening the Apollo hatch and pulling on the panel with a long pole. The crew then docked with Skylab, and entered the overheated station the next day. Once inside, they pushed a parasol-like sunshade through one of the scientific airlocks to compensate for the thermal shield that had been ripped off at launch. The temporary sunshade brought down the temperature, but normal operations were still prevented by a severe shortage of electrical power. To rectify this, the crew carried out a 3.5-hour spacewalk, during which they used a pair of wire cutters on an 8m-long pole to cut the strap that was holding the solar panel shut. This allowed the panel to be pulled open and power began to flow. The astronauts were then able to carry out their experimental programme, during a 28-day flight which set a new space endurance record.

Skylab was originally expected to remain in space until well into the 1980s and it was intended that, once the Space Shuttle entered service, a special mission should be mounted to rendezvous with it. During this flight a robot craft called a Teleoperator Retrieval System would be docked with Skylab and used either to boost the derelict space station into a higher orbit or to de-orbit it so that it fell harmlessly into the ocean. However, the Space Shuttle fell behind schedule and unexpectedly high atmospheric drag caused Skylab to re-enter on 11 July 1979, before a rescue mission could be mounted. Most of the station fell into the Indian Ocean, but some fragments landed in Australia.

Skylab was originally known by the rather prosaic name of the Apollo Applications Program.

The most famous non-human passengers on Skylab were two common cross spiders called Arabella and Anita. The two spiders were flown to investigate how well they could spin webs in the weightless conditions of space. The experiment, suggested by Judith S Miles of Lexington, Massachusetts, was one of 25 student experiments carried on Skylab. The two spiders were launched with the second Skylab crew and Arabella was allowed to spin a web soon after reaching space. Her first attempts were poor but after about two days of acclimatization she began to spin almost normal webs. The second spider, Anita, was allowed to begin spinning a web only after several days in space and she began to spin normal webs almost immediately.

Regular operations

The second and third Skylab crews had more-or-less routine missions, punctuated by occasional crises as equipment malfunctioned from time to time. The second crew constructed a rather more substantial sunshade than the parasol deployed on the first mission then settled down to a 59-day stay. During the flight they conducted scientific experiments including solar and stellar astronomy, Earth observation, materials science and space medicine. The crew also made several planned spacewalks to recover film from the ATM.

The third and final visit to Skylab got off to a rather bad start when the crew concealed from mission control the fact that they had suffered from a bout of space sickness. For a time relationships between the astronauts and ground control were strained, but these problems were solved and the crew went on to carry out a very productive mission that was extended to 84 days. Scientific highlights included studies of comet Kohoutek and of a giant solar flare as well as continued medical experiments to determine the effects of the record-making flight on the astronauts' bodies.

Skylab mission log

Skylab 2: Launched 25 May 1973

Charles Conrad	Orbits:	404
Joseph Kerwin	Duration:	28 days 50 min
Paul Weitz	Splashdown:	22 June 1973

Skylab 3: Launched 28 July 1973

Alan Bean	Orbits:	858
Owen Garriott	Duration:	59 days 11 hours 9 min
Jack Lousma	Splashdown:	25 September 1973

Skylab 4: Launched 16 November 1973

Gerald Carr	Orbits:	1214
Edward Gibson	Duration:	84 days 1 hour 15 min
William Pogue	Splashdown:	8 February 1974

This picture of Skylab was taken by the last crew as they prepared to return to Earth. The emergency Sun shade is clearly visible

Solar sails

Regatta 1994?

Riding on the power of light

Gossamer-thin solar sails may prove to be a key space propulsion system for the next century.

The idea of using the pressure of sunlight for propulsion is not new: like many aspects of space travel it was foreseen by the Soviet space pioneer Konstantin Tsiolkovski in the 1920s. Solar sailing was later taken up by science fiction writers and was considered in technical papers published in the late 1950s and early 1960s. NASA conducted a number of solar sail studies during the mid 1960s, but the budget cutbacks which followed the Apollo programme meant that nothing came of these. The concept eventually re-emerged during the late 1970s when it was suggested that NASA send a solar sail propelled spacecraft to explore Halley's comet. This proposal led to a year-long study under the direction of the NASA Jet Propulsion Laboratory, during which the feasibility of the concept was confirmed. However, NASA felt that there was insufficient time to develop and test the necessary technology and the idea was dropped. Even so, the study led to a resurgence of interest in solar sailing and several new ideas, some of which may one day be put into practise, have emerged.

The principle of solar sailing

The solar sail utilizes the pressure of sunlight to propel itself through space. This is best understood by remembering that although light is often considered as a continuous wave, physicists regard it as a form of electromagnetic radiation comprised of a stream of tiny packets of energy called photons. Each photon travels at the speed of light (about 300 000 km per second) and, although they are rather ephemeral things, photons can be thought of as tiny particles. When a beam of light, that is, a stream of photons, reflects off something then the momentum in the photons is transferred to the reflecting surface. The momentum of a single photon is very small, but if enough photons strike a reflector then the combined effect can be significant. A crude analogy is that using the momentum of photons to propel a spacecraft is rather like trying to move a large object by throwing grains of rice at it. It can only be done if a very large number of rice grains can be thrown and even then, the acceleration of the object will be very slow.

An effective solar sail for space propulsion must be very large, so that it will reflect as many photons as possible, and yet have low mass so that the tiny force of sunlight can produce a reasonable acceleration. These requirements present considerable challenges to space engineers, but if these can be overcome then solar sails do possess important advantages: sunlight is free, it arrives continuously and it is never likely to run out. So, although the thrust provided by a solar sail is small, it can be applied continuously over very long periods without the need to carry fuel.

Navigating a solar sail

It is important to appreciate that because of their size and fragility solar sails can only be used in space, and therefore must first be lifted into Earth orbit by conventional rockets. Once in orbit and assembled the sail can then be orientated so that the pressure of sunlight either pushes the craft along in the direction of its orbit, increasing its velocity and causing it to spiral outwards from the Earth, or against the direction of the orbit, decelerating it and causing it to spiral downwards. Similarly, once the sail has escaped into an orbit around the Sun, the pressure of sunlight can be used to slow it down and send it in towards Venus, or speed it up and push it slowly out towards Mars.

The inherent disadvantage of solar sails is that the rate at which they can change their velocity is very small. When the mass of the sail material and its support structure are taken into account, a realistic design for a solar sail might be able to accelerate (or decelerate) at about 1mm per second^2, or about 0 to 10km per hour in an hour. This is very small compared to the acceleration produced by firing a rocket motor but, as already remarked, it can continue for days or weeks. This means that the velocity change required to send a spacecraft from Earth orbit to Mars (about 8km per second) can be built up over a period of about three months of solar sailing. To send a solar sail to Mars would take about 400 days, somewhat longer than a trip using a conventional rocket, but for missions in which time is not important (such as sending a cache of supplies ahead to await the arrival of a manned mission) sails could prove to be useful.

Solar sail design

The most likely form of solar sail would be a plastic sheet, not unlike the clear plastic used to wrap food, coated with a layer of a reflective material such as aluminium. To make the sail as light as possible the film must be very thin, perhaps only 2 to 4 microns thick. This material must then be spread out over a huge area, perhaps covering many square kilometres, and must keep its shape to ensure maximum performance. This presents some interesting design problems.

One possibility is that the sail could be supported by rigid poles and bracing wires like the sails on a ship. The shape of such a sail can be maintained quite well and it would be quite easy to turn the sail to different angles relative to the sunlight for manoeuvring, but the extra mass of spars and wires would reduce performance. An alternative solution is to make the sail circular and to spin it so that centrifugal force keeps the spinning sheet outstretched like a lasso whirled by a cowboy. This does not require any extra structure, but manoeuvring the sail is more difficult. Another possibility is to make the sail out of long thin strips and rotate them like the blades of a helicopter. This scheme, called a heliogyro, could be manoeuvred by tilting the angles of the blades as the sail slowly cartwheels across space.

Despite the simplicity of the concepts, the practical difficulties of such designs must not be underestimated. It has been calculated that to send a mass of 6000kg between planets would require a square sail 2km on a side, or a heliogyro with 12 sails each 10m wide and over 30km long. This gossamer-thin structure must be folded into some sort of container that can be launched by a rocket, and then must be deployed to the correct shape once in space. This would be a formidable undertaking even for the most imaginative space engineer.

The concept of the solar sail is so elegant that several space groups are trying to persuade a major space agency such as NASA or ESA (European Space Agency) to attempt to launch one for test purposes. It has even been proposed that a race is held in which the winner would be the first solar sail to travel from Earth orbit to beyond the orbit of the Moon. It has also been reported that the USSR was possibly planning to launch a solar sail propelled spacecraft called Regatta in about 1994.

It is important to realize that solar sails do not derive their thrust from the solar wind, the stream of atomic particles blowing off the Sun. The momentum carried by the particles of the solar wind is only a tiny fraction of that carried by the photons of sunlight.

Soviet launch centres

Orbital launches: Baikanour 1957, Kapustin Yar 1962, Plesetsk 1966

The cosmodromes

The USSR has launched about 100 satellites a year from its three cosmodromes.

The USSR has three sites which are used for orbital launches and each has a more-or-less direct equivalent in the US space programme.

Kapustin Yar

The oldest of the Soviet cosmodromes is located about 200km south-east of Volgagrad near the town of Kapustin Yar and was first used to launch V2 rockets captured from the Germans at the end of World War II. The base remained in service between 1947 and about 1952 and then fell into disuse until 1962 when it was reactivated for launching small satellites, in a similar way to NASA's Wallops Island facility.

Most launches from Kapustin Yar are of small rockets such as the B-1 and C-1 (see p179) which orbit scientific satellites in the Cosmos and Intercosmos series. The cosmodrome has also been used to launch military satellites. The B-1 rocket was eventually replaced by the C-1 launcher, which continues to be launched from Kapustin Yar. Beginning in 1982 the cosmodrome was also used to launch a series of unmanned spaceplanes used during the development of the Soviet space shuttle Buran.

Tyuratam/Baikanour

When the USSR launched the first Sputnik in 1957 it was announced that the satellite had been launched from the Baikanour cosmodrome. However, the cosmodrome is not at the town of Baikanour, but 370km to the south-west near Leninsk, which has itself grown up around an old town known as Tyuratam. The reason for this deception, which lasted until satellite photographs revealed the true location of the launch complexes, was bound up with the politics of the Cold War. The rocket which launched Sputnik, and the pad from which it was launched, formed part of the Soviet intercontinental ballistic missile system and the USSR was anxious to preserve as many of its military secrets as possible.

Work on the cosmodrome began in about 1955 when it was decided to move rocket research away from Kapustin Yar, which lay uncomfortably close to US radar stations and U2 spy plane bases in Turkey. It has since grown to a huge Y-shaped complex which extends over 100km from west to east and about 50km from north to south. The city of Leninsk lies to the south at the base of the Y. The climate is hot and dry during the summer and yet very cold, with deep and lasting snow, during the winter. Because of this, much of the pre-launch assembly and checking of rockets is done inside special buildings and the completed rocket is only moved to the launch pad a few days before the launch date. The rockets are assembled horizontally, and moved to the launch pad on flat railway wagons, before being erected and fuelled on the launch pad itself. Unlike its US counterpart at Cape Canaveral, which is essentially a launching base for rockets built throughout the USA, the Baikanour cosmodrome also provides facilities to actually assemble rockets from components brought in from elsewhere.

At the western end of the complex are silos and buildings which have not changed since the 1970s and are probably abandoned missile sites. About 25km east of these are a number of launch pads and underground silos which form part of a complex used to test large military missiles.

At the centre of the cosmodrome are several launch pads used for rockets in the A series (SL-1, 2, 3, 4 and 6). These are variants of the original rocket used to launch the first Sputnik and are today used to launch both Soyuz spacecraft and a wide variety of unmanned satellites. The launch complexes comprise a final assembly building connected by rail to a large concrete launch pad. The pad is supported over a large flame trench and has a hole about 16m in diameter which allows the rocket exhaust to enter the flame trench, from where it is deflected harmlessly off to one side. When a rocket arrives at the pad from the assembly hanger, it is first brought to the vertical and then surrounded by a number of towers which rise up from the ground. Some of the towers are equipped with platforms that close around the rocket to provide working areas from which technicians can perform final checks before launch. Others are support arms used to stop the rocket falling over. The rocket is filled with fuel a few hours before launch and then the large access towers are swung down, leaving only the smaller arms to keep the rocket upright. Finally, as the motors ignite and the thrust builds up, the support arms are retracted and the rocket begins its climb into space.

The central area of the cosmodrome is also the site of the launch facilities originally developed for the Soviet moonrocket and now used for Energia and the space shuttle Buran. There are three launch pads and a number of large buildings, which are described on page 20. The runway on which Buran lands when returning from space is located about 12km from the launch complex.

In the eastern arm of the Y-shaped cosmodrome are a group of launch pads which provide further facilities for both A-type rockets and the smaller rockets used to launch Cosmos satellites. There is also an industrial area. At the end of this eastern arm are pads for the Proton launch vehicle. These rockets are also assembled horizontally and moved to the launch pad by rail. Each pad has two large towers which support floodlights and television cameras as well as a gantry which wraps around the rocket to provide access to the rocket and its cargo once it is on the launch pad.

Plesetsk

The secret northern cosmodrome at Plesetsk is located about 300km south of Archangel and lies just outside the Arctic Circle. Plesetsk is mostly used for launches of military satellites and for test firing of long-range ballistic missiles. This makes its function similar to the US launch centre at Vandenberg Air Force Base in California. The existence of the cosmodrome was not admitted by the USSR until 1987, although its location was deduced in 1966 following the launch of the Cosmos 112 satellite. Tradition has it that the discovery was made by the schoolboy satellite trackers of Kettering Grammar School in England, under the direction of their physics teacher Mr Perry, but no doubt the military space observers of the major powers were also aware that Cosmos 112 could not have been launched from either Baikanour or Kapustin Yar.

Plesetsk has launch pads for the SL-3, the SL-4 Vostok and SL-6 Molniya launch vehicles as well as the smaller SL-7, SL-8, SL-11 and SL-14 rockets. There is also a memorial to 48 technicians killed in 1980 during an accident which occurred while loading a rocket with propellants. Launches from Plesetsk are frequent and in some years they occur more than once a week. Many of these launches are used to place Earth-observation satellites, which may be military or civilian in nature, into polar orbits. The location of Plesetsk means that such orbits can be reached more easily than from the two more southerly cosmodromes.

The coordinates of the cosmodromes are:

Plesetsk	60°43′N	40°18′E
Tyuratam/Baikanour	45°36′N	63°24′E
Kapustin Yar	48°30′N	45°46′E

Kapustin Yar means 'little cabbage patch'.

Soviet launch vehicles

Sputnik 1957, Energia 1987

An evolutionary approach

At the beginning of the 1990s the USSR was still using a version of the rocket which launched Sputnik 1 in 1957.

The USSR produced a number of space launch vehicles but, unlike the USA which developed and then abandoned various kinds of rockets, the USSR adopted an evolutionary approach, often building upon its existing designs. One result of this has been that the type of rocket used to launch Soyuz spacecraft at the beginning of the 1990s was basically an improved version of the one used by Yuri Gagarin in 1961. Overall, this philosophy has been highly successful but Soviet rocket development suffered a setback in the 1960s and early 1970s when attempts to build a large moonrocket floundered. The design teams eventually recovered from this debacle and recently several new launchers were developed, culminating in Energia, the most powerful launch vehicle in the world.

Names and codes

The first Soviet launch vehicles were based on military missiles, so not surprisingly the history of Soviet rocket development was clouded in secrecy for many years. As a result, various schemes have sprung up to identify particular rockets and, since the same rocket can have several different designations, this can be rather confusing. The most widely used names used to be derived from major satellites orbited, for example Vostok or Soyuz, but more recently specific names, such as Tsyklon (Cyclone) and Energia, were announced by the Soviets themselves. In parallel to these names the US Department of Defense classifies Soviet rockets using a system consisting of the letters SL and a number. According to this scheme the Soyuz rocket is known as the SL-4. Another scheme was introduced by the late Dr Charles Sheldon of the US Library of Congress. Sheldon assigned each basic type of rocket a capital letter, then added numbers to denote the number of extra stages carried. In the case of launchers which used an upper stage with a special purpose, for example to boost a probe out of parking orbit, small letters such as e (escape), m (manoeuvrable) and r (recoverable) were added. For example Sheldon called the Soyuz rocket an A-2, but it became an A-2-e if an upper (escape) stage was added on top. Finally, launchers based on missiles may also have a NATO code name for the missile version. An example is the Cosmos rocket, also known as SL-8 or C-1, which is based on the missile with the NATO codename Skean.

The Sputnik rocket (SL-1, SL-2, A, Sapwood, R-7)

The rocket which launched Sputnik 1,2 and 3 was developed from the ballistic missile known to NATO as Sapwood and to the Soviets as the R-7. It comprised a central core containing four RD-107 rockets and four small (vernier) rockets. Clustered around the core were four, tapered, booster stages each containing four RD-108 rockets plus two verniers. All the motors used liquid oxygen and kerosene propellants. The four boosters were jettisoned during the launch, leaving the core to place the Sputnik in orbit. The Soviet authorities usually referred to this as a two-stage rocket, regarding the four strap-on boosters as the first stage, but in the West it has been regarded as a single stage plus strap-ons or a 1.5-stage rocket.

The Vostok rocket (SL-3, A-1)

This was basically the Sputnik rocket, to which was added an upper stage supported by a lattice-work arrangement. The upper stage had a single RD-7 engine. This combination could place about 4 700kg into low Earth orbit. It was used to launch the manned Vostok spacecraft and unmanned satellites in the Luna, Electron and Meteor series. The SL-5 is an obscure variant of this rocket.

The Soyuz rocket (SL-4, A-2)

The Soyuz rocket is a more powerful variant of the original Sputnik rocket with a second stage carrying two engines and able to lift over 7 200kg into low Earth orbit. This was used in the Voskhod programme and remains in use today to send Soyuz spacecraft to the Mir space station.

The Molniya rocket (SL-6, A-2-e)

This is a Soyuz rocket with an additional upper stage and is used to send probes to the Moon and planets (hence -e for escape). It is also used to launch Molniya communications satellites into highly eccentric Earth orbits.

The Cosmos rocket (SL-7, B-1)

Based on the missile known to NATO as Sandal, this was a small two-stage rocket which was used to launch satellites in the Cosmos and Intercosmos series until 1977 when it was phased out. The first stage had a single RD-214 motor and the second stage had a single RD-119.

The Cosmos rocket (SL-8, C-1)

This is another small launcher based on a missile, in this case the Skean medium-range ballistic missile with its RD-216 motor. To convert the Skean into the Cosmos rocket a restartable second stage was added. This rocket remains in use launching small satellites and can place about 1 500kg into low Earth orbit. It is the only Soviet rocket type that has been launched from all three Soviet cosmodromes.

The SL-10 and 11 (F-1-r, F-1-m)

This is a two-stage launcher based on the Scarp missile. A third stage was added for special operations in orbit. The F-1-r version was used in tests of the Fractional Orbit Bombardment System and the F-1-e was used for tests of the anti-satellite system. (see p39).

Proton (SL-9, 12 & 13, D, D-1, D-1-e)

This launcher has been used in a number of variants all of which have been known as Proton, possibly because details of the design were kept hidden for many years. Although first launched in 1965, it was not until 1985 that the true layout of the rocket was revealed when detailed photographs showing the launch of the Vega probes to Halley's comet were released. Proton comprises a six-engine first stage (the D version or SL-9) to which a second stage (D-1 or SL-13) and escape stages can be added (D-1-e or SL-12). The layout is rather unusual, comprising a central fuel tank with the six RD-253 rocket engines in what appear to be, but are not, jettisonable boosters clustered around the base. Proton can place place about 20 tonnes into low Earth orbit or send about 5 tonnes to the inner planets. An improved version of Proton, called the Proton-KM, is expected to enter service in 1995.

Tsyklon (SL-14, F-2)

Tsyklon is a new generation launcher closely related to the SL-11 and introduced in 1977. It has three stages all using nitrogen tetroxide and unsymmetrical dimethylhydrazine propellants. Tsyklon can launch about 4 000kg into low Earth orbit.

SL-15 (TT-5, G, Webb's Giant, N-1)

This was the giant Soviet moonrocket abandoned in 1972 (see p183).

Zenit (SL-16, J-1)

Zenit is a new, three-stage, booster intermediate in capability between the Soyuz and

Proton rockets. It was first launched in 1985. The first stage is also used as a strap-on booster for the Energia rocket and uses the same RD-170 engines as the Energia booster variant. It can lift about 13 700kg into low orbit.

Energia (SL-17)

Following the total failure of their moonrocket during the 1960s and 1970s Soviet designers decided to try again with a totally new design. The result was Energia, a 2000-tonne rocket able to place 100 tonnes into low Earth orbit. Energia comprises a central core to which four large boosters, regarded as the first stage, are fitted. These boosters have RD-170 engines and use kerosene and liquid oxygen propellants. The second, or core stage is equipped with four engines using liquid hydrogen and liquid oxygen (a first for a Soviet rocket) and satellite payloads are mounted in a special canister on the side of the core stage. When Energia is being used to launch satellites a small third stage is added to inject the payload into orbit. Energia can also launch the Soviet space shuttle Buran, in which case no third stage is required since Buran has its own small rocket engines.

The first launch of Energia was on 14 May 1987 and although, due to a failure in the third stage, its test payload fell back to Earth, the flight was considered a success. The first launch with the Buran space shuttle was on 15 November 1988, but the dissolution of the USSR in 1991 has left the future of the Energia programme in doubt. Since there seems little interest in using the Buran space shuttle on a regular basis, and there do not appear be any other payloads which require a rocket as large as Energia, then the programme may be scrapped. However, despite this uncertainty, there have been reports that a new, scaled down, version called the Engeria-M (which will have about half the lifting capability of the standard Energia) is to be developed.

The possibility of launching the Soviet Zenit rocket from a proposed commercial launch site at Cape York, Australia was discussed in 1990 but by 1991 the plan had run into serious financial problems.

Unlike US launch vehicles, whose abandoned stages fall into the oceans, the location of the main Soviet launch sites mean that their rocket stages fall on land. The arctic tundra around the northern cosmodrome in Plesetsk is littered with tonnes of debris called 'metal from heaven' by the local inhabitants. In a similar vein, residents of Dzhezkazgan, which lies close to the flightpath of rockets from the Baikanour cosmodrome, have complained that toxic fuel from crashed rocket stages has contaminated the soil and made it impossible to graze livestock.

A Proton launcher

Soviet Moon programme

Cancelled 1972

Moon race cover-up

After years of denying that they planned a manned lunar mission, Soviet space engineers and cosmonauts recently began to reveal the details of a secret programme intended to beat Apollo to the Moon.

During the period from 1962 to 1964 the Soviet chief designer Sergei Korolov was planning what became known as the 'Soyuz complex', a new programme intended to comprise three different spacecraft:

Soyuz-A. A manned spacecraft comprising a cylindrical equipment module, a bell-shaped descent module and a roughly cylindrical orbital module. This evolved into the Soyuz craft in use today.

Soyuz-B. An unmanned craft equipped with a powerful rocket motor, but which would be launched with almost no propellants.

Soyuz-V. An unmanned tanker which could carry propellants to the Soyuz-B.

The mission of the Soyuz complex seems to have involved the launch of a Soyuz-B rocket block followed by a series of Soyuz-V tanker craft. The tankers would transfer propellants to the Soyuz-B before a manned Soyuz-A would fly to Soyuz-B and dock with it. Western analysts believe that this combination would have been able to send the Soyuz-A on a flight around the Moon, although a landing would not have been possible. We may never know if this was the intended mission of the Soyuz complex because the Soyuz-B and Soyuz-V versions were cancelled in about 1964 and plans for a Moon flight using the more powerful Proton rocket were made instead.

Zond

Between 1968 and 1970 a series of unmanned Zond missions (Zond 4-8) were flown. Several of these passed close to the Moon and this led many Western observers to believe that these flights were tests of a manned lunar programme, a suspicion recently confirmed by the Soviet authorities.

The Zond 4-8 spacecraft (Zond 1-3 were unrelated) were based on a Soyuz-A spacecraft without an orbital module. In this configuration, known to the Soviets as L-1, a single cosmonaut could be launched towards the Moon by the D-1-e version of the Proton rocket (see p179). The Zond's own rockets were not powerful enough to brake the craft into lunar orbit, but Zond could swing around the Moon and then return to Earth, allowing the USSR to claim the first manned flight into deep space.

The unmanned Zond 4 spacecraft was launched on 21 November 1967 and boosted into space in the opposite direction to the Moon. Since Zond 4 was a test flight, it may have been deliberately sent away from the Moon to simplify the navigation problems associated with a deep space flight. Its fate is uncertain; it was thought to have gone into orbit around the Sun, but it has been claimed recently that Zond 4 crashed in the USSR during an attempt to recover it.

Zond 5, which carried a cargo of turtles and other biological specimens, was launched on 14 September 1968. It looped behind the Moon on 18 September, passing within 2000km of the lunar surface, then returned to Earth. Zond 5 was intended to land in the USSR, but instead it splashed down in the Indian Ocean due to a problem with its guidance system. Zond 6, launched on 10 November 1968, followed a similar course, but used a different re-entry technique. Instead of making a direct plunge into the atmosphere, Zond 6 made a partial re-entry then skipped back into space and made a second re-entry a few minutes later. This allowed a relatively gentle re-entry, more suitable for a manned flight.

By this time, the US Apollo programme was well under way and a manned Zond mission to beat the Americans to the Moon was anticipated in late 1968. Warned of this by US intelligence, NASA decided to change the flight plan of Apollo 8 and to send it into lunar orbit on Christmas Eve 1968. According to cosmonaut Oleg Makarov, who claims to have worked on the manned Zond project, a manned Zond mission after Zond 6 was delayed by problems with the Proton booster. By the time these problems were resolved, the American Apollo 11 mission had already landed on the Moon, so the Zond flight was abandoned.

There were two other unmanned Zond missions. Zond 7 was launched on 7 August 1969 and Zond 8 on 20 October 1970. Both looped around the Moon and landed safely in the USSR.

The Moon landing programme

Full details of the Soviet manned lunar landing programme are still vague, but they appear to have hinged on the development of a new super rocket known to the Soviets as the N-booster. This huge booster had five stages, the first using a cluster of 30 separate rocket engines, the second stage using eight and the third stage four. Two further stages, each with a single engine, were required for the lunar mission. Above these would have been a shroud containing a lunar landing craft, known as the lunar cabin (Russian acronym LOK), then a two-man Soyuz spacecraft. The lunar cabin, first revealed to Western scientists in 1989, had a spherical upper stage based on the Soyuz orbital module. The lower stage had four landing legs. Unmanned tests of the craft, and of related hardware, were carried out in Earth orbit during 1970 as part of the Cosmos programme.

As described by designer Vasili Mishin, the mission called for the first three stages to place the rest of the N-1 into Earth orbit and for the fourth stage to send the fifth stage towards the Moon. The fifth stage would be used to enter lunar orbit. Once in lunar orbit, a single cosmonaut would make a spacewalk from the Soyuz spacecraft and enter the LOK lander. (The spacewalks made between Soyuz 4 and 5 in 1969 may have been rehearsals for this complicated and dangerous procedure.) The Soyuz would then pull free, leaving the lander attached to the fifth stage. This would be fired again to start the descent towards the Moon before the lunar lander separated and used its own engines to make a soft landing. This might have been a rather dangerous manoeuvre since cosmonaut Alexei Leonov has said that there was only enough fuel to hover for about 25 seconds while searching for a safe landing spot. After touchdown, the cosmonaut would conduct a brief moonwalk and return to his landing craft. The upper portion of the lunar lander would then blast off and rendezvous and dock with the orbiting Soyuz. The two cosmonauts, plus their lunar samples, would then return to Earth using the skip re-entry technique tested during the Zond missions.

Interestingly, Mishin's description is similar to the mission profiles which Western space experts had deduced from the scanty information released by the Soviets over the years. Some of these schemes envisaged that the manned Soyuz would be launched by a separate Proton rocket and would dock with the rest of the complex in Earth orbit before leaving for the Moon. It is possible that this was a plan for another, more complicated, mission that would land the entire Soyuz on the Moon and allow a longer stay on the lunar surface. We may never know exactly what was intended.

Unfortunately, there were problems with the N-1 rocket and it failed four times between 1969 and 1972. These failures, combined with the success of the Apollo programme, led to the cancellation of the Soviet lunar programme in 1972.

If all had gone well, who would have been the first cosmonaut on the Moon? Alexi Leonov (the first spacewalker) is one of several men who are believed to have been considered for the lunar landing.

Soyuz

Soyuz 1 1967, Soyuz T 1979, Soyuz TM 1986

A remarkably adaptable spaceship

Designed to go to the Moon, the USSR's Soyuz has always been used in Earth orbit.

The history of the Soyuz spacecraft starts in 1961 when Soviet chief designer Sergei Korolov conceived what became known as the 'Soyuz complex' (see p182). This consisted of unmanned booster modules, orbiting fuel tankers and a manned spacecraft which would all rendezvous in Earth orbit before leaving for the Moon. However, the Soviet plans changed and only the manned version, known as Soyuz A, was ever built. Despite its origin as a lunar spaceship, the Soyuz design proved remarkably successful and was developed into a number of different versions for use in Earth orbit.

Soyuz 1-9, The first generation

The original Soyuz consisted of three modules. At the rear was a cylindrical instrument section 2.7m wide and 2.3m long. This contained electronics, rocket engines and propellants. The main engine, and its backup, each delivered a thrust of about 3.9kNewtons. Next came the descent module, a headlamp-shaped capsule about 2.2m in diameter and 2.2m long which could carry up to three men. It had a curved heatshield at the rear and was the only section of the Soyuz that could return to Earth intact. The descent module had two small portholes and, since the view forward was blocked, a periscope which projected below the cabin. The forward section of the Soyuz was an egg-shaped orbital module about 2.2m x 2.65m which provided extra working space for the cosmonauts once they were in space. Power was provided by a pair of solar arrays on either side of the instrument module which were folded during launch and spanned 8.3m when deployed. The total mass of a Soyuz was just under 7 tonnes and the combined length of the three modules was about 7m, depending on the type of docking system carried.

The descent module was so cramped that the crew could not wear spacesuits, but wore cloth flying suits instead. Once in space the cosmonauts could leave their seats and move into the orbital module to conduct experiments. The orbital module could be depressurized so that it could serve as an airlock allowing the cosmonauts to leave the Soyuz to perform spacewalks. Although the forward section of the orbital module did have provision for docking to another Soyuz, there was no connecting hatch and so crew exchanges could only be accomplished by spacewalking. This apparent oversight stems from the fact that the early Soyuz craft were connected with the Soviet Moon landing programme and were not intended to dock to each other or to orbiting space stations.

At the conclusion of the flight the crew returned to the descent module and fired the main engine to begin the return to Earth. Once this was done the three modules were separated and the descent module, protected by its heatshield, reentered. A single large parachute lowered the craft to the ground. Just before touchdown the heat shield was released and four solid rocket motors fired to cushion the landing.

After several unmanned tests, the first Soyuz was launched in April 1967. Unfortunately it crashed, killing its single occupant, when its landing parachute failed to open properly (see p192). Subsequent flights related to the planned lunar mission included one-, two- and three-man crews and one automatic mission (Soyuz 2). Since then, the role of Soyuz has changed, but the basic three-module layout of the spacecraft has been retained and is still in use today.

Soyuz 10 and 11

When the USSR gave up its plans to fly to the Moon, it was decided to remodel Soyuz to send crews to and from orbiting space stations. The main modifications were the removal of a large fuel tank at the rear of the instrument module and the provision of a new docking assembly at the front of the orbital module. The new docking system incorporated a hatch so that cosmonauts could transfer from Soyuz to a space station without the need for a spacewalk. Only two of these versions ever carried men into space. Soyuz 10 docked briefly with the Salyut 1 space station and Soyuz 11 docked with the station normally but its crew perished during re-entry when a valve opened suddenly and their air supply rushed out (see p193).

Soyuz 12-40, two-man Soyuz

Following the Soyuz 11 disaster, it was decided that each cosmonaut must wear a spacesuit during dangerous manoeuvres such as launch, docking and re-entry. Unfortunately, the Soyuz descent module was so cramped that the extra equipment could only be carried if one of the seats was removed, and so Soyuz became a two-man spacecraft. Since by then it had been planned that Soyuz was to be used to travel to and from Salyut space stations, it was decided to remove the solar panels and to rely on batteries for electrical power during the brief period of independent flight. A consequence of this was that if the Soyuz failed to dock on schedule it would be forced to return to Earth quickly before its batteries ran down. The first of these new craft was Soyuz 12, launched on a two-day test flight on 27 September 1973.

Four of the two-man Soyuz craft were used for independent flights and these carried solar panels so that they could remain in orbit for more than the two-day maximum of the battery powered versions. The flights involved were the Apollo–Soyuz international mission in 1975 (Soyuz 19) and its rehearsal (Soyuz 16). There were also two scientific missions, Soyuz 13 and Soyuz 22.

Soyuz T

The next Soyuz variant was the Soyuz T, or Transport, which first flew in December 1979. Soyuz T was also intended as a space station ferry, but it carried two solar panels each 1.4m wide and 4.14m long. This meant it was capable of an independent flight of up to four days in the event of a failure to dock with a Salyut station. Soyuz T also carried new, smaller electronics which meant that the descent module could now accommodate up to three cosmonauts in spacesuits. In addition, the propulsion system was modified to allow fuel to be transferred between the main rocket motor and the smaller attitude control rockets.

Soyuz T also introduced a refined operating procedure for re-entry. The orbital module of a Soyuz T was cast off before firing the main engine to begin the return to Earth. By reducing the mass which had to be decelerated to begin re-entry, the fuel required for the de-orbit manoeuvre was decreased, meaning that more fuel was available during the early phases of the mission.

Soyuz TM

The latest Soyuz is the Soyuz TM, first flown in 1986. This is a modernized version of Soyuz T with improved power supplies, new parachutes and extra space for equipment. It can carry an extra 200kg into orbit and return up to 150kg of cargo to Earth, 100kg more than a Soyuz T. The TM version is expected to remain in service for several years and will probably still be in use three decades after the launch of Soyuz 1.

> *Soyuz* means 'union', an appropriate name for a spacecraft which makes a docking on almost every mission.

Soyuz programme

Soyuz 1   23 April 1967

25 years of space travel

Soyuz began as an element in the Soviet lunar programme and evolved into a ferry for cosmonauts travelling to the Salyut and Mir space stations.

The Soyuz spacecraft is described on page 184 and was originally intended for a Soviet lunar mission. The proposed Moon flight (see p182) involved a rendezvous in lunar orbit and required a cosmonaut to transfer to and from the lunar landing module by means of a spacewalk. Like the Americans, the Soviets decided to practice these complicated manoeuvres in Earth orbit before attempting them in deep space.

Soyuz 1, launched in 1967, was intended as a spectacular mission which would accomplish all of these objectives in a single flight and restore the propaganda lead which had been lost during the US Gemini programme. According to the plan the launch of Soyuz 1, with a single cosmonaut, was to have been followed the next day by the launch of a three-man crew in Soyuz 2. The two craft were intended to dock so that two men could transfer from Soyuz 2 to Soyuz 1 and then return to Earth. However, problems developed with Soyuz 1 soon after launch and Soyuz 2 was cancelled as mission controllers strove to return Soyuz 1 to Earth. Despite their efforts, Soyuz 1 crashed, killing the cosmonaut Vladimir Komorov (see p192) and delaying the next Soyuz flight for over a year.

Soyuz 2 was launched unmanned and was followed into space by a single cosmonaut aboard Soyuz 3. The two craft rendezvoused, but did not dock, and both then returned to Earth. The Soyuz 4/5 mission in 1969 carried out the mission originally intended for Soyuz 1 and 2, with a docking and crew transfer taking place. Described at the time as the first step in building an orbiting laboratory and in proving the Soviets' ability to perform a space rescue, the mission may also have been a rehearsal of the operations required during the proposed lunar flight.

The next Soyuz flights did not take place until after the successful Apollo 11 mission to the Moon and marked a shift in emphasis for the Soviet programme. Soyuz 6,7 and 8 were launched on consecutive days and performed a series of joint manoeuvres in Earth orbit. No docking or spacewalks took place and it is likely that an attempt to dock Soyuz 7 and 8 ran into technical problems and was cancelled. Soyuz 6 carried out welding experiments and was probably launched in order to secure a propaganda point by having three manned spacecraft in orbit simultaneously. Soyuz 9 was a solo flight intended to develop Soviet experience of long duration spaceflight. Its crew of two set a space endurance record of 18 days.

After the launch of the Salyut 1 space station, almost all Soyuz missions were simply ferry flights to and from orbiting space stations. Soyuz 10 and 11 flew to Salyut 1, but as described on page 185, the basic Soyuz had to be modified following the fatal accident to Soyuz 11. The modified Soyuz became a two-man spacecraft with only limited capabilities for independent flight and this resulted in a number of Soyuz missions being aborted when the crew failed to dock with a space station and were forced to return to Earth. The basic Soyuz was later upgraded into the two- or three-man Soyuz T and then the improved Soyuz TM.

As the USSR developed experience in long duration flights, it became necessary to divide Soyuz missions into two types.

Long-stay crews docked with a space station and remained onboard for progressively longer flights. Since it was considered unsafe to return to Earth in a Soyuz that had been docked to a space station for more than about three (and later six months), visiting crews were used to fly 'fresh' spacecraft to the station and then return to Earth in the old one about a week later. These visiting crews sometimes included foreign guest cosmonauts or paying passengers. During the later stages of the Salyut programme partial or total crew exchanges took place from time to time, so that listing cosmonauts by Soyuz mission becomes rather complicated. The table below lists the crews of each Soyuz as it was launched. Frequently cosmonauts returned to Earth in different Soyuz capsules from the one in which they were launched and sometimes with different companions from those with whom they entered space.

Apart from failed rendezvous missions, five independent Soyuz flights were flown during the 1970s. Soyuz 12 was a two-day flight to check out the modifications made after the Soyuz 11 accident. Soyuz 13 had a battery of telescopes fitted to the front of the orbital module in place of a docking mechanism and was devoted to astronomical research. Soyuz 16 was a practice for the international Apollo–Soyuz Test Project (ASTP) which was flown by Soyuz 19, and Soyuz 22 was an Earth observation mission with a large camera mounted in the orbital module.

Soyuz mission summary

Mission	Launched	Crew	Notes
Soyuz 1	23 Apr 67	Komarov	Komarov killed in crash landing
Soyuz 2	25 Oct 68	—	Unmanned target for Soyuz 3
Soyuz 3	26 Oct 68	Beregovoi	Rendezvous with Soyuz 2
Soyuz 4	14 Jan 69	Shatalov	Docked to Soyuz 5
Soyuz 5	15 Jan 69	Volynov, Khrunov, Yeliseyev	Khrunov and Yeliseyev spacewalk to Soyuz 4 and land with Shatalov
Soyuz 6	11 Oct 69	Shonin, Kubasov	Multiple rendezvous, no docking
Soyuz 7	12 Oct 69	Filipchenko, Volkov, Gorbatko	Rendezvous with Soyuz 6 and 8
Soyuz 8	13 Oct 69	Shatalov,Yeliseyev	Rendezvous with Soyuz 6 and 7
Soyuz 9	1 Jun 70	Nikolayev, Sevastyanov	18-day mission
Soyuz 10	23 Apr 71	Shatalov, Yeliseyev, Rukhavishnikov	Docked to, but did not enter, Salyut 1
Soyuz 11	6 Jun 71	Dobrovolsky, Volkov, Patseyev	Docked to Salyut 1, crew died on re-entry
Soyuz 12	27 Sep 73	Lazarev, Makarov	Test flight of modified Soyuz
Soyuz 13	18 Dec 73	Klimuk, Lebedev	Solo flight, astronomy
Soyuz 14	3 Jul 74	Popovitch, Artyukhin	Docked to Salyut 3
Soyuz 15	26 Aug 74	Sarafanov, Demin	Failed to dock with Salyut 3
Soyuz 16	2 Dec 74	Filipchenko, Rukhavishnikov	Solo test for US/USSR ASTP mission
Soyuz 17	11 Jan 75	Gubarev, Grechko	Docked with Salyut 4
Soyuz -	5 Apr 75	Lazarev, Makarov	Launch failure, sub-orbital abort
Soyuz 18	24 May 75	Klimuk, Sevastyanov	Docked with Salyut 4
Soyuz 19	15 Jul 75	Leonov, Kubasov	Docked with Apollo in ASTP
Soyuz 20	17 Nov 75	—	Docked with Salyut 4, test of systems for Progress tanker
Soyuz 21	6 Jul 76	Volynov, Zholobov	Docked with Salyut 5
Soyuz 22	15 Sep 76	Bykovsky, Aksyonov	Solo flight, Earth observations
Soyuz 23	14 Oct 76	Zudov, Rozhdestvensky	Failed to dock to Salyut 5, night landing in icy lake, crew rescued
Soyuz 24	7 Feb 77	Gorbatko, Glazhkov	Docked with Salyut 5
Soyuz 25	9 Oct 77	Kovalyonok, Ryumin	Failed to dock with Salyut 6
Soyuz 26	10 Dec 77	Romanenko, Grechko	Docked to Salyut 6, 96-day mission
Soyuz 27	10 Jan 78	Dzhanibekov, Makarov	Visitors to Salyut 6

Soyuz 28	2 Mar 78	Gubarev, Remek (Czechoslovakia)	Visitors to Salyut 6
Soyuz 29	15 Jun 78	Kovalyonok, Ivanchenkov	Docked to Salyut 6, 140-day mission
Soyuz 30	27 Jun 78	Klimuk, Hermasziewski (Poland)	Visitors to Salyut 6
Soyuz 31	26 Aug 78	Bykovsky, Jahn (DDR)	Visitors to Salyut 6
Soyuz 32	25 Feb 79	Lyakhov, Ryumin	Docked with Salyut 6, 175-day mission
Soyuz 33	10 Apr 79	Rukhavishnikov, Ivanov(Bulgaria)	Failed to dock to Salyut 6
Soyuz 34	6 Jun 79	—	Launched unmanned, returned crew of Soyuz 32 from Salyut 6
Soyuz T1	16 Dec 79	—	Unmanned test of new Soyuz, docked to Salyut 6
Soyuz 35	9 Apr 80	Popov, Ryumin	Docked with Salyut 6, 185-day mission
Soyuz 36	26 May 80	Kubasov, Farkas (Hungary)	Visitors to Salyut 6
Soyuz T2	5 Jun 80	Malashev, Aksyonov	Visitors to Salyut 6
Soyuz 37	23 Jul 80	Gorbatko, Tuan (North Vietnam)	Visitors to Salyut 6
Soyuz 38	18 Aug 80	Romanenko, Mendez(Cuba)	Visitors to Salyut 6
Soyuz T3	27 Nov 80	Kizim, Makarov, Strekalov	Docked to Salyut 6 for short repair mission
Soyuz T4	12 Mar 81	Kovalyonok, Savinykh	Docked with Salyut 6
Soyuz 39	22 Mar 81	Dzhanibekov, Gurragcha (Mongolia)	Visitors to Salyut 6
Soyuz 40	15 May 81	Popov, Prunariu (Rumania)	Visitors to Salyut 6
Soyuz T5	13 May 82	Berezovoi, Lebedev	Docked with Salyut 7, 211-day mission
Soyuz T6	24 Jun 82	Dzhanibekov, Ivanchenkov, Chretien (France)	Visitors to Salyut 7
Soyuz T7	19 Aug 82	Popov, Serebrov, Savitskaya	Visitors to Salyut 7
Soyuz T8	20 Apr 83	Titov, Strekalov, Serebrov	Failed to dock with Salyut 7
Soyuz T9	27 Jun 83	Lyakhov, Alexandrov	Docked to Salyut 7, 149-day mission, no visitors.
Soyuz —	27 Sep 83	Titov, Strekalov	Abort off launch pad, crew saved
Soyuz T10	8 Feb 84	Kzim, Solovyov,Atkov	Docked with Salyut, 237-day mission
Soyuz T11	3 Apr 84	Malashev, Strekalov, Sharma (India)	Visitors to Salyut 7
Soyuz T12	17 Jul 84	Dzhanibekov, Saviskaya, Volk	Visitors to Salyut 7
Soyuz T13	6 Jun 85	Dzhanibekov, Savinykh	Docked to Salyut 7, crew rotation with Soyuz T14
Soyuz T14	17 Sep 85	Vasyutin, Volkov, Grechko	Docked with Salyut 7, mission ended early, Vasyutin ill
Soyuz T15	13 Mar 86	Kizim, Solovyev	Docked with Mir and Salyut 7
Soyuz TM1	21 May 86	—	Test flight of improved Soyuz, docked with Mir
Soyuz TM2	5 Feb 87	Romanenko, Laveikin	Docked with Mir
Soyuz TM3	22 Jul 87	Viktorenko, Alexandrov, Faris (Syria)	Docked with Mir, partial crew exchange with TM2, Romanenko, in space for 326 days
Soyuz TM4	21 Dec 87	Titov, Manarov, Levchenko	Docked to Mir, Titov and Manarov in orbit for one year
Soyuz TM5	7 Jun 88	Solovyev, Savinikh, Alexandrov	Visited Mir, returned in TM4

Soyuz TM6	29 Aug 88	Lyakhov, Poliakov, Mohmand (Afghanistan)	Visited Mir, Lyakhov and Mohmand returned in TM5, Poliakov made medical checks on Manarov and Titov
Soyuz TM7	26 Nov 88	Volkov, Krikalyev, Chretien (France)	Docked to Mir, TM6 returned Titov and Manarov with Chretien. TM7 later returned Volkov, Krikalyev and Poliakov leaving Mir empty.
Soyuz TM8	5 Sep 89	Viktorenko, Serebrov	Reactivated Mir for 6-month mission
Soyuz TM9	11 Feb 90	Solovyov, Balandin	Docked to Mir for 6-month mission, crew rotation
Soyuz TM10	1 Aug 90	Manakov, Strekalov	Docked to Mir, crew rotation
Soyuz TM11	2 Dec 90	Afanasyev, Manarov, Akiyama (Japan)	Docked to Mir, crew rotation, carried Japanese journalist
Soyuz TM12	18 May 91	Artsebarsky, Krikalyev, Sharman (UK)	Docked to Mir, crew rotation
Soyuz TM13	2 Oct 91	Volkov, Aubakirov, Vietiboeck (Austria)	Docked to Mir, partial crew rotation
Soyuz TM14	17 Mar 92	Viktorenko, Kaleri Flade (Germany)	Docked to Mir, crew rotation (Volkov, Krikalyev down)

Soyuz takes off

Space debris

Cosmos 1275 destroyed 1981

Pollution in orbit

Debris in orbit around the Earth is now presenting a serious hazard to satellites and astronauts.

Before astronauts ventured into space one of the biggest unknowns was the danger posed by meteoroids (orbiting dust grains) which might strike a spacecraft with catastrophic consequences. It soon became clear that the risk from meteoroids was very small, but a new hazard is now manifesting itself: more than three decades of space exploration have left some regions of space dangerously cluttered with debris.

The danger from debris

Since 1957 there have been over 3 000 rocket launches which, between them, have placed almost 4 000 satellites in orbit. Most of these satellites are no longer operating, and many have already re-entered the atmosphere and burned up, but hundreds are still in space. However, these are just the tip of the iceberg because complete satellites represent only a few per cent of the objects in orbit. Military radars can detect objects as small as a few centimetres across and these systems have revealed that there are thousands of man-made objects in orbit. Most of these tiny satellites are pieces of space junk, burned-out rocket stages and the like, but each one presents a hazard to other satellites and so must be catalogued and tracked.

The problem of space debris is not just that two satellites might one day collide; even in the most congested region of low Earth orbit the likelihood of a collision between two catalogued satellites is very small. The overriding concern is that for every object large enough to be tracked from the ground there are probably many more which are too small to be detected, but which are large enough to do serious damage if they should strike another satellite. At orbital speeds the effect of an impact by an object only a few centimetres across may be similar to that produced by a hand grenade: quite enough to ruin an expensive satellite and potentially lethal if a manned spacecraft were to be hit.

The source of the debris

Space debris arises from a number of causes. The most well publicized, although the least important, examples are objects left in space by astronauts. Probably the most famous of these are the glove which floated out of the Gemini 4 spacecraft during the first US spacewalk in 1964 and the Hasselblad camera lost by Michael Collins during the Gemini 10 mission. Less well known are what appear to be bags of rubbish jettisoned from Soviet space stations. Another source of debris is panels and covers used to protect sensitive parts of satellites during launch and cast off once the satellite was in orbit. However, even these are relatively unimportant compared to the huge numbers of fragments produced by satellites which have disintegrated in space.

The destruction of satellites, including discarded rocket stages that have entered stable orbits, may occur by accident or design. Occasionally, during the firing of a rocket motor a major fault develops which causes an explosion. Cosmos 1423, which blew up in December 1982, may have met its end in this way. Fortunately, such events are rare. More common are the unplanned explosions of rocket stages after they have reached orbit and their payload has been released. For example, during the 1970s the abandoned upper stages of seven Delta rockets exploded in orbit. Between them these explosions produced over 1 200 fragments large enough to be tracked and catalogued from the ground. It was eventually discovered that the explosions were caused by the accidental mixing of propellants left over when the rocket shut down. To prevent this recurring, new procedures to burn off all the remaining propellants were introduced and no further explosions of Deltas have occurred since. However, the Delta rocket is by no means the only culprit. US Titan and Agena stages have also disintegrated in orbit and the upper stage of a European Ariane rocket exploded in November 1986.

The deliberate destruction of satellites is another major source of space debris. Some military satellites are destroyed to prevent them falling into the hands of another nation should they re-enter over foreign soil. The USSR in particular had a history of exploding its spy satellites once they had completed their missions. The USSR also conducted tests of killer satellites that home in on a target and then explode, adding to the number of fragments in orbit.

Another, and more worrying, cause of satellite fragmentation are high-speed collisions with pieces of space debris. In 1981 the Soviet Cosmos 1275 satellite broke into hundreds of pieces for no apparent reason, and it is believed that a collision with an uncatalogued piece of space junk may have been responsible. If this was the case, then it is easy to see that the problem may already be getting out of control. The destruction of Cosmos 1275 produced almost 300 fragments large enough to be tracked. Any of these fragments could probably destroy another satellite, generating another shower of lethal debris and hastening the destruction of more satellites. This could lead to a chain reaction that might, quite quickly, surround our planet with a blizzard of satellite fragments.

Clearing up space

Fortunately, much space debris eventually re-enters the atmosphere and burns up. Objects in low Earth orbit experience a tiny amount of drag from the tenuous upper atmosphere and this is often enough to cause them to spiral down to destruction. The Sun's 11-year cycle has a powerful cleaning action, for at times of high solar activity the Earth's atmosphere expands and this helps to bring down debris in low orbits. Objects in higher orbits, such as elliptical geostationary transfer orbits, are not so immediately vulnerable to air drag as objects in low circular orbits, but the gravitational influences of the Sun and Moon can subtly alter these orbits so that they move into regions where air drag is significant.

Despite the effects of air drag, some satellites are in orbits that will be stable for centuries and, at present, little can be done about these. Although the Space Shuttle has returned several satellites to Earth for repairs, it is not practical to use astronauts as celestial road sweepers to tidy up space. Other, more exotic, methods of removing space debris are being considered, but for the present the main priority must be to stop the production of more debris before it is too late. Obvious preventative measures that could be implemented immediately include deliberately de-orbiting satellites and empty rocket stages that have finished their missions and stopping the destruction of military spacecraft. If such measures are not taken, mankind may soon find itself trapped in an orbiting cocoon of its own making.

> Space debris is already proving expensive. A Space Shuttle window, cracked by a tiny impact that occurred in orbit, cost $50 000 to replace.

> Scientists estimate that there is a 1% chance that the $1 500 million Hubble Space Telescope will be seriously damaged by a collision with a piece of space junk.

> Some panels from the Solar Maximum Mission satellite were brought back to Earth by Space Shuttle astronauts in 1984. Of the 186 tiny craters found on these panels, 20 were due to meteoroids and the other 166 contained traces of spacecraft paint.

Space disasters

Apollo 1 1967, Soyuz 1 1967, Soyuz 11 1971, Challenger 1986

Paying the price

Although space travel has proved to be relatively safe, there have been tragedies.

Before humans ventured into space, science fiction writers assumed that fatal accidents would be relatively frequent and often made passing references to lost astronauts in their stories. However, despite many spectacular failures of unmanned rockets, the USA's Mercury and Gemini, and the USSR's Vostok and Voskhod programmes were completed without an astronaut or cosmonaut being killed as a direct result of a spaceflight (although some died in training accidents such as aircraft crashes). This blemish-free record was lost in 1967 when the space programmes of both nations suffered their first fatal accidents.

Apollo 1

On 27 January 1967 the crew assigned to the first Apollo mission, Virgil Grissom, Edward White and Roger Chaffee entered their spacecraft for a routine countdown test. Apollo 1 was installed on top of a Saturn 1B rocket on Pad 34 at Cape Canaveral and, because the rocket was empty, the test was not considered hazardous. Just as it would be during the actual countdown the Apollo command module was filled with pure oxygen at slightly more than atmospheric pressure. The test proceeded slowly, but by just after 6:30pm the simulated countdown had reached T-10 minutes. Suddenly Virgil Grissom shouted 'Fire in the spacecraft' and seconds later Roger Chaffee called 'We've got a bad fire'. A few more words were heard and the astronauts could be seen moving about inside Apollo 1, but within seconds all movement ceased. Moments later the capsule split open and flames rushed out. Technicians tried to rescue the astronauts, but heat, smoke and flames drove them back. In fact, there was nothing they could do, the three astronauts were already beyond help.

After the charred bodies had been removed from the capsule, post-mortems showed that although burned, the astronauts had died from inhaling smoke. The accident, occurring as it did not in space but on the ground, threw the USA into a state of shock and disbelief. The three astronauts were buried with full military honours, Grissom and Chaffee in the national cemetery at Arlington on the outskirts of Washington and White at the US Air Force Academy at West Point.

An investigation of the accident reported that the cause of the fire could not be established for certain, but was probably an electrical arc between poorly insulated wires somewhere below Grissom's couch. The Apollo capsule contained a large amount of flammable material and this burned with unusual ferocity in the pure oxygen atmosphere. The hatch, the only escape route for the astronauts, took about 1.5 minutes to open, but the crew were unconscious within seconds. The inquiry also found a huge catalogue of shortcomings, poor workmanship and bad design throughout the Apollo spacecraft. It was clear that there were other potentially fatal problems and a major programme of safety improvements was begun. These modifications delayed the Apollo programme by about 18 months, but the result was a much safer and more reliable spacecraft. It was little comfort, but the death of three men on the ground almost certainly saved others from dying in space.

Soyuz 1

On 23 April 1967, a few months after the tragedy at Cape Kennedy, the USSR launched Vladimir Komarov aboard the first of their new Soyuz spacecraft. The mission was expected to involve a link-up with another Soyuz spacecraft and possibly a transfer of crews in orbit. However, although the Soviet authorities were vague about the details, the mission went badly wrong from the start. It is believed that at least one of Soyuz's two solar panels failed to deploy properly, depriving Komarov of vital electrical power. There were also problems with either the spacecraft's thrusters or with the sensors required for stabilization, and these made Soyuz difficult to control. The other Soyuz launch was abandoned and the efforts of the Soviet mission controllers were concentrated on solving the problems with Soyuz and bringing Komarov back. Re-entry attempts were called off at the last minute on orbits 16 and 17, probably because the spacecraft could not be stabilized in the correct attitude to fire its retro-rockets, and re-entry did not start until orbit 18. At first all went well, but then a problem developed with the capsule's landing parachute, which became tangled. Komarov was killed when his capsule smashed into the ground near a town called Orsk, close to the Urals. A monument marks the spot where he died.

Soyuz 11

On 29 June 1971, after spending a record 24 days aboard the Salyut 1 space station, the crew of Soyuz 11 prepared to return to Earth. Commander Georgi Dobrovolsky, engineer Vladislav Volkov and researcher Victor Patseyev sealed themselves into their Soyuz ferry and undocked from Salyut. As was normal practice at the time, they were not wearing spacesuits. A few hours later they fired the Soyuz retro-rockets to begin their descent to Earth. Shortly afterwards, following the standard procedure, the Soyuz re-entry module was separated from the remainder of the craft. At this point a valve, intended to open in order to equalize the pressure between the capsule and the air outside just before touchdown, was unexpectedly jerked open. At once, air began to rush out of the capsule into the vacuum of space. Victor Patseyev seems to have tried to close the valve by hand, but it was impossible, the valve took a full minute to close and by then the crew, unprotected by spacesuits, had suffocated. The re-entry continued normally and Soyuz 11 landed in full view of a recovery helicopter. The recovery team opened the Soyuz hatch and gazed in horror at the three dead cosmonauts.

At first it was feared that the Soyuz 11 crew had perished because they had remained in space too long, causing irreversible changes in their bodies which meant that they could not survive the forces of re-entry, but the true cause of the tragedy was soon established. As a result, it was decided that in future Soviet cosmonauts must always wear pressure suits during critical phases of their missions and Soyuz was redesigned to accommodate two spacesuited cosmonauts instead of three in overalls.

After a state funeral the Soyuz 11 crew were cremated and on 2 July 1971 their ashes joined those of Vladimir Komorov and Yuri Gagarin, who had died in an aircraft accident, in the Kremlin wall.

Challenger

The twenty-fifth mission of the Space Shuttle had been postponed and delayed four times before its crew of seven rode out to the launch pad on the morning of 28 January 1986. It had been a cold night at Cape Canaveral and the launch tower was festooned with icicles, but it was decided that these would not be dangerous and that the launch could go ahead. Accompanying commander Dick Scobee and pilot Mike Smith were mission specialists Ellison Onizuka, Judy Resnik and Ron McNair, and 'passengers' Gregory Jarvis, from the Hughes Aircraft Company, and teacher Christa McAuliffe. McAuliffe, selected from 11 000 applicants to be the first teacher in space, was expected to broadcast several lessons from orbit directly into US schools.

Challenger lifted off at 11:38 Eastern Standard Time and, although unnoticed at the time, several puffs of smoke emerged from the right-hand solid rocket booster as Challenger climbed away from the launch pad. The puffs were evidence of a leak between two sections of the booster which are bolted together and sealed by a combination of putty and rubber O-rings. The leak apparently sealed itself a few seconds after launch and for almost the next minute Challenger continued to climb normally. Then, 58 seconds after launch, a small flame appeared from the same joint. Within seconds it had formed a burning torch which melted a hole in the side of the huge external fuel tank and began to weaken the lower strut which attached the solid rocket to the tank. Seventy-two seconds after launch the strut burned through and the solid booster swung around its top strut and crashed into the external tank which began to rupture. Within another second the Space Shuttle was enveloped in a huge fireball of exploding propellants and the Challenger was torn into several large pieces by aerodynamic forces. The solid boosters continued firing until they were destroyed by remote control. The fragments of Challenger, including the crew compartment, continued upwards for a few moments then fell back into the Atlantic Ocean. All seven astronauts were killed.

A presidential commission soon identified the cause of the accident as a failure of the two O-ring seals at the joint between two sections of the solid rocket booster. This had been brought on by the low temperatures to which the Shuttle had been exposed during the preceding night and which had caused the rubber O-rings to lose their flexibility, preventing them from sealing properly when the boosters were ignited. The investigation also revealed that similar failures had occurred in the past, but without such appalling consequences. More alarmingly, the commission also found numerous other potentially dangerous problems, such as shortcomings in the braking system used when the Space Shuttle lands.

Although the commission recognized that it was impossible to escape from the Space Shuttle while the solid rocket boosters were firing, Space Shuttle launches were suspended for over two years while various other safety issues were attended to and new, safer operational procedures defined. Amongst the changes which followed the Challenger explosion were the fitting of a third O-ring seal, the use of pressure suits by the crew during launch and the provision of an escape system so the astronauts could parachute to safety at any time while the Space Shuttle was in controlled gliding flight.

The Apollo 15 crew left this memorial to 14 deceased astronauts and cosmonauts in a small lunar crater

Space emergencies

Gemini 8 1966, Apollo 13 1970, Soyuz T10 1987

Dicing with death

Space is a hostile environment and it is the skill of the crews, and mission controllers, which must prevent an emergency becoming a disaster.

Many spaceflights have had moments of high tension when it appeared that things were going wrong. Sometimes, as when it was feared that the heatshield of US astronaut John Glenn's Mercury capsule was loose, the problem is a faulty sensor giving a false alarm. At other times, such as when the Apollo 11 computer overloaded minutes before the first landing on the Moon, the problem is real, but can be circumvented by advice from the ground. Occasionally, things go so wrong, so quickly, that astronauts have been very close to death.

Gemini 8

When the USA's Gemini 8 astronauts Neil Armstrong and David Scott were launched on 16 March 1966 they were looking forward to a three-day mission which would involve a docking to an unmanned Agena target rocket and a spacewalk by Scott. At first, all went according to plan and seven hours after launch Armstrong eased the nose of Gemini 8 into the Agena to complete the first ever manned docking in space. About half an hour later the combined spacecraft began to roll to the left for no obvious reason. Suspecting a problem with the Agena rocket's control system, the astronauts turned it off, but the rolling continued. After a few minutes the tumbling had become so violent that it was becoming dangerous, threatening to damage the docking system, and Armstrong decided to undock from the Agena. This did not help, indeed the tumbling rapidly got worse in both roll and yaw. Scott reported 'We have serious problems here ... we're tumbling end over end'. Scott did not exaggerate, the Gemini was spinning once per second and within moments the astronauts could expect to become dizzy and suffer blurred vision, ending all hope of regaining control. Furthermore, the Gemini itself might be damaged if the wild gyrations could not be stopped. Realizing that one of their 16 attitude control thrusters was firing continuously, but unable to tell which one, the astronauts switched off the whole system and activated the separate re-entry thrusters. This system, normally only used during the last few minutes of the mission, worked perfectly and the astronauts soon regained control.

Unfortunately, the mission rules demanded that if the re-entry thrusters were activated then the spacecraft must return to Earth without delay. So, on its seventh orbit, Gemini 8 fired its retro-rockets and made an emergency splashdown in the Pacific Ocean. The flight had lasted just 10 hours. The fault which had caused the problem was never positively identified, but was most probably a short circuit in the spacecraft's electrical system.

Apollo 13

James Lovell, Fred Haise and Jack Swigert were launched aboard the USA's Apollo 13 at 14:13 local time on 11 April 1970. They planned to land on the Moon near the crater Fra Mauro, but they never reached their destination. On 13 April, almost 56 hours into the flight, the crew performed a routine operation to stir up the liquid oxygen in the tanks within the Apollo service module. Unknown to anyone at the time, the wiring in the tank had been damaged weeks before at Cape Kennedy and the astronauts' actions started an electrical arc inside the tank. This set fire to some teflon insulation and the resulting heat started boiling the liquid oxygen and caused the pressure in the tank to rise rapidly. Within a few seconds, unable to contain the pressure, the oxygen tank ruptured. Gas flooded out into the service module and less than half a second later the sudden rise in pressure inside the bay blew a complete panel off the side of the service module. Jarred by the explosion, Apollo 13 lurched suddenly and alarms began to go off inside the command module warning that the fuel cells, which used oxygen from the ruptured tank to produce electricity, were not operating correctly, and Jack Swigert radioed 'OK Houston, we've had a problem here'.

In fact, the problem was much worse than a failed electrical system. The astronauts soon noticed gas blowing off into space and realized that they were in grave danger. When the tank exploded, it had damaged the only other oxygen tank as well and the astronauts vital oxygen was boiling off into space before their eyes. Despite frantic efforts there was nothing they could do to stop the gas escaping or to restore power to the crippled service module. They had only one chance: the lunar module was still attached and had both oxygen and batteries which could supply a limited amount of electrical power. Hurriedly, the astronauts began to switch on the lunar module's equipment so that they could use it as a lifeboat.

The events of the next few days were a triumph for the mission controllers and their industrial contractors. They were also a test of endurance for the astronauts. Ground controllers improvised new procedures to use the lunar module's engine to bring Apollo 13 back to Earth and to eke out its limited supplies of power. Since all unnecessary systems were switched off, Apollo 13 became very cold and in addition the astronauts suffered from a shortage of drinking water (water was usually produced as a by-product of the now defunct fuel cells). After a firing of the lunar module's descent engine, Apollo 13 rounded the Moon and headed back towards Earth. Shortly before splashdown the astronauts returned to the command module, jettisoned the useless service module and the lifesaving lunar module and began re-entry. On 17 April, almost 143 hours after lift-off, Apollo 13 splashed down in the Pacific Ocean.

Soyuz T10

The narrow escape of the crew of Soyuz 18 in 1974 is described on page 164, but this was not the only Soviet launch failure. In September 1983 the Soviet space station Salyut 7 developed problems with its fuel system and its electrical power supply. On 27 September a repair crew consisting of Vladimir Titov and Gennadi Strekhalov boarded Soyuz T10. Their mission was to fly to Salyut 7, install new solar panels and investigate the leaky fuel system. The Soyuz countdown proceeded normally until about a minute from launch when a fire developed at the base of the Soyuz rocket. At this point the escape tower, a small rocket atop Soyuz designed to pull the spacecraft clear from the rocket beneath, should have been triggered. However, the fire damaged the wiring which should have activated the escape system and the Soyuz remained attached to its burning rocket. A disastrous explosion was only seconds away.

According to some Western sources, the only remaining way to trigger the escape system was by a radio command from ground controllers, but this required simultaneous countdowns in two different rooms. Another 10 seconds elapsed before this procedure could be completed and then, just as the rocket exploded, the capsule containing Titov and Strekhalov was blasted clear by the escape tower. The Soyuz was dragged more than 1km into the air by the escape tower and seconds later the descent module, with its shaken and bruised, but otherwise uninjured crew was seen dangling below its parachute. The descent module landed 4km from the wrecked launch pad.

The Apollo 13 astronauts never did make the legendary remark 'Houston, we have a problem'. Jack Swigert said 'Hey, we've had a problem here' and Lovell repeated moments later 'Houston we've had a problem here'. The phrase 'we have a problem' originated with HAL, the recalcitrant computer in Arthur C Clarke's novel *2001 A Space Odyssey* (1968).

The Apollo 13 astronauts took this photograph during their perilous journey back to Earth

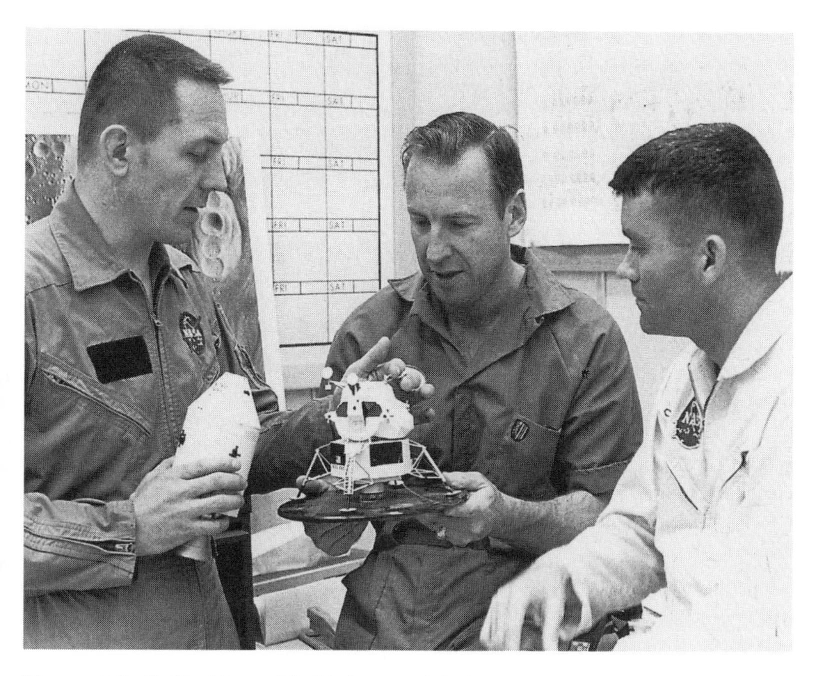

The crew of Apollo 13 with a model of the lunar module that was to save their lives

Space habitats

Gerard O'Neill 1973

Cities in orbit

The concept of huge space stations holding thousands of people came into prominence during the 1970s.

Although first considered by prophetic writers like Konstatin Tsiolkovski and J D Bernal, it was not until the early 1970s that the idea of self-sufficient space stations accommodating tens of thousands of people began to receive serious consideration. The sudden surge of interest probably arose because of an interesting coincidence of technological and social factors. The USA's Apollo Moon landings had demonstrated that mankind could achieve almost any realistic technological objective but, at the same time, concerns about pollution and a series of energy crises seemed to threaten the stability of Western technological societies. To some, most notably a professor of physics from Princetown University named Gerard K O'Neill, the solution was obvious: the surface of planet was not the best place for a technological civilization. Instead, O'Neill and co-workers began to develop the idea of moving society out into space.

O'Neill's vision was staggering in its audacity, yet it was cleverly thought out and only required modest extensions of existing technology. The key was that if a space station could be made large enough, then it could be self sufficient. O'Neill imagined that fields of crops and grazing animals would provide food for the occupants, who would drink recycled water and breath air that had been cleaned of carbon dioxide by plants growing inside the station. Unlimited energy could be obtained from the Sun, and the climate, and even the gravity, inside the station could be controlled. The industrial processes required to maintain the station, and to provide it with products which it could export, would be carried out in nearby, but independent, space factories so that toxic wastes did not accumulate in the space station itself. The result would be a closed ecology that would not require constant resupply from Earth. To distinguish them from ordinary space stations like the Soviet Mir, these self-sustaining mini-worlds are called space habitats.

The proponents of space habitats believed that although the cost of the first example would be huge, the occupants could develop a viable economy by building and operating huge solar power satellites. These orbiting power stations would harness sunlight to produce electricity and then beam it down to Earth as a cheap, pollution-free source of energy. The income from this would repay the initial investment and provide a profit which could be reinvested in the building of new, larger habitats. To those who argued that it would be impossible to launch into orbit sufficient material to build a space city, O'Neill had an easy answer — don't bring material up from Earth, mine the Moon instead.

Mining the moon

Mining the Moon makes economic sense if large quantities of material are required to build space habitats. The Moon's weaker gravity means that less energy is required to launch material from the Moon than from the Earth and for most locations in the Earth–Moon system, including Earth orbit, it requires less energy to collect material from the Moon than it does to launch it from Earth. Furthermore, the Moon has no atmosphere, so payloads can be accelerated to escape velocity at ground level without experiencing atmospheric drag. To exploit this, the lunar miners would not use rockets to export their products, but would probably employ electromagnetic catapults called mass drivers.

A mass driver uses linear induction motors to suspend a truck a fraction of a centimetre above a guide rail. Powerful electric and magnetic fields are used to accelerate the truck along the rail and, at the same time, stop it from touching the rail itself. Such devices have already been demonstrated on Earth. The lunar mass driver would be used to accelerate containers of lunar material to escape velocity, releasing them at just the right moment to send them towards an orbital construction site. Power for the mass driver, and for the mining operations, would be obtained from solar panels laid out across the Moon.

Mining the Moon like this sounds like fantasy but, provided a suitable lunar base exists, there are no obvious reasons why it should not be done. As with other aspects of space habitats, there are enormous economies of scale once the large initial costs have been met.

Island 1

Island 1 was one of O'Neill's ideas for a space habitat. It was to be a sphere about 460m in diameter with windows that would admit sunlight directed in by giant mirrors. The sphere would be constructed from metal as thick as battleship armour and the occupants would be shielded from dangerous solar and cosmic radiation by a few metres of lunar soil packed around the outside. About 10 000 people could live on the inside of the sphere, in apartments no smaller than they might demand on Earth. The sphere would be rotated twice a minute so that, in the sphere's equatorial zone, centrifugal force would produce an artificial gravity about equal to that of the Earth. Gravity would reduce to about one-third of normal at about 45°of north and south latitude and the poles would be regions of zero gravity, ideally suited to all kind of sports and recreations.

O'Neill envisaged an idyllic habitat, with private homes and public areas such as shops, parks and even small rivers laid out exactly as the inhabitants chose. For convenience, the accommodation could be arranged as three communities each living in a time zone eight hours different from the others. This would allow around-the-clock work to go on in the maintenance departments, and in the satellite power station assembly areas, without anyone having to work a night shift.

So that it would not need rockets to maintain itself in the correct orbit, Island 1 would probably be built at one of the five gravitationally stable regions in the Earth–Moon system. These regions are known as Lagrangian points after the mathematician who showed that an object placed in one would tend to be kept there by the combined gravitation fields of the Earth and Moon. The most stable positions, known as L4 and L5, are at the same distance from the Earth as the Moon, but are 60° ahead or behind the Moon itself.

Other designs

The spherical Island 1 was not the only concept that sprang from the fertile imaginations of O'Neill and his followers. Other versions considered were long cylinders with domed ends and with three long, equally spaced, rectangular windows admitting sunlight to a strip of 'land' on the opposite side of the cylinder. Each cylinder would spin around its long axis to produce artificial gravity and two cylinders would probably be joined together in parallel. A number of different-sized models were studied so that realistic estimates of how to build various designs could be made. The largest model was for a cylinder 32km long and 6.2km in diameter intended to support 20 million people. Another version, known as a Stanford Torus, resulted from a NASA-sponsored university summer school in 1975. This was a structure shaped like a spinning wheel, about 1.5km in diameter, which received its sunlight via a circular mirror hovering above the spin axis.

O'Neill originally referred to his concepts as 'space colonies' but this name has fallen from favour and 'space habitats' is now generally preferred.

Spacelab

First flight 10 November 1983

A reusable space laboratory

A modular laboratory that fits into the USA's Space Shuttle, Spacelab was Europe's first venture into manned spaceflight.

The seeds of the Spacelab system were sown in the late 1960s when NASA invited international cooperation in its post-Apollo programme, which it hoped would include a space station and manned missions to Mars. NASA's original plans turned out to be too ambitious and it was eventually decided that the USA would build a reusable Space Shuttle and that Europe would develop a laboratory module which could be carried into orbit and used to conduct scientific experiments. The laboratory was initially called a 'sortie module' and later named 'Spacelab'. The agreement was sealed when a memorandum of understanding, which stated that the European Space Agency (ESA) would provide one complete Spacelab system to NASA free of charge and that NASA would then purchase a second system from the ESA, was initialled by NASA and European representatives on 14 August 1973. The first Spacelab flight was to be a joint NASA/ESA mission after which each Spacelab user, whether ESA or an individual nation, would have to reimburse NASA for the costs of launching and operating Spacelab.

Development of Spacelab was an optional programme within the ESA system and Germany elected to pay the major share (53.3%) of the development costs. The other major participants were Italy (18%), France (10%) and the UK (6.3%).

The design

Spacelab is not a spacecraft, but a flexible system of modules that can be installed in the cargo bay of the Space Shuttle to provide facilities for scientific research. In each configuration Spacelab remains attached to the Shuttle and is dependent on it for power and other vital supplies. The most obvious Spacelab elements are the pressurized modules, which provide a shirt-sleeve environment in which scientist-astronauts can work. Each module is built from cylindrical aluminium segments 2.7m long and 4m in diameter. The outside of each segment is coated with layers of thermal insulation and removable experiment racks can be installed inside. One segment always contains controls and monitoring equipment as well as research equipment and this can be operated alone as a 'short module'. If a rear segment containing extra research equipment is added then the combination is called a 'long module'. The pressurized modules are linked to the main cabin of the Space Shuttle by a tunnel that connects to the door of the Space Shuttle airlock in the cargo bay. The astronauts live in the Space Shuttle, but move into Spacelab to conduct their research programmes.

A second part of the system are the Spacelab pallets. These are U-shaped platforms that can be used to mount experiments that do not need to be inside the pressurized laboratory. These might include automatic telescopes or cameras, or equipment that needs to be exposed to the vacuum of space. A single pallet weighs 1 200kg and is 4m across and 3m long.

Another element of Spacelab is the Instrument Pointing System or IPS. This is a movable platform, attached to a pallet, which is able to aim and stabilize instruments with great precision. The main role of the IPS is for astronomical missions which must point telescopes either at selected regions of the Sun or at particular stars. The IPS can be pointed either by commands from the ground or by astronauts using a joystick controller inside the Space Shuttle.

Many different combinations of these units are possible. A Space Shuttle can carry just a single pallet with a few instruments attached or a complete train of five pallets which completely fills the payload bay. If a pressurized module is carried then it can be short or long and can be combined with either three or two pallets respectively. All the elements of the system are reusable and are intended to make many flights.

Early Spacelab missions

Spacelab-1, the joint NASA–ESA mission was launched on 28 November 1983 aboard the Space Shuttle Columbia. It comprised a long module with a single pallet and carried dozens of experiments covering a wide range of scientific disciplines. The six-member crew included the first European astronaut, Dr Ulf Merbold. The mission lasted 10 days and the Spacelab system performed well.

The next mission was Spacelab-3 which was a seven-day flight launched aboard Challenger on 29 April 1985. The crew of seven used a long module plus one pallet configuration to conduct experiments in life sciences and materials science.

Spacelab-2 was an astronomy mission devoted to studies of both the Sun and stars. The launch was delayed because of a series of problems with the IPS and was eventually overtaken by Spacelab-3. After further delays on the launch pad, and a near abort when one engine of the Space Shuttle was shut down during ascent, Spacelab-2 and its crew of seven finally got into orbit on 29 July 1985. There were further problems with the IPS and some difficulties with a large infrared telescope, but eventually good results were obtained from both the solar and astronomical experiments.

Spacelab D-1 was a German national mission for which NASA received compensation from the German space agency. The configuration was a long module plus a pallet and the mission carried an eight-person crew which included two German scientists, an ESA astronaut and five Americans. Launched aboard the Challenger on 30 October 1985, the flight lasted seven days.

The future of Spacelab

The destruction of Challenger in January 1986 caused many Spacelab flights to be delayed or cancelled. However, even before the loss of the Challenger it was becoming clear that Shuttle launches were so costly that Spacelab flights would not be as frequent as originally hoped. A few missions are being flown in the 1990s and the first was the Spacelab Life Sciences mission SLS-1 in 1991. Other flights include another ESA/NASA mission and a Japanese flight (Spacelab-J). There will also be missions connected with the development of the international space station Freedom. Sadly, the problems with the Space Shuttle mean that Spacelab will probably never really achieve its full scientific potential.

> Despite the international nature of Spacelab, it was never possible to compromise in one problem area: the Shuttle was built in imperial units whilst Spacelab uses the metric system. Great care was required when making sections that had to be joined together.

Spacelab 1 in the payload bay of Space Shuttle Columbia during 1983

Space pioneers

1890–1940

Dreamers and achievers

A few visionaries have secured a certain place in the history of astronautics.

Wernher von Braun (1912–1977)

As a teenager, von Braun joined the German society for space travel (VfR–Verein für Raumschiffahrt), a private organization which carried out rocket experiments during the 1930s. When the work of the VfR was taken over by the Army, von Braun continued to develop rockets, first at Kummersdorf near Berlin and later at Peenemunde on the Baltic coast. Although he was once arrested by the Gestapo and accused of being more interested in space travel than military rockets, von Braun was responsible for the V2 ballistic missiles that were launched against England and Holland during the closing stages of World War II.

In 1945, to avoid capture by the Soviet Army, von Braun and many of his team headed south to surrender to the Americans, reasoning that the USA offered them the best chance of being allowed to continue their rocket research. Von Braun and many of his engineers were employed by the US Army, first at the White Sands missile range in New Mexico and later at the Redstone Arsenal in Huntsville, Alabama. Von Braun developed a series of missiles for the US Army and in 1957 he organized the launch of Explorer 1, the USA's first satellite. A tireless proponent of manned spaceflight, he masterminded the development of the Saturn rockets used to send men to the Moon. Disappointed by the USA's failure to follow up the Moon programme with a mission to Mars, von Braun resigned from NASA in 1972 and worked in industry before succumbing to cancer in 1977.

Valentin Petrovich Glushko (1908–1988)

One of the pioneers of Soviet rocketry, Glushko was born in Odessa and began to study astronomy as a child. His enthusiasm for space travel was encouraged when he wrote to Konstantin Tsiolkovski and eventually Glushko entered the Physics Department of Leningrad University to obtain a suitable technical education. Glushko then moved to the Gas Dynamics Laboratory (GDL) in Leningrad to carry out research in rocketry. During is work at the GDL during the early 1930s Glushko designed the ORM-1, the first liquid-fuelled rocket engine fired in the USSR. Glushko's group went on to develop a series of other rocket motors for various military purposes.

The staff of the GDL maintained contacts with the unofficial rocket societies such as the GIRDs and in 1932 Glushko met future chief designer Sergei Korolov. The two men worked together for a time at the Jet Propulsion Research Institute (Russian acronym RNII) on various rocket projects before Korolov was arrested and removed from the RNII on what proved to be trumped-up charges.

During the Second World War the two men worked together again and, once the war was over, went on to develop the R-7 or Sputnik rocket. Glushko designed the RD-107 and RD-108 engines for this rocket and was later involved in work on the rocket motors used in a variety of Soviet launch vehicles although the precise nature of his work is shrouded in secrecy.

Robert Hutchings Goddard (1882–1945)

Goddard was a professor of physics at Clark University in Worcester, Massachusetts when he began experimenting with rockets, and, beginning in 1914, he took out a string of patents concerning rocket technology. In 1916 he published a paper entitled 'A Method of Reaching Extreme Altitudes' which, although mainly concerned with using rockets for atmospheric research, did point out the possibility of sending a small rocket to the Moon. This aspect of his paper attracted considerable uninformed criticism and, as a result, Goddard began to conduct much of his later work in near secrecy.

On 16 March 1926 Goddard launched the world's first liquid-fuelled rocket which, using liquid oxygen and gasoline propellants, rose to a height of 12.5m. The aviator Charles Lindbergh persuaded the Daniel Guggenheim foundation to provide extra funds for Goddard's work and so Goddard moved to a farm in Roswell, New Mexico where he could experiment with larger and more powerful rockets. In 1936 he published details of his work in a monograph entitled *Liquid Propellant Rocket Development.* With the help of a few assistants, Goddard eventually launched rockets over 6 metres long and weighing 200kg. During World War II Goddard was employed on military work, mostly concerned with small rockets used to shorten the take-off run of heavily loaded aircraft. He died before he could resume his civilian research. The NASA Goddard Spaceflight Centre near Washington is named after him.

Theodore von Karmen (1882–1963)

Theodore von Karmen was born in Budapest, Hungary and moved to Germany in order to obtain his higher education. He remained in Germany doing research into aerodynamics until 1930, when the rise of Nazism caused him to emigrate to the USA. He soon became a professor at the Guggenhiem Aeronautical Laboratory of the California Institute of Technology where he did pioneering work on aeronautics and rocketry. In 1936 students of von Karmen fired a liquid-fuelled rocket in the Arroyo Seco, a dried-up river bed near Pasadena, north of Los Angeles, and this eventually led to a request from the US War Department to develop rockets able to shorten the take-off run of heavily loaded military aircraft. To conduct this programme, some land near Pasadena was leased as a testing site and over the years this area has developed into the Jet Propulsion Laboratory, now one of the most important space facilities in the world.

Nikolai Ivanovich Kibalchich (1853–1881)

A Russian anarchist who, while awaiting execution for killing Tsar Alexander II, is said to have sketched a man carrying rocket which was powered by the explosion of gunpowder cartridges.

Sergei Pavlovich Korolov (1907–1966)

His identity masked under the pseudonym of 'Chief Designer of Carrier Rockets and Spacecraft', Korolov was the driving force behind the Soviet space programme until his death during routine surgery. Korolov was a member of the Moscow Group for the Study of Rocket Propulsion which was formed in 1931. Within this group, which merged with a similar society from Leningrad in 1933, Korolov helped to develop the first liquid-fuelled rockets launched in the USSR. He was imprisoned by Stalin in 1938 and forced to work in a scientific labour camp during World War II. Once freed from prison, Korolov continued his work on rockets, beginning by developing improved versions of the German V2 missile and moving on to more powerful military rockets. Korolov designed the first Soviet intercontinental ballistic missile and later adapted this to form the launch vehicle for the first Sputnik.

After the success of Sputnik, Korolov went on to mastermind the development of the Vostok and Voskhod spacecraft and a series of early unmanned missions to the planets. He also planned the Soyuz Complex, a trio of spacecraft intended to send Soviet cosmonauts to the Moon. Improved versions of his ballistic missile design were recently used to launch Soviet spacecraft and cosmonauts still travel in Soyuz spacecraft whose basic design Korolov sketched in 1961.

Hermann Julius Oberth (1894–1990)

The German-born Oberth was one of the founding fathers of modern rocketry, but was basically a theoretician. In 1923 he published a book entitled *The Rocket into Interplanetary Space* which included details of manned orbiting space stations. During the 1930s Oberth worked with the German society for space travel (VfR–Verein für Raumschiffahrt), which fostered rocket development in Germany until research was taken over by the German Army. After the war Oberth joined the German rocket engineers working under von Braun in the USA, but he soon left to return to his native country. He survived to see the Apollo 11 mission which sent the first men to the Moon.

Eugen Sanger (1905–1964)

An Austrian engineer who started work on rockets at Vienna University in 1931, Sanger was interested in winged rockets. From 1936 to 1945 he directed rocket research for the German Luftwaffe and proposed a winged rocket bomber that would skip in and out of the atmosphere to increase its range. In some senses, this was a forerunner of the Space Shuttle.

Friedrich Arturovich Tsander (1887–1933)

Tsander was a Soviet rocket pioneer involved in the development of liquid-fuelled rocket engines. In 1931 he became head of the Moscow Group for the Study of Rocket Propulsion and designed several rockets before his early death from typhoid fever.

Konstantin Eduardovich Tsiolkovski (1857–1935)

Tsiolkovski never built a rocket, but is today regarded as the father of Soviet rocketry. Rendered deaf by disease at the age of 10, Tsiolkovski qualified as a teacher, but spent much of his spare time dreaming about spaceflight. He published at least two books concerning space travel before the end of the nineteenth century and in 1903 published *The Exploration of Space Using Reaction Devices* which described a rocket fuelled by liquid hydrogen and liquid oxygen. His other work included descriptions of multistage rockets (which he called rocket trains), proposals to use gyroscopes and tilting rocket nozzles to control a rocket in flight, and the description of life-support systems, airlocks and spacesuits to sustain life outside the atmosphere. Almost unknown in pre-revolutionary Russia, Tsiolkovski was recognized as a true visionary later in his life and received many awards from the new Soviet government.

Max Valier (1895–1930)

A German inventor interested in rocket propulsion, Valier believed that the best way to develop rockets for space travel was first to prove them in terrestrial applications. He did considerable experimental work, including producing a series of rocket-propelled cars, before being killed by a laboratory explosion while testing a rocket motor on 17 May 1930.

Johannes Winkler (1887–1947)

A German rocket pioneer, Winkler edited the journal *Die Rakete* which was produced by the German Society for space travel (VfR–Verein für Raumschiffahrt), and carried out experiments with solid rockets in 1928. He went on to develop liquid-fuelled rockets and was probably the first European to launch such a rocket when his liquid methane and liquid oxygen powered rocket was launched on 21 February 1931.

> Konstantin Tsiolkovski is credited with the famous saying 'The Earth is the cradle of mankind, but you cannot live in the cradle forever'.

Spaceplanes

HOTOL 1984

Single stage to orbit?

The 1980s saw a resurgence of interest in fully reusable spaceplanes when Britain, the USA and Germany all put forward new proposals.

Most of the mass of a space launch vehicle is comprised of its propellants and the tanks to hold them. Lifting these structures all the way into orbit is inefficient, which is why rockets are built in stages. By dropping off fuel tanks as they are emptied, the amount of dead weight carried into orbit can be reduced and the payload correspondingly increased. None the less, multi-stage rockets are complicated and expensive so engineers have long sought a design which can reach orbit without using several stages; such a vehicle is known as a Single Stage To Orbit (SSTO) launcher.

By using ultra-lightweight materials, and small but powerful rocket motors, it may be possible to build an SSTO launcher which looks similar to a conventional rocket, but recent developments in engine technology suggest that there may also be an alternative. A rocket motor burns a mixture of fuel and oxidiser, and a conventional rocket carries a complete supply of both of these at the moment it takes off. However, for part of its flight the rocket is in the Earth's atmosphere so it could, in principle, extract oxygen from the air, reducing the total mass that must be launched. (This is how automobile and aircraft engines work, they carry only fuel, petrol or kerosene, and draw in oxygen from the air.) A normal rocket ascends quickly through the lower atmosphere into the near vacuum of space, so extracting oxygen from the air is not very practical, but a winged spaceplane could use a different technique, taking off horizontally like an aeroplane and drawing in oxygen from the air as it climbed. Once at high altitude the vehicle could accelerate in level flight, still using atmospheric oxygen, until it had achieved considerable velocity. At this point it could zoom into space, using oxidiser from a small tank in the spaceplane for the final rocket burn. At the end of its mission the spaceplane would re-enter and glide in to land like a Space Shuttle, ready for reuse. Since only a small oxidiser tank would be needed, the spaceplane could be smaller, and so lighter, than a conventional rocket and might be a practical SSTO vehicle.

HOTOL

The British Aerospace HOTOL (Horizontal Take-off and Landing) unmanned satellite launcher concept was announced in 1984. The HOTOL design is for an SSTO spaceplane intended to place 7 000–8 000kg into low Earth orbit. The spaceplane, which would have a sausage shaped fuselage 72m long and a stubby, double-delta wing 20m across, would weigh about 240 tonnes. The satellite payload would be carried in a small bay in the fuselage and be released into space via cargo bay doors which would open for deployment. For manned missions a pressurized cabin would be fitted in the cargo bay in place of the satellite.

HOTOL is designed to take off from a conventional aircraft runway, using a detachable trolley in place of a heavy-duty undercarriage, and then climb to 26km using an air-breathing engine. At this point, when HOTOL would be travelling at five times the speed of sound, it would switch to an internal supply of liquid oxygen and climb out of the atmosphere into space. For landing, HOTOL would return to a normal runway and touchdown on a lightweight undercarriage. By using normal airport-type facilities it is intended that many HOTOL flights could be made on a regular schedule, thus reducing the operating costs per mission.

It was hoped that Europe would adopt the HOTOL idea for further development, but little official interest was shown, perhaps because of worries that the cost of developing the new technology engines for HOTOL would be too great. In response to this, an interim HOTOL, using conventional rockets instead of the original air-breathing engines, has been proposed. The interim HOTOL, which has a shorter, fatter fuselage, would be carried to high altitude on the back of the huge Soviet-developed Antonov An-225 transport aircraft and then released. Once clear of the aircraft, HOTOL would fire its rocket motors to climb the rest of the way into orbit. If development were to be authorized it is believed that the first flight of the interim HOTOL could be in 2005.

NASP

At about the same time as HOTOL was proposed, the USA announced that it would develop a similar vehicle called the National Aerospace Plane (NASP). The NASP is sometimes referred to as the Orient Express, since it might be used for long-range passenger flights as well as for satellite launches. The NASP might also be used for high-speed military reconnaissance.

Initial studies for NASP are being done with wind tunnels and supercomputer simulations, but it is hoped that it will be possible to build a prototype experimental aircraft called the X-30 towards the middle of the 1990s. The X-30 will be used to verify the results of the wind tunnel tests and confirm that the advanced aerodynamics required for the NASP are well enough understood for development of a full-scale vehicle.

Sanger

In 1985 West Germany proposed an air-breathing spaceplane concept which they named after the famous German engineer Eugene Sanger (1905–1964). Similar to a proposal first advanced in the 1960s, the new Sanger would be a two-stage system in which a delta-winged carrier vehicle would lift a smaller, upper stage to an altitude of 35km and a speed of seven times the speed of sound. The upper stage, which could be a manned shuttle-type spacecraft called HORUS (Hypersonic Orbital Upper Stage) or an unmanned cargo carrier called CARGUS, would then use conventional rocket engines to boost itself into orbit. The carrier aircraft would then return to its base ready for the next mission. The CARGUS module would not be recovered, but the HORUS shuttle would return to Earth for reuse after it had completed its mission.

> If HOTOL is ever built, then British Aerospace will have pioneered a fighter plane that takes off vertically and a rocket that takes off horizontally.

An artist's impression of the interim HOTOL being launched from an Antanov 225 cargo aircraft

Space Shuttle

First flight 12 April 1981

A remarkable flying machine

The USA's Space Shuttle was the world's first reusable manned spacecraft.

The concept of reusable spacecraft is not new. Most early space writers assumed that manned spaceships would be recovered intact and used many times and they envisaged that winged spaceplanes would be used to build a space station from which a Moon flight could begin. However, President John F Kennedy's decision to declare a race to the Moon disrupted these carefully considered plans; the need to win at almost any cost led to the rapid production of giant rockets rather than the progressive development of reusable spaceplanes. Only once the Moon race had been won did NASA seek approval to develop a more economical Space Transportation System (STS). The STS was to have been an integrated system of reusable shuttles and space tugs servicing a manned space station, but financial cutbacks in the early 1970s caused the space tugs to be cancelled and the space station to be delayed. Only the Space Shuttle remained.

The Space Shuttle's original purpose was to service the, now delayed, space station, so NASA managers were forced to seek a new justification for its development. They argued that the Space Shuttle was not just a delivery vehicle, but that it could serve as a flying laboratory, a service to repair or recover faulty satellites, an economic launch vehicle for commercial satellites and could serve the needs of national defence. They proposed a fully reusable system in which a manned, jumbo-jet-sized, first stage carried a smaller orbiter to high altitude. The first stage would then release the orbiter and return to its launch site while the orbiter continued into space under its own power. It soon became clear that this system would cost more than the US Congress would approve and so the manned first stage was replaced by a partially reusable system of solid rockets and a disposable fuel tank. The project was approved in 1972 and the first flight was expected in 1979.

The original timetable proved unrealistic since many new technologies were needed for the Space Shuttle and the project's initial budget was inadequate to develop them all in time. Although an engineless prototype called Enterprise made a number of approach and landing tests in 1977, delays associated with the main engines and the heat-resistant tiles used for protection during re-entry, pushed the first launch back until 1981.

The system

The Space Shuttle comprises four main components, two reusable Solid Rocket Boosters (SRBs), a disposable External Tank (ET) and a winged Orbiter.

The Orbiter is about the size of a DC-9 jet airliner and can carry up to eight people (10 in an emergency) in a two-level crew compartment in the vehicle's nose. The upper level comprises the flight deck, from which the pilots fly the craft, and a rear workstation from which astronauts can control the deployment of satellites, operate equipment in the payload bay and control the robot arm (officially called the Remote Manipulator System or RMS). The lower deck is mainly used for sleeping and eating, but also features an airlock which opens out into the payload bay. The payload bay occupies most of the Orbiter's fuselage and is 18.3m long and 4.6m in diameter. Large doors over its entire length can be opened to deploy satellites or to expose scientific instruments to space. The inside of the doors also supports radiators which are used to dispose of the waste heat produced by the electrical equipment and the astronauts' bodies. The doors must be opened on every flight to prevent the shuttle from overheating. At the rear of the Orbiter, on either side of the tailplane, are pods containing the Orbital Manoeuvring System (OMS). This comprises two rocket engines used for orbital adjustments and sets of smaller thrusters used, in conjunction with others in the nose, for attitude control. The entire Orbiter (apart from the windows) is covered with various forms of insulation to protect the aluminium structure from the heat of re-entry. On the underside, where the re-entry heating is greatest, a large number of individual silica fibre tiles (which can be replaced as required) are fitted. Less critical areas are protected by flexible blankets of heat-resistant material.

At the very rear of the Orbiter are the three Space Shuttle main engines. These are the most advanced liquid-fuelled rocket engines ever built and can be used many times. Each engine develops 2 090kNewtons of thrust at its nominal 100% power, but can be operated at up to 109% thrust if required. The engines use liquid hydrogen and liquid oxygen propellants which are carried in the External Tank and sucked into the Orbiter by a special piping system.

The External Tank (ET) is the only part of the Shuttle which is not reused. It carries the 633 tonnes of liquid oxygen and 106 tonnes of liquid hydrogen for the Orbiter's three main engines. To keep the supercold propellants from boiling away, and to prevent ice formation on the tank before launch, the entire tank is coated with spray-on insulation. Extra ablative insulation is applied where aerodynamic heating during flight is likely, or where radiant heat from the main engines may impinge on the tank. The ET also serves as the structural backbone of the fully assembled Space Shuttle, holding together the Orbiter and the two large Solid Rocket Boosters.

The Solid Rocket Boosters (SRBs), the largest ever developed, provide additional thrust during the first few minutes of flight. Each SRB is made from a number of segments that are filled at the manufacture's facilities and then shipped to Cape Canaveral to be joined together during final assembly of the Space Shuttle. Each complete booster contains about 500 tonnes of propellant and develops a thrust of 12.9 million Newtons at launch. The nozzle of each booster can be tipped while the booster is firing to assist in steering during flight.

After separation, the SRBs are parachuted into the Atlantic Ocean and float until they are collected by recovery ships which tow them back to Cape Canaveral. The SRBs are then dismantled and the segments are returned to the manufacturer for refurbishment and refilling ready for reuse.

Mission profile

Each Space Shuttle flight is different, but a typical mission is as follows. A few seconds before launch the three main engines are ignited, brought up to full power and checked. If any problems are detected then the engines are immediately shut down and the launch is cancelled. If all is well, the two SRBs are ignited and the bolts which hold the Space Shuttle to the launch pad are severed by small explosive charges. One minute later, at an altitude of about 10km, the Space Shuttle approaches the point of maximum aerodynamic pressure and the main engines are throttled back for a few seconds to reduce the stresses on the vehicle. Once this point is passed, the engines are returned to full power and by about two minutes after launch the Space Shuttle is 44km high and the thrust from the SRBs is beginning to decay. At this point the SRBs are jettisoned.

The Space Shuttle continues to climb under the power of its three main engines until it has reached a velocity of about 7 800m per second and an altitude of about 110km. The main engines are then switched off and the External Tank is released. Valves in the ET open, allowing any remaining liquid oxygen to escape and causing the tank to tumble so that it breaks up during re-entry. At this point the Orbiter is not usually travelling fast enough to remain in space and so its OMS engines are fired to provide the final push into orbit. The final orbit varies, but is typically about 250km high.

A standard Orbiter can remain in space for up to 8 days (plus emergency supplies for a further 2 days) although a mission of 5 days is typical. If extra supplies for the crew and for the electrical power generation system are carried, then flights of up to 30 days are possible.

When the mission is over, the payload bay doors are closed and the OMS engines are fired to reduce the Orbiter's velocity by about 100m per second. This is sufficient to begin re-entry since atmospheric drag soon slows the Orbiter and causes it to plunge back into the lower atmosphere. As the air gets thicker, the aerodynamic controls become effective and the Orbiter glides down to a landing. Touchdown, which takes place at about 360km per hour, is usually at the Edwards Air Force Base in California or on the specially built runway at Cape Canaveral.

The Space Shuttle programme

The first four Space Shuttle missions were intended as development flights, after which an aggressive policy of satellite launches began to dominate Shuttle operations. NASA, anxious to prove the flexibility of its new system, also allowed a series of ambitious, and highly successful, satellite rescue missions. However, fundemental flaws in the Shuttle's design, combined with an unintentional fall in safety standards brought on by the pressures of maintaining a high launch rate, led to the fatal explosion of the Challenger in January 1986 (see p193). After a two-year pause for modifications to a number of critical systems, operations were restarted with a new, more conservative, policy. This decreed that the Space Shuttle would only be used where the presence of astronauts was essential, and routine satellite launches were removed from the Space Shuttle programme and transferred back to unmanned rockets.

Mission designations

The first flights were simply numbered consecutively but as the schedule became more complicated, with missions being cancelled and payloads switched from one flight to another, a code system was introduced. This used the last numeral of the US fiscal year in which the mission was expected, a launch site designator (1 = Cape Kennedy, 2 = Vandenberg Air Force Base) and a letter indicating the expected sequence of launches in that fiscal year. The system was abandoned from January 1986, after which each mission was assigned an STS number irrespective of the actual order in which the launches took place.

Endeavour, NASA's newest Space Shuttle, is named after the ship of British explorer Captain Cook. Costing two billion dollars to build, Endeavour is similar to earlier Shuttles but has improved electronics and is fitted with a tail parachute designed to shorten the Shuttle's landing run after touchdown. Endeavour's first mission, STS-49, was launched on 7 May 1992 and was completed on 16 May. The highlight of the flight was a series of complex and dangerous spacewalks to attach a motor to an INTELSAT-6 satellite which had been stranded in the wrong orbit by a failure of its Titan launch vehicle in March 1990. After two failures, a record breaking three-person spacewalk succeeded in attaching the motor allowing INTELSAT to be successfully moved to its correct orbit.

Space Shuttle flights: the first 10 years

Flight	Launch	Landing	Commander/Pilot/ Number of other crew	Payload/Mission highlights
STS 1 (C)	12 Apr 81	14 Apr 81	Young/Crippen/0	First Flight
STS 2 (C)	12 Nov 81	14 Nov 81	Engle/Truly/0	OSTA-1, Earth observation
STS 3 (C)	22 Mar 82	30 Mar 82	Lousma/Fullerton/0	OSS-1, Space sciences
STS 4 (C)	27 Jun 82	4 Jul 82	Mattingly/Hartsfield/0	Dept of Defense
STS 5 (C)	11 Nov 82	16 Nov 82	Brand/Overmyer/2	SBS-C,Anik C-3
STS 6 (Ch)	4 Apr 83	9 Apr 83	Weitz/Bobko/2	TDRS-A/ two-man EVA
STS 7 (Ch)	18 Jun 83	24 Jun 83	Crippen/Hauck/3	Anik C-2, Palapa B-1, SPAS,OSTA-2.
STS 8 (Ch)	30 Aug 83	5 Sep 83	Truly/Brandenstein/3	INSAT 1B
STS 9 (C)	28 Nov 83	8 Dec 83	Young/Shaw/4	Spacelab 1
STS 41B (Ch)	3 Feb 84	11 Feb 84	Brand/Gibson/3	Westar 6, Palapa B-2/ MMU flight
STS 41C (Ch)	6 Apr 84	13 Apr 84	Crippen/Scobee/3	LDEF/ SMM repair
STS 41D (D)	30 Aug 84	5 Sep 84	Hartsfield/Coates/4	SBS-D, Telstar 3D, Syncom IV-2
STS 41G (Ch)	5 Oct 84	13 Oct 84	Crippen/McBride/5	ERBS, OSTA-3
STS 51A (D)	8 Nov 84	16 Nov 84	Hauck/Walker/3	Telesat-H, Syncom IV-1 & 2 rescued
STS 51C (D)	24 Jan 85	27 Jan 85	Mattingly/Schriver/3	Dept of Defense
STS 51D (D)	12 Apr 85	19 Apr 85	Bobko/Williams/5	Telesat 01,Syncom IV-3
STS 51B (Ch)	29 Apr 85	6 May 85	Overmyer/Gregory/5	Spacelab 3
STS 51G (D)	17 Jun 85	24 Jun 85	Brandenstein/Creighton/5	Arabsat 1, Morelos 1, Telstar 3D
STS 51F (Ch)	29 Jul 85	6 Aug 85	Fullerton/Bridges/5	Spacelab 2
STS 51I (D)	27 Aug 85	3 Sep 85	Engle/Covey/3	AUSSAT-1, ASC-1, Syncom IV-4
STS 51J (A)	3 Oct 85	7 Oct 85	Bobko/Grabe/3	Dept of Defense
STS 61A (Ch)	30 Oct 85	6 Nov 85	Hartsfield/Nagel/6	Spacelab D-1
STS 61B (A)	26 Nov 85	3 Dec 85	Shaw/O'Connor/5	Morelos-B, Aussat-2, Satcom Ku 2
STS 61C (C)	12 Jan 86	18 Jan 86	Gibson/Bolden/5	Satcom Ku 1
STS 51L (Ch)	28 Jan 86	Exploded	Scobee/Smith/5	Fatal explosion
STS 26 (D)	29 Sep 88	3 Oct 88	Hauck/Covey/3	TDRS-C
STS 27 (A)	2 Dec 88	6 Dec 88	Gibson/Gardner/3	Dept of Defense
STS 29 (D)	13 Mar 89	18 Mar 89	Coates/Blaha/3	TDRS-D
STS 30 (A)	4 May 89	8 May 89	Walker/Grabe/3	Magellan
STS 28 (C)	8 Aug 89	13 Aug 89	Shaw/Richards/3	Dept of Defense
STS 34 (A)	18 Oct 89	23 Oct 89	Williams/McCulley/3	Galileo
STS 33 (D)	23 Nov 8	28 Nov 89	Gregory/Blaha/3	Dept of Defense
STS 32 (C)	9 Jan 90	20 Jan 90	Brandenstein/Wetherbee/3	Syncom IV-5,LDEF recovery
STS 36 (A)	28 Feb 90	4 Mar 90	Creighton/Casper/3	Dept of Defense
STS 31 (D)	25 Apr 90	29 Apr 90	Shriver/Bolden/3	HST
STS 41 (D)	6 Oct 90	10 Oct 90	Richards/Cabana/3	Ulysses
STS 38 (A)	16 Nov 90	20 Nov 90	Covey/Culbertson/3	Dept of Defense
STS 35 (C)	Dec 90	10 Dec 90	Brand/Gardner/3	Astro-1
STS 37 (A)	5 Apr 91	11 Apr 91	Nagel/Cameron/3	GRO
STS 40 (C)	5 Jun 91	14 Jun 91	O'Conner/Gutierrez/5	Spacelab Life Sciences 1
STS 43 (A)	2 Aug 91	11 Aug 91	Blaha/Baker/3	TDRS-E
STS 48 (D)	13 Sep 91	18 Sep 91	Creighton/Reightler/3	UARS
STS 44 (A)	24 Nov 91	1 Dec 91	Gregory/Henricks/4	Dept of Defense

Key: C = Columbia, Ch = Challenger, D = Discovery, A = Atlantis

The Space Shuttle Challenger in orbit

Spacesuits

Willy Post 1935

A spaceship you can wear

An astronaut's spacesuit is a natural extension of the pressure suits worn by high-flying pilots.

Pilots and astronauts need pressure suits because the boiling point of a liquid decreases as the atmospheric pressure is reduced. At altitudes above about 20 000m the air pressure is so low that fluids such as blood will boil just from the natural warmth of the human body. Because of this, jet pilots and astronauts must wear full pressure suits to survive. The first pressure suits was built in 1933 by the B F Goodrich company for pioneer US aviator Willy Post. Post was killed in 1935 and interest in pressure suits died with him until the 1940s, when military aviators began to fly so high that they had to use pressure suits on a regular basis.

Into space

When the USA began its Mercury programme it was realized that the astronaut, although sealed inside a pressurized capsule, would need protection in case his capsule's walls were punctured. Naturally, NASA turned to the pressure suits already available and so the Mercury spacesuit was a modified version of the suits worn by US Navy jet pilots.

The USA's next space programme, Gemini, called for astronauts to leave their spacecraft and 'walk' in space, and this presented special problems. What was needed was a spacesuit which was airtight but comfortable for long periods, would not balloon up when inflated, was strong enough to protect against meteorites and accidental tearing and yet flexible enough for the astronaut to move about. This was (and remains) a complex problem, since some human joints, such as the shoulder, can perform a wide variety of possible motions.

The design chosen for Gemini was based on a Mercury suit which had been adapted to allow for spacewalks. The Gemini spacesuit comprised an airtight pressure bladder followed by a layer of a fishnet-like woven fabric called Link Net used to prevent the bladder from ballooning when pressurized. Next came a layer of felt, seven layers of insulation to protect against temperature extremes and an outer nylon cover. The suit was pressurized at one-quarter of normal atmospheric pressure and oxygen was piped in from the Gemini spacecraft's life-support system through an umbilical cord which plugged into the suit at the astronaut's midriff. An emergency oxygen supply was carried in a pack on the astronaut's chest. The helmet and gloves were clipped on to the suit via metal rings and were normally removed when the astronauts were inside the spacecraft. The helmet had a clear visor that could open and another over-visor which served as a sunshield during spacewalks. The astronaut put on the suit by climbing in through a slit in the back which was then sealed with a heavy duty, airtight zip.

The lunar suit

A new suit design was used during the Apollo programme. Called the AL7B the Apollo suit incorporated special joints and bellows at the wrists, elbows, shoulders, waist, ankles and thighs which made it easier for the astronaut to move while working on the Moon (the command module pilot who remained in orbit had a slightly different version without a flexible waist). Gemini astronauts had overheated during strenuous activity, so the inner layer of a lunar spacesuit was a liquid-cooled garment, worn next to the skin, which comprised a network of polyvinyl tubing through which cold water was circulated. Next came the pressure garment, which included a linen comfort layer, a rubber-coated nylon bladder, and a nylon restraint layer to prevent ballooning. Above this came a thermal and meteoroid protection garment comprising a layer of rubber-coated nylon followed by several alternate layers of aluminized material separated by an insulating spacer fabric. In the vacuum of space, these layers acted like a thermos flask to prevent heat leaking in or out of the suit. The outermost layer provided protection against fire and accidental damage. The plastic, fishbowl-type helmet was attached rigidly to the suit, but the astronaut could move his head freely within it. The removable helmet visor assembly incorporated two visors, the outermost of which was gold coated to provide protection against sunlight. The astronaut's lunar boots pulled on over the suit and provided insulation from the lunar surface. They had an outer layer of woven metal fabric to protect against abrasion.

Oxygen, power and cooling water for the lunar suit were supplied from a Portable Life Support System (PLSS) or backpack which was connected to the suit via umbilicals at the astronaut's waist. The PLSS could sustain an astronaut for about seven hours and could be recharged from the lunar module's supplies. A separate 30-minute emergency supply of oxygen was carried in a small pack mounted above the main backpack.

A modified Apollo suit, which used umbilicals to supply oxygen, power and cooling water, was used during the Skylab programme.

Space Shuttle suits

To protect them against emergencies during launch and landing, Space Shuttle astronauts wear orange pressure suits, but a quite different suit is used for spacewalking. Unlike earlier spacesuits which were tailored to each astronaut, the Space Shuttle suit comprises a number of separate components which are made in different sizes and which can be put together to fit almost any particular person, male or female. Each suit has supplies for a 6.5 hour spacewalk plus a 30-minute reserve. It is pressurized to just under one-third of atmospheric pressure and weighs about 110kg.

To don the suit the astronaut first puts on a liquid-cooled undergarment (which has 100m of plastic tubing) then steps into the lower body assembly which is already fitted with boots. The flexible lower torso comprises several layers, beginning with a bladder of urethane-coated nylon, a restraining layer of dacron, a thermal protection garment of nylon and aluminized mylar and an outer protective layer. The hard upper torso, which includes a built-in backpack containing oxygen, water, fans and a radio, is made of fibreglass and is fixed to the wall of the Space Shuttle when not in use. The astronaut squats under this and then slides up into it, slipping the sleeves on and popping his or her head into the helmet. The two sections are then joined by a connecting ring at the waist.

Soviet space suits

The pressure suit worn by Vostok cosmonauts was hidden under an orange coverall and the Voskhod 1 crew flew without spacesuits at all. Voskhod 2 cosmonaut Alexi Leonov wore a special suit for his historic spacewalk in 1965. Leonov drew his supplies from a backpack, suggesting that his may have been a suit designed for use on the Moon. Four years later when the crew of Soyuz 4 transferred to Soyuz 5 they had a modified suit with no backpack, but with air supplies attached to their legs.

After the Soyuz 11 disaster (see p193) all Soviet cosmonauts wore pressure suits during launch, docking and landing, but special suits have been used for spacewalks. These suits are one-piece garments with a built-in backpack which swings to one side revealing a hatch through which the cosmonaut enters the suit. Closing the backpack seals the suit ready for use.

> The official name for a NASA spacesuit is Extravehicular Mobility Unit, or EMU.

Space travellers

Yuri Gagarin 1961

A high-flying profession

Over 200 men and women have now flown in space, many on more than one occasion.

Aldrin, Edwin E (20 Jan 1930–) Edwin 'Buzz' Aldrin flew on Gemini 12 (1966) during which he performed three spacewalks. As lunar module pilot of Apollo 11 (1969) he was the second man on the Moon.
Armstrong, Neil A (5 Aug 1930–) After flying with the US Navy in Korea, Neil Armstrong became a civilian test pilot for NASA. Selected as an astronaut in 1962 he commanded the aborted Gemini 8 mission (1966) and went on to command Apollo 11 in 1969. He was the first man on the Moon.
Bean, Alan L (15 Mar 1932–) The lunar module pilot of Apollo 12 (1969) and so the fourth man on the Moon, Bean went on to command the second manned Skylab mission (1973). He retired from NASA and has since become an artist.
Borman, Frank (14 Mar 1928–) Selected as an astronaut in 1962 Borman commanded the Gemini 7 mission (1965) which set a new space endurance record. He also commanded Apollo 8 (1968) which made the first flight in lunar orbit. Borman later became president of Eastern Airlines.
Cernan, Eugene A (14 Mar 1934–) A US Navy pilot, Cernan became an astronaut in 1963 and performed a spacewalk from Gemini 9 (1966). As lunar module pilot on Apollo 10 (1969) he descended to within 15km of the Moon and as commander of Apollo 17 (1972) he was the eleventh man on the Moon.
Chaffee, Roger B (15 Feb 1935–27 Jan 1967) Selected as a US astronaut in 1963, Chaffee died in the Apollo 1 fire without ever flying in space.
Chretien, Jean-Loup (20 Aug 1938–) The first French spationaut he flew to Salyut 7 as a guest cosmonaut on Soyuz T6 (1982). He flew to the Mir space station aboard Soyuz TM7 (1989) and became the first European to perform a spacewalk.
Collins, Michael (31 Oct 1930–) Pilot of Gemini 10 (1966) and command module pilot of Apollo 11, Collins remained in lunar orbit during the first Moon landing.
Conrad, Charles (2 Jun 1930–) Selected as a US astronaut in 1962 Charles 'Pete' Conrad flew as pilot of Gemini 5 in 1965 and commanded Gemini 11 in 1966. As commander of the Apollo 12 mission in 1969 Conrad was the third man on the Moon. His final flight was the first manned Skylab mission in 1973.
Crippen, Robert L (11 Sep 1937–) Selected as a US astronaut in 1969 Crippen waited until 1981 before flying as co-pilot of the first Space Shuttle mission. He commanded Space Shuttle missions STS 7 (1983), STS 41C (1984) and STS 41G (1984) before taking a senior management post at NASA HQ. He became director of the Kennedy Space Center in 1992.
Dobrovolsky, Georgi T (1 Jun 1925–29 Jun 71) Probably a member of the team for the abandoned Soviet lunar programme Georgi Dobrovolsky commanded the Soyuz 11/Salyut 1 mission (1971) and died when Soyuz 11 lost pressure during re-entry.
Duke, Charles M (3 Oct 1935–) Duke was the lunar module pilot of the USA's Apollo 16 (1972) and the tenth man on the Moon.
Gagarin, Yuri A (9 Mar 1934–27 Mar 1968) As pilot of the USSR's Vostok 1 (1961) Gagarin was the first man in space. He was killed in an air crash while reportedly training for a Soyuz mission.
Glenn, John H (18 Jul 1921–) John Glenn was the first American in orbit. His Mercury capsule 'Friendship 7' made three orbits of the Earth in 1962. Glenn left NASA and became the US Senator for Ohio.

Grissom, Virgil I (3 Apr 1926–27 Jan 1967) As pilot of Liberty Bell 7 (1961), Virgil 'Gus' Grissom was the second American in space. In 1965 he commanded Gemini 3, the first manned Gemini mission. He died in the Apollo 1 fire.

Irwin, James B (17 Mar 1930–8 Aug 1991) Irwin was the lunar module pilot of the USA's Apollo 15 (1971) and the eighth man on the Moon. After leaving NASA for religious work he died of a heart attack, the first moonwalker to die.

Komarov, Vladimir M (16 Mar 1927–24 April 1967) Komarov was the commander of the USSR's three-man Voskhod 1 (1964) and the sole pilot of Soyuz 1 (1967). He was killed when the parachutes of Soyuz 1 became tangled and the spacecraft crashed on landing.

Leonov, Alexei A (30 May 1934–) Leonov made the world's first spacewalk from the Voskhod 2 spacecraft in 1965 and was probably part of the Soviet lunar team in the late 1960s. When this programme was abandoned he trained for a Salyut mission, but eventually commanded the Soviet half of the Apollo–Soyuz Test Project in 1975.

Lovell, James A (25 Mar 1928–) A US Navy pilot, Lovell flew on Gemini 7 (1965) and commanded Gemini 12 (1966). During Apollo 8 (1968) he was one of the first men to orbit the Moon, but his chance of being the fifth man on the Moon was lost when Apollo 13, of which he was the commander, suffered a near catastrophic explosion in space.

Manarov, Musa K (22 Mar 1951–) Musa Manarov set a world space endurance record of 366 days aboard the USSR's Mir space station after his launch on 21 December 1987 in Soyuz TM 4. In 1991 he completed another six-month visit to Mir.

Merbold, Ulf (20 Jun 1941–) A West German national, Merbold was the first European Space Agency (ESA) astronaut to be assigned to a mission. He flew aboard the STS 9 mission as a payload specialist in the European Spacelab.

Mitchell, Edgar D (17 Sep 1930–) Selected as a US astronaut in 1966 Ed Mitchell was the lunar module pilot of Apollo 14 (1971) and the sixth man on the Moon. He is often remembered for carrying out unauthorized ESP experiments during his mission.

Patseyev, Victor I (19 Jun 1922–29 Jun 1971) The research engineer on the USSR's Soyuz 11/Salyut 1 mission, Patseyev died when Soyuz 11 lost pressure during re-entry.

Remek, Vladimir (26 Sep 1948–) The first non-American/non-Soviet in space, Remek (a Czech) flew to the Salyut 6 space station as a guest cosmonaut on Soyuz 28 (1978).

Ride, Sally K (26 May 1951–) The USA's first woman astronaut, Sally Ride flew on Shuttle missions STS 7 (1983) and STS 41G (1984).

Savitskaya, Svetlana Y (8 Aug 1948–) The second woman in space, on the USSR's Soyuz T7/Salyut 7 mission (1982), Savitskaya was the first woman to walk in space, on the Soyuz T12/Salyut 7 mission (1984).

Schirra, Walter M (12 Mar 1923–) 'Wally' Schirra was the third American in orbit, in the Mercury capsule Sigma 7 (1962). He commanded Gemini 7 (1965) which made the first space rendezvous and later commanded the first manned Apollo mission, Apollo 7 (1968). He was the only man to fly in Mercury, Gemini and Apollo spacecraft.

Schmitt, Harrison H (3 Jul 1953–) A professional geologist, 'Jack' Schmitt was the lunar module pilot of the USA's Apollo 17 (1972) and the twelfth man on the Moon. After leaving NASA he served for a time in the US Senate.

Scott, David R (6 Jun 1932–) A US Air Force pilot, Scott was selected as an astronaut in 1963 and flew in Gemini 8 (1966) and the Earth orbiting Apollo 9 test flight (1969). As commander of Apollo 15 in 1971 he was the seventh man on the Moon.

Shatalov, Vladimir A (8 Dec 1927–) The commander of the USSR's Soyuz 4 (1969), a mission which featured the first exchange of crews in space, Shatalov also commanded Soyuz 8 (1969). As commander of Soyuz 10 (1971) he docked with the Salyut space station, but was unable to enter the station because of technical problems. He later became head of cosmonaut training.

Shepard, Alan B (18 Nov 1923–) The first American in space (1961), Shepard was grounded for some time for medical reasons. When these were resolved he commanded the Apollo 14 mission (1971) and was the fifth man on the Moon.

Slayton, Donald K (1 Mar 1924–) Selected as one of the original seven US Mercury astronauts 'Deke' Slayton was suddenly grounded because of a minor heart irregularity. He finally flew on the Apollo–Soyuz Test Project in 1975.

Stafford, Thomas P (17 Sep 1930–) Stafford flew on the USA's Gemini 6 (1965) and commanded Gemini 10 (1966). He then commanded Apollo 10 (1969) and flew to within 15km of the lunar surface in a dress rehearsal of the lunar landing mission. He was commander of the US part of the Apollo–Soyuz Test project mission in 1975.

Tereshkova, Valentina V (6 Mar 1937–). As pilot of the USSR's Vostok 6, Tereshkova was the first woman in space. She married cosmonaut Adrian Nikolayev.

Titov, Gherman S (11 Sep 1935–) Titov was the pilot of the USSR's Vostok 2 (1961) and the second man in space.

Truly, Richard H (12 Nov 1937–) Truly was involved in the glide tests of the US Space Shuttle Enterprise in 1977 and then flew on Shuttle missions STS 2 (1981) and STS 8 (1983). He later became head of NASA.

Volkov, Vladislav N (23 Nov 1935–29 Jun 1971) Volkov was flight engineer on the USSR's Soyuz 7 (1969) and then on the Soyuz 11/Salyut 1 mission (1971). He died when Soyuz 11 lost pressure during re-entry.

White, Edward H (14 Nov 1930– 27 Jan 1967) Selected as an astronaut in 1962, Ed White performed the first US spacewalk, during the Gemini 4 mission (1965). He died in the Apollo 1 fire.

Young, John W (24 Sept 1930-) John Young flew as pilot of the USA's Gemini 3 (1965) and commanded Gemini 10 (1966). As command module pilot of Apollo 10 (1969) he circled the moon during a rehearsal for the lunar landing and went on to command Apollo 16 (1972) and become the ninth man on the Moon. Young commanded the first flight of the Space Shuttle (1981) and the STS 9 mission (1983) which featured the first flight of the European Spacelab.

Astronaut means 'star voyager'.

Astronauts R Crippen and J Young at the Kennedy Space Center in 1979

Spies in the sky

Discoverer 1 2 February 1959

Watching and listening

Although some military satellites can be used for peaceful purposes such as navigation, others have purely military uses.

Photographic reconnaissance

Just as aircraft replaced hot-air balloons, satellites have replaced aircraft as vehicles from which to photograph potentially hostile forces. The US Discoverer programme, which began in the late 1950s, was used to develop the techniques required to take photographs from orbit and then return the exposed film to Earth. The first 12 missions failed to return capsules, but finally, on 11 August 1960, the film capsule from Discoverer 13 (which had been launched the previous day) was recovered from the Pacific Ocean. The next flight was also successful, with the returning capsule being snatched from the air by an aircraft towing a special trapeze-like snare. The Discoverer programme continued for some time before launches in the series were classified.

SAMOS, the Satellite and Missile Observation System, was similar to Discoverer and consisted of satellites launched into polar orbits. The SAMOS satellites usually returned their data by means of television systems which recorded pictures as they passed over foreign territory and then replayed them as the satellite flew over the USA. Some SAMOS satellites could also return film images using recoverable capsules. Launches after SAMOS 3 were secret, but the series is believed to have continued, with successively improved versions of satellite, until the early 1970s.

The SAMOS series was replaced by a new generation of satellites called Big Bird. These large satellites, launched by a Titan 3 rocket, combined the function of searching wide areas and taking close-up images of regions of particular interest, tasks previously assigned to different types of satellite. They were based on the Agena rocket stage and were equipped with high-resolution cameras able to detect objects as small as 30cm. Although much of the data obtained by these satellites was transmitted via a television-type system, six capsules were carried to return exposed film of targets for which the best possible images were required. The Big Birds were placed in very low orbits and had to use their own rocket engines from time to time to prevent atmospheric drag causing them to re-enter. The first Big Bird was launched on 15 June 1971 from Vandenberg Air Force Base in California and one or two were launched each year from then on. The Big Birds were replaced from 1976 on by satellites called KH-11s which were similar in size, but which operated from a much higher orbit. The KH-11 satellites return their images by digital data transmission instead of recoverable film capsules.

Soviet reconnaissance satellites were based on modified versions of manned spacecraft in which the re-entry module is used to carry the camera system and to return it, and its film, to Earth at the end of each mission. These satellites usually remain in orbit for relatively short periods.

France is planning to develop a military observation satellite named Helios which will be based on the civilian SPOT Earth observation satellite.

Ocean reconnaissance

The growth of Soviet naval power led the US Navy to develop ocean reconnaissance satellites under a project called Whitecloud.

The first of these, NOSS-1, launched in 1976 is believed to have carried a millimetre wave radar able to track surface ships even if they were hidden by cloud. The NOSS series may also be able to search for submerged submarines by detecting the warm water they leave in their wakes. Several NOSS satellites have released smaller sub-satellites once in orbit. The USSR also developed radar satellites (known as RORSATs) and these carry nuclear reactors to generate the large quantities of electricity required to operate the radar.

Early warning

In the early 1960s, the USA decided to develop satellites which could detect nuclear missile launches. The first of these was the Missile Defence Alarm System (MIDAS) which used infrared sensors, mounted on an Agena rocket stage, to detect the heat from the exhaust of ballistic missiles. After a few early failures, a series of MIDAS satellites were launched into 3 000km-high polar orbits. Unfortunately, the problem of missile detection proved more difficult than expected (for example false alarms were caused by sunlight reflecting off clouds) and MIDAS never went into full operation. Instead, a series of experimental missions were flown under the covername Program 461.

In the late 1960s four early warning satellites were placed into geosynchronous orbits which traced a figure-of-eight path in sky over the USSR. These were in turn replaced by a new class of satellites, the Integrated Missile Early Warning System (IMEWS), which carried large infrared and optical telescopes to detect and photograph Soviet missiles within seconds of them taking off. The first IMEWS was launched in 1971 and, although it failed to reach its correct orbit, a number of other craft in the series have been launched successfully. The Defense Support Programme (DSP) satellites are related to the IMEWS series and are placed in geosynchronous orbits, from where they look down with infrared telescopes to search for both land and submarine launched ballistic missiles.

Nuclear explosion monitoring

To detect infringements of the nuclear test ban treaties the USA developed satellites under a programme called Vela Hotel. These craft carried equipment that could detect the intense pulse of X-rays, gamma rays and electromagnetic radiation produced by a nuclear explosion. Each launch placed two satellites into circular orbits about 100 000km high, positioned so that the two satellites were on the opposite side of the Earth. The first six Velas were 20-sided polyhedra about 1.25m across and weighed 230kg. The first pair were orbited on 16 October 1963 with further pairs being launched on 17 July 1964 and 20 July 1965. The remaining six craft in the series were of a more advanced design with 26 sides and a mass of 260kg. They were launched on 28 April 1967, 23 May 1969 and 8 April 1970. Several of these satellites remained in operation for many years, before being replaced by equipment carried on the IMEWS and DSP satellites.

Electronic Intelligence (ELINT)

ELINT involves monitoring radio transmissions, which can provide information on enemy communications, radar and other electronic warfare systems. Such monitoring is often carried out from satellites known as 'ferrets'. What is thought to have been the first in a series of US ferrets was launched on 15 May 1962 by a Thor-Agena B rocket. Up to 17 of these satellites were launched before their work was transferred to the Rhyolite series, the first of which was launched in 1973. Rhyolite satellites were placed into a geosynchronous orbit over the USSR to monitor Soviet missile tests. Other satellites with similar missions were codenamed Argus and Chalet. The latest ELINT craft is Aquacade, a 4-tonne satellite launched from the Space Shuttle. The first example was placed into geosynchronous orbit inclined at 28.5° in January 1985.

> Although the Velas never unambiguously detected a clandestine nuclear test, scientists using data from the Velas discovered the first gamma ray burster, a type of astronomical objects which is still not fully understood.

Sputnik and the first satellites

Sputnik 1 1957, Explorer 1 1958

Starting the space race

When the USSR launched the world's first artificial satellite it triggered a space race against the USA.

Although the idea of an artificial satellite of the Earth can be traced back for over 100 years, it was not until the advent of ballistic missiles that satellites became a serious possibility. During World War II Wernher von Braun and his V2 missile team dreamed of using rockets for space travel and in 1945 British writer Arthur C Clarke proposed the use of satellites as radio relay stations. In the same year the US Navy began studies of possible satellites and, on 2 May 1946, the US Air Force and the RAND (Research and Development) corporation published a report which indicated that a 225kg satellite could be launched by 1951. US interest in satellite development then languished, although in 1953 Professor Fred Singer proposed Project MOUSE (Minimum Orbital Unmanned Satellites of Earth) and in 1954 a possible joint US Army–Navy project called Orbiter emerged from a meeting in Washington. Soon after this meeting Wernher von Braun showed how the US Army could launch a small satellite using existing military hardware, but this proposal was quashed in favour of a US Navy project called Vanguard. Vanguard, which would use a launch vehicle based on rockets developed for scientific research, was felt to be more politically acceptable than a project based on US Army missiles. Project Vanguard was initiated on 9 September 1955 with the aim of launching a satellite during International Geophysical Year in 1957–8.

Sputnik

Unknown to the USA, the USSR had been considering launching satellites since the early 1950s. On 15 April 1955 the Soviet Academy of Sciences set up a commission to develop a satellite and a design team was formed. Shortly after President Eisenhower announced the start of Project Vanguard, the Soviet Union announced that it too would launch a satellite. Few people believed this at the time.

In fact, the guiding light behind the Soviet programme, Sergei Korolov, was thinking along the same lines as von Braun and planned to use the USSR's new ballistic missile as a launch vehicle. However, unlike Project Orbiter, Korolov was not hamstrung by political sensitivities about having a civilian space programme and so he was able to use the most powerful rocket available. This meant that his team could afford to design much heavier satellites than those being planned in the USA.

The first Soviet satellite was simple in concept, little more than a steel ball containing a radio transmitter and batteries. Four whip aerials streamed out from the 83.6kg sphere which was called Sputnik, or 'fellow traveller'. Sputnik 1 was launched by an A-1 rocket from the Baikanour cosmodrome on 4 October 1957. It went into an orbit of 228km × 947km and 90 minutes later it flew back over Baikanour sending a steady bleep-bleep-bleep from its radio transmitter. The space age had begun and the Soviet launch team was ecstatic. However, in the USA Sputnik was seen as a blow to national prestige and a sign of military weakness.

As the shock waves from the launch of Sputnik 1 echoed around the world Korolov's team prepared another, more sophisticated satellite for launch. Sputnik 2 consisted of two compartments; one contained instruments for measuring temperatures, pressures and so on and the other was a capsule containing a dog named Laika. The 508kg satellite was launched into a 225km × 1 671km orbit on 3 November 1957. Laika seemed to suffer no ill effects from her launch into space, nor from weightlessness once in orbit, but Sputnik 2 had no means of returning the animal to Earth and she died a week later when her air supply gave out. Laika's body was incinerated when Sputnik 2 re-entered in April 1958.

The US response

When they heard about Sputnik 1, von Braun and his superior, General John Medaris, strove to restart Project Orbiter, but permission was refused since it was still hoped that Vanguard would soon even the score. By 6 December 1957 Vanguard was ready and the first rocket, called TV3 (Test Vehicle 3), stood on the launch pad at Cape Canaveral. One second after lift-off the single engine of the first stage lost power and the rocket subsided gently back on to the launch pad before toppling over in a dramatic and humiliating fireball. The satellite, thrown clear of the explosion, was found nearby with its radio beacon still transmitting. The press called it 'Flopnik' and 'Kaputnik'. Five days later Project Orbiter was restarted and von Braun promised to launch a satellite within 90 days.

Von Braun's planned to use a four-stage rocket called Juno 1. It was a modified version of a rocket called Jupiter C (C for Composite) which had already been launched three times during a test programme connected with the development of ballistic missiles. The first stage of the Jupiter C was a Redstone ballistic missile with enlarged fuel tanks. Stage 2 was a drum containing 11 Sergeant solid-fuelled rockets and stage 3, which nested inside the drum, comprised a further 3 Sergeants. The fourth stage, added to convert Jupiter C into a satellite launcher, was another Sergeant with a scientific payload attached. Including the final stage, the satellite was 2m long, 15cm in diameter and weighed 14kg. The main scientific instrument was a Geiger-Mueller counter provided by James van Allen from Iowa University, who hoped to use it to detect cosmic rays.

The satellite was named Explorer 1 and was launched into a 360km × 2 534km orbit on 31 January 1958. As well as restoring US prestige Explorer 1 detected an intense zone of radiation trapped by the Earth's magnetic field. These results were confirmed by later satellites and the zones are now known as the Van Allen belts.

Mixed fortunes

A few days after the triumph of Explorer 1, a second attempt was made to launch a Vanguard rocket. This time the rocket climbed away from the launch pad successfully, but veered off course and broke up 57 seconds later. Explorer 2, launched on 5 March 1958, also failed to get into orbit when the final stage of its Jupiter C did not perform as planned. The Vanguard programme finally achieved some success on 17 March 1958 when it placed a satellite into a 186km × 2 799km orbit. Although derided by the USSR as a grapefruit because it was only 16cm in diameter and weighed 1.47kg, the satellite continued to operate until the middle of 1964. Vanguard 1 was followed into orbit by Explorer 3, launched by another Jupiter C on 26 March, but another failure followed on 28 April when the third stage of Vanguard TV5 failed. Two more Vanguards failed in the following months, but by then the USSR had already launched another massive satellite.

Sputnik 3

On 15 May 1958 the USSR launched Sputnik 3 into a 226km × 1 880km orbit. This 1 500kg orbiting laboratory was a polished metallic cone with solar cells mounted along its sides and festooned with aerials. During its 691 days in orbit it returned data on micrometeorites, the van Allen belts and cosmic rays. The size of Sputnik 3 demonstrated, beyond any doubt, that the USSR had rockets far more powerful than those of the USA and gave the impression of overwhelming Soviet superiority in space.

During its development, Sputnik 1 was known as PS or Preliminary Satellite, but many of its design team called it SP in unofficial tribute to Korolov's first names, Sergei Pavlovich.

The final record for the Vanguard rocket was 3 successes in 11 attempts. Jupiter C achieved 3 successes in 6 attempts.

Model of Sputnik 2

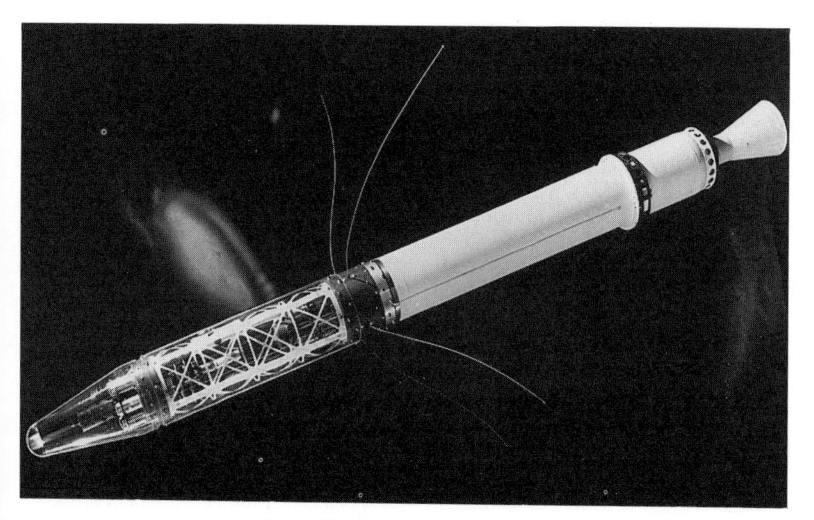

A model of Explorer 1, the first US satellite

Surveyor

Surveyor 1 1966, Surveyor 7 1968

A robot on the Moon

NASA's Surveyor spacecraft was a pathfinder for the manned Apollo landings on the Moon.

The Surveyor programme was started by NASA in 1961. Overall responsibility for the project was given to the Jet Propulsion Laboratory in Pasadena, California and the main industrial contractor was the Hughes Aircraft Company. The programme's primary purpose was to serve as a pathfinder for the manned Apollo missions by achieving a soft landing on the Moon, returning data in support of Apollo and conducting scientific experiments. Although these objectives now sound rather modest, it is important to realize that in 1961 no spacecraft had ever made a controlled landing on the Moon and that there was considerable disagreement about the conditions on the lunar surface. For example, at least one astronomer believed that the Moon was covered with a layer of fine dust deep enough to bury a spacecraft as soon as it landed. Ideas like this, and other less dramatic possibilities, had to be checked before the detailed plans for the Apollo lunar landings could be finalized.

The spacecraft

The Surveyor spacecraft weighed about 1 000kg at launch. The basic structure was a triangular aluminium frame with a landing leg at each corner. Each leg had shock absorbers and ended in a circular footpad made from aluminium honeycomb. The legs were folded for launch and sprang out and locked into place once en route to the Moon. The aluminium frame supported two main electronic boxes containing radios and other equipment, two low-gain antennae, a television camera and scientific equipment. The precise details of the scientific instruments varied from mission to mission, with the later Surveyors being somewhat more lavishly equipped. Three small liquid-fuelled rocket engines, called verniers, were provided both for course corrections and for use during the final phase of the landing on the Moon. A large, solid-fuelled retro-rocket, which amounted to 60% of the total mass of the spacecraft, was mounted in the centre of the base of the framework. Projecting above the main structure of Surveyor was a mast which supported a flat, almost square, high-gain antenna and a single panel of solar cells.

The flight plan for a typical Surveyor mission was as follows. The spacecraft was launched from the Kennedy Space Centre by an Atlas-Centaur rocket and placed on course for the Moon. Next, the Surveyor separated from its launch vehicle and used its own Sun and star sensors to find its orientation in space. Once this was done Surveyor used small nitrogen jets mounted near the footpads to stabilize itself in the correct attitude for its 60-hour cruise to the Moon. During the flight the vernier rockets were fired when required to make course corrections and steer Surveyor towards its chosen landing site. As the spacecraft approached the Moon it was positioned so that its retro-rocket was pointed along the direction of flight, ready to brake the spacecraft for landing. Next, when Surveyor was about 325km high, the landing radar system was turned on and at about 100km from its intended touchdown point the solid retro-rocket and the vernier rockets were fired. The solid motor fired for about 40 seconds and when the motor firing stopped the Surveyor was about 40km above the Moon and travelling at about 400km per hour. The empty solid rocket motor casing was jettisoned and Surveyor's onboard control system used data from the radar to adjust the thrust of the verniers so that the spacecraft was travelling at only a few kilometres per hour when it arrived just above the Moon's surface. The verniers were then shut down to avoid disturbing the dust around the landing site and the spacecraft fell the last 3m to the surface, using its shock absorbers to cushion the final impact. Once Surveyor was safely on the Moon, controllers back on Earth sent messages to operate the television camera and other scientific instruments by remote control.

The missions

Surveyor 1 was launched on 30 May 1966 and made a perfect touchdown after a flight lasting 63 hours, 36 minutes and 35 seconds. The landing occurred near the crater Flamsteed in the region of the Moon called the Ocean of Storms and was about 15km away from the pre-planned landing site. Surveyor 1 was basically an engineering test and did not carry any scientific instruments other than a television camera. The camera was highly successful and returned 10 150 images during Surveyor's first lunar day. Surveyor 1 survived the lunar night and was reactivated in early July when it returned another 1 000 pictures. Contact was maintained occasionally until January 1967, but no further pictures were taken.

Surveyor 2 was launched on 20 September 1966, but problems developed when one of the verniers failed to fire during a course correction. This put Surveyor into a spin which could not be corrected by its small nitrogen thrusters. Contact was maintained for a time, but the spacecraft eventually crashed near the crater Copernicus. Surveyor 3 followed on 17 April 1967 and landed in the Ocean of Storms on 19 April after a couple of small bounces across the lunar surface. Surveyor 3 was equipped with a robot scoop which was used to dig a series of trenches and conduct tests to determine the bearing strength of the lunar surface. In addition, 6315 high-quality television pictures were returned. Although contact could not be regained after Surveyor's first night on the Moon, this was not the last heard of Surveyor 3; in 1969 the Apollo 12 astronauts landed near Surveyor 3 and returned several of its components, including the television camera, to Earth so that they could be studied to determine how well various materials had survived their stay on the Moon.

Although Surveyor 4, launched on 14 July 1967, crashed on the Moon when contact was lost about 2.5 minutes before touchdown, the remaining three missions were successful. Surveyor 5 was launched on 8 September 1967 and, although it developed problems with its vernier engines en route to the Moon, these were overcome by mission controllers who quickly developed a new landing procedure. This worked perfectly and, after landing in the Sea of Tranquillity, Surveyor 5 returned over 19 000 pictures. Surveyor 5 also carried an instrument which was dropped on to the surface to determine the chemical composition of the lunar soil. This was done by bombarding the surface with alpha particles from a radioactive source and measuring the nuclear particles that were bounced back. From the number and type of atomic particles that bounced back it was possible to get some idea of the bulk chemical composition of the lunar surface. This turned out to be a basalt-type rock, confirming theories that the smooth lunar plains were formed by outpouring of lava from below the lunar surface.

Surveyor 6 was launched on 6 November 1967 and landed normally only a few kilometres from its target in the Sinus Medii (Central Bay). Over 30 000 pictures were returned and the chemical composition of the lunar soil was again tested. Surveyor 6 also achieved the first lift-off and controlled flight from the Moon by re-igniting its vernier rockets to perform a 6.5-second flight that moved it about 2.5m west of its original landing point. The maximum altitude during the hop was about 3.5m. Surveyor 7, the last in the series, was launched on 6 January 1968 and bounced to a landing near the lunar crater Tycho in the Moon's southern hemisphere. Surveyor 7 had both a mechanical scoop and a soil composition experiment, which was fortunate because the soil experiment did not drop to the surface as planned and had to be pushed down using the mechanical scoop. Surveyor 7 returned over 21 000 images during its first lunar day. Although contact was re-established after the lunar night, the spacecraft was not operating properly and contact was lost again before the end of the second lunar day.

Despite the failure of Surveyor 2 and 4 the programme had been a tremendous success and had achieved all that had been planned for it. The next soft landing on the Moon was made by Apollo astronauts.

The mechanical scoop used on the later Surveyor spacecraft was a triumph of lightweight engineering. Developed in a period of only a few months the scoop was used to dig trenches, turn over stones and reposition the lunar surface composition experiment. On Surveyor 7 the scoop responded to 12639 commands, dug 8 trenches, performed 15 tests of the bearing strength of the lunar surface and pushed, lifted and even broke lunar rocks. The scoop, officially described as the Surveyor Soil Mechanics Surface Sampler, was also known as the shovel with a handle 400000km long.

In 1968 the Surveyor Programme team was awarded the Robert J Collier trophy. The trophy was established in 1912 by Collier, an aviation enthusiast, and is awarded annually 'for the greatest achievement in aeronautics or astronautics in America, with respect to improving the performance, efficiency or safety of air or space vehicles, the value of which has been thoroughly demonstrated by actual use during the preceding year'.

Surveyor landing sites

Surveyor 1	Landed	2° 27'S	43° 19'W
Surveyor 2	Crashed	5° 30'N	12° 00'W
Surveyor 3	Landed	2° 56'S	23° 20'W
Surveyor 4	Crashed	0° 26'N	1° 30'W
Surveyor 5	Landed	1° 30'N	23° 11'E
Surveyor 6	Landed	0° 28'N	1° 29'W
Surveyor 7	Landed	40° 53'N	11° 26'W

The unmanned Surveyor 3 spacecraft is visited by Apollo 12 astronauts 2½ years after it landed on the Moon

Ultraviolet astronomy satellites

OAO-1 1966, IUE 1978

Hot stars and cold space

The ultraviolet region of the electromagnetic spectrum has been extensively observed from space.

Ultraviolet light is a form of electromagnetic radiation with wavelengths shorter than those of visible light. It is emitted by high temperature material such as the atmospheres of massive stars and hot gas around supernovae. As well as providing an opportunity to observe very hot objects, ultraviolet light is of special interest to astronomers because heavy atoms such as carbon, nitrogen and oxygen have important spectral lines which occur at ultraviolet wavelengths and this means that ultraviolet observations can be used to study the chemical composition of all sorts of astronomical objects. Unfortunately for astronomers, (but fortunately for life on Earth since ultraviolet radiation can be harmful) most of the ultraviolet radiation from stars is absorbed in the ozone layer and so ultraviolet astronomy is only possible from space.

The first attempts to detect extraterrestrial ultraviolet radiation were made in the 1920s from balloons, but balloons cannot rise high enough to penetrate the ozone layer and it was not until 1946 that a spectrometer launched on a captured V-2 rocket detected ultraviolet radiation from the Sun. The first ultraviolet observations of stars were made from rockets during the late 1950s and these early results were followed by small experiments on satellites such as 1964-83C and the USSR's Cosmos 51. The success of these experiments led to the development of the NASA Orbiting Astronomical Observatories (OAO) which were large satellites devoted to ultraviolet astronomy.

The OAOs

Each OAO spacecraft was comprised of an octagonal structure 3m long and 2m in diameter. The satellite's main sub-systems were grouped around the inside walls of the body, leaving a cylinder about 1.25m in diameter down the centre which was used to carry one or more telescopes sensitive to ultraviolet radiation. Power was supplied by solar panels which folded out on opposite sides of the body once in orbit. The OAOs were large and sophisticated satellites for their time, with a mass of about 2 000kg and an ability to be pointed and stabilized with great accuracy.

The first OAO was launched on 8 April 1966 by an Atlas-Agena D rocket. Problems with the satellite's electrical system developed soon after launch and the mission was abandoned after only three days.

The second OAO carried two experiments, one looking out of each end of the spacecraft body. It was launched by an Atlas-Centaur rocket on 7 December 1968 and went into a 760km-high orbit. This time all went well and OAO-2 made ultraviolet observations of thousands of stars. It also made special observations of objects such as a comet, a nova in the constellation Serpens and some galaxies beyond our Milky Way. OAO-2 was finally switched off in February 1973.

The next OAO, known as OAO-B, was lost in a launch failure on 30 November 1970 and it was not until 21 August 1972 that OAO-3 was placed in an orbit about 730km high. OAO-3 weighed 2 200kg and was the heaviest NASA payload launched up to that time. The main experiment was an 80cm Cassegrain telescope equipped with an ultraviolet spectrograph. In addition, a small X-ray telescope was carried as a secondary experiment. After its launch the satellite was named Copernicus, to commemorate the 500th anniversary of the birth of this famous astronomer. OAO-3 remained in service until 1981 when its capabilities were overtaken by newer satellites with more sensitive detectors.

Other early missions

The TD-1A satellite was one of the first major projects of the European Space Research Organisation. It was designed to survey the sky and produce a catalogue of all the stars detectable in the ultraviolet. The satellite comprised a box 1m square and 2.16m long equipped with two solar panels. TD-1A was placed in a polar orbit on 12 March 1972 (11 March in the USA) and operated for two periods of six months separated by a four-month hibernation. Although problems developed with the satellite's tape recorders early in the mission, much useful data was collected.

The ANS (Astronomishe Nederlandse Satelliet), which was launched on 30 August 1974 by a US Scout rocket, was the first Dutch national satellite. The ANS satellite carried a 22cm telescope coupled to an ultraviolet photometer able to measure the brightness of selected stars. Although not placed in the correct orbit, the satellite worked well until it re-entered in 1977. Another small national satellite was D2B-Aura, a French mission used for ultraviolet astronomy. It was launched by a Diamant rocket on 27 September 1975.

Ultraviolet experiments also featured on manned space missions during the 1970s. The Apollo 16 astronauts took an ultraviolet camera to the Moon in 1972 and used it to take a series of photographs from the lunar surface. Three small ultraviolet experiments, one based on the Apollo 16 camera, were flown on the US Skylab space station during 1973–4. The Soviet Orion-1 experiment was carried on the ill-fated Salyut 1/Soyuz 11 flight and a similar package, Orion-2, was carried on Soyuz 13 in December 1973.

The International Ultraviolet Explorer (IUE)

Probably the most famous ultraviolet astronomy satellite has been the IUE, a joint mission of NASA, ESA and the UK, which was launched on 28 January 1978. The satellite was built by NASA, ESA provided the solar panels and a European ground station and the UK built the ultraviolet detectors for the satellite's 45cm telescope. The observing time was divided up between US and European astronomers with about two-thirds of the time going to the USA.

Unlike most astronomy satellites, in which observations are collected automatically and sent to each astronomer days or weeks later, the IUE is operated like a normal telescope. Astronomers using the satellite visit a control station and make the observations with the help of local staff. The scientists can usually see their results immediately and change their observing plan if they need to. To make this possible there are two ground stations, one in the USA and one in Europe, and the IUE orbit of 26 000km × 45 000km is arranged so that one or other of these is always in contact with the satellite.

Despite some technical problems the IUE has been a tremendous success and in 1991 celebrated its thirteenth year of operations. During this time it has been used to make observations of a huge variety of astronomical objects, ranging from comets to very distant galaxies.

Astron 1

Another, less publicized ultraviolet observatory was the Soviet Astron 1 mission, which was launched on 23 March 1983. The 4 000kg spacecraft carried a French ultraviolet telescope and was placed into a very eccentric orbit ranging from 1 950km to 201 000km from the Earth. It remained in service for several years.

Extreme Ultraviolet (EUV) astronomy is the study of even shorter wavelengths and uses techniques similar to those used by X-ray astronomers. The first orbital EUV experiment was carried on the Apollo–Soyuz mission in 1975. In 1990 the German ROSAT X-ray mission carried an EUV telescope which was used to explore the sky at these hitherto unexplored wavelengths.

Astro

The USA's Astro is a set of three ultraviolet telescopes designed to be carried in the payload bay of the Space Shuttle. The three telescopes are fitted to a special pointing system attached to the Shuttle and can be operated by the astronauts or by ground controllers. The first mission was originally intended to be launched in early 1986, but was delayed until December 1990 by the explosion of the Space Shuttle Challenger. A second flight is expected in about 1994.

The OAO-3 Copernicus satellite

Ulysses

Launched 1990

To the Sun via Jupiter

After surviving almost as many perils as its mythical namesake, the joint ESA/NASA Ulysses mission will provide a unique perspective on our Sun.

Every spacecraft which has flown beyond the Earth–Moon system has always remained close to the ecliptic, the thin disk in which the orbits of all the major planets lie. The reasons for this are bound up both with the missions of these spacecraft and with the laws of orbital dynamics. Spacecraft destined for other planets must travel by the shortest practical route and can gain nothing by climbing above the ecliptic plane after launch only to drop down again as they approach their destination. In the case of spacecraft intended to study interplanetary space, such as Pioneer 5–9 (see p150), orbital dynamics decree that considerable extra rocket power is required to move out of the ecliptic plane and since the use of a powerful rocket is expensive, most such spacecraft have remained close to the ecliptic plane. As a result, even though spacecraft have travelled to the very edge of the solar system, almost all the data on the solar wind and the Sun's magnetic field has come only from a narrow region of space within a few degrees of the Sun's equator. Little is known about the poles of the Sun, or about the nature of the solar wind and solar magnetic field at high ecliptic latitudes. This limited perspective is a major disadvantage for physicists trying to understand the full, three-dimensional, properties of the Sun and the solar wind and is the major motivation for a mission to the solar poles.

A history of Ulysses

The first detailed consideration of a mission to fly towards the poles of the Sun was probably the European Out-of-Ecliptic mission study performed in 1971–2. In 1973–4 NASA joined in these studies and in 1977 the European Space Agency (ESA) decided to go ahead with the project. In March 1979 NASA and ESA signed a Memorandum of Understanding for a joint mission to be known as the International Solar Polar Mission (ISPM). Under this agreement NASA and ESA would each build one spacecraft and the two would be launched together in 1983. The plan was that one spacecraft would fly over the Sun's north pole and one would swoop below the south pole, providing a stereoscopic view of the complex solar phenomena. These observations could be combined with data from other craft in the ecliptic plane to provide the first three-dimensional picture of the Sun's influence on the rest of the solar system.

In keeping with the international nature of the project each ISPM spacecraft would carry a mixture of European and US experiments although each partner was to design and finance its own spacecraft. NASA also agreed to provide a Space Shuttle launch, a Radioisotope Thermoelectric Generator nuclear power generator and tracking services for the European spacecraft. In 1981 NASA, faced with one of its now familiar budget crises, unilaterally cancelled its spacecraft in order to save money needed for its own planetary projects Galileo and Magellan. This caused enormous anger within Europe and culminated in formal protests to the US government. In the ensuing negotiations aimed at salvaging something from the debacle, ESA offered to build a copy of its spacecraft and sell it to NASA for the bargain price of $40 million, but this proposal was declined by the US authorities. Eventually it was agreed that NASA would still provide the nuclear generators for the European spacecraft and pay for the launch and tracking as promised, but some US experiments would have to remain on the European probe. As a result of these changes many experiments were deleted and, more importantly, the ability to make simultaneous observations from above and below the Sun was lost.

ISPM, which was renamed Ulysses in 1984, then suffered further delays due to uncertainties about which booster would be used to fire it from Earth orbit into deep space once it had been deployed from the Space Shuttle. These were eventually resolved and the launch was then set for the summer of 1986. The explosion of the Space Shuttle Challenger postponed the launch again and it was not until 6 October 1990 that Ulysses was finally launched aboard the Space Shuttle Discovery.

The spacecraft

Ulysses is a relatively small spacecraft, comprising a box about 3m square topped by a fixed dish antenna 1.65m in diameter. A single Radioactive Thermoelectric Generator projects from one side of the box. Several deployable booms, one 72m long, carry a variety of sensors which must be removed as far as possible from the electrical and magnetic fields of the spacecraft to sample the weak interplanetary magnetic and electric fields. Ulysses cruises through space with its dish antenna aimed permanently back towards Earth and is stabilized by spinning five times per minute. This spinning motion has the added advantage that the scientific instruments scan across the sky without needing a movable platform. The total mass of the spacecraft is 366.6kg of which 33.5kg is hydrazine fuel for the thrusters used for course adjustments.

Nine scientific experiments are carried on the spacecraft, and two other research projects, which do not require specific equipment onboard Ulysses, are carried out by carefully monitoring the radio transmissions from the spacecraft. The mission's objectives are to study the interplanetary medium at high solar latitudes and determine the effects of sunspots and solar flares on conditions in interplanetary space; characterize solar wind particles and interplanetary dust in previously unexplored regions of the solar system; and observe the Sun's electric and magnetic fields over a wide range of solar latitudes.

The mission profile

After deployment from the Space Shuttle, Ulysses was not aimed at the Sun, but in the opposite direction, outwards towards the giant planet Jupiter. This is because even the most powerful upper stages which can be used with the Space Shuttle (in this case an IUS plus a PAM-S booster) are not powerful enough to impart the 42km per second boost needed to put the probe directly into its solar polar orbit. Ulysses spent 16 months cruising towards Jupiter along a fairly typical interplanetary trajectory in the ecliptic plane. On arrival at Jupiter in early 1992 the probe used Jupiter's gravity to swing it around and southwards so that it will return to the vicinity of the Sun in 1994. Although it never approaches the Sun closely, always remaining outside the orbit of the Earth, Ulysses will spend a considerable time above 70° South solar latitude, providing scientists with their first real information on this region of space. The Sun's gravity will then pull Ulysses northwards so that it crosses the solar equator and then rises up to look down on the northern regions of the Sun in the summer of 1995. By then Ulysses will be moving away from the Sun back out towards Jupiter's orbit. This marks the official end of the mission although, if the spacecraft is still operating normally, it is possible that the mission may be extended for as long as useful scientific data can be obtained.

By the time it was launched, Ulysses had been tested and approved for flight three times, dismantled and put into storage twice and had made three trips across the Atlantic.

After launch Ulysses was the fastest man-made object in the Universe as it hurtled away from the Earth at 11.4km per second.

US launch centres

Wallops Island 1945, Cape Canaveral 1949,

Vandenberg Air Force Base 1958

Spaceport USA

NASA has three space launch centres, by far the most famous of which is the Kennedy Space Center at Cape Canaveral.

The Kennedy Space Center is NASA's main launch site. It is part of a collection of launch pads and other facilities which lie around Cape Canaveral on the Atlantic Coast of Florida. NASA operates the northern portion of the complex, on Merrit Island, and the southern area forms the US Department of Defense's Cape Canaveral Air Force Station. It is usual to consider all the facilities together, since NASA and commercial launch companies frequently use launch pads in the US Air Force section.

The site was selected as a missile range soon after World War II when it became clear that new rockets were becoming too powerful to use inland ranges such as the one at White Sands in New Mexico. Cape Canaveral offered good weather and a string of islands to the east suitable for tracking stations from which to monitor missiles fired out over the Atlantic Ocean. The range was set up in 1949 and the first launch, a V2 rocket with a WAC Corporal second stage, was made on 24 July 1950. The range was then used for test flights of a whole series of winged missiles including the Naveho, Snark and Matador as well as for launches of the Redstone and Jupiter C ballistic missiles.

When the USA decided to develop an artificial satellite, Cape Canaveral was the obvious launch site. Not only were suitable facilities already in place, but the rocket could take advantage of the extra velocity imparted by the centrifugal force of the Earth, which is greatest when rockets are launched eastwards from a site near the equator. Attempts to launch a satellite with the Vanguard rocket failed disastrously in 1957 and it was a Jupiter C launched from Complex 26 which orbited the USA's first satellite on 31 January 1958 (see p222). Since then a number of specialized launch complexes have been built to support the many different types of rockets launched from Cape Canaveral.

At the southern end of the Cape are pads used to test fire missiles such as the Trident and Minuteman. There are also pads for small and medium-sized unmanned satellite launchers such as Scout, Delta (Complex 17) and Atlas-Centaur (Complex 36A and 36B). In the centre of the site is what was called ICBM row, a chain of a dozen major launch pads used for missiles such as Redstone, Atlas and Titan. Since the rockets used to launch the first US astronauts were versions of these missiles, some of these pads were also used for the Mercury and Gemini spacecraft. At the northern end of ICBM row are Complex 37, used for Saturn 1 rockets, and Complex 34, the site of the fire which took the lives of three astronauts in 1967. Many of the pads, including Complex 14 from which John Glenn was launched, have now been demolished. Others, such as the site of the two-man Gemini launches at Complex 19, have been abandoned and are rusting away in the salt air.

The northern section of Cape Canaveral comprises Complexes 39, 40 and 41. These are used to launch the largest US rockets and operate in a different way to the smaller pads to the south. Instead of assembling and testing each rocket on the launch pad itself, the rockets are assembled and checked inside special buildings and then moved to the launch pad when the day of the launch approaches. The central cores of Titan rockets destined for Complexes 40 and 41 are assembled in a special Vertical Integration

Building large enough to hold four rockets side by side. Each assembled rocket is then moved on railway tracks to another building where its two huge solid rocket boosters are attached. Only then is the rocket moved to the launch pad from which it will begin its journey into space.

Launch Complex 39, the one originally used for the Saturn 5 moonrocket, is the most northerly at Cape Canaveral. The Saturn 5's were assembled in the Vehicle Assembly Building (VAB), a 52-storey building large enough to hold four Saturn rockets at a time. Each Saturn 5 was assembled on a mobile launch platform, a huge square structure with a 100m-high tower at one end. The tower had nine arms which swung across to allow access to the rocket and supported the pipes through which fuel was loaded during the last hours of the countdown. The whole assembly, weighing over 5000 tonnes, was moved by a huge vehicle known as a 'crawler transporter' to one of two identical launch pads (39A and 39B) about 5km away. The crawler was then used to collect a separate mobile service structure which was positioned around the Saturn 5 so that technicians could work on the rocket in the weeks before launch. The mobile service structure was removed about 12 hours before lift-off.

After 1975, Complex 39 was modified for use by the Space Shuttle. The main change was to introduce a fixed launch tower on the side of the pad in place of the one on the mobile launcher. This new tower incorporated a rotating service structure which could be swung into place around the Space Shuttle cargo bay, removing the need for the mobile service structure. The Space Shuttle is still joined to its booster rockets and external tank in the VAB and then moved to the pad by a crawler a few weeks before launch. Other specialized buildings are used to check the Space Shuttle after it returns from orbit and to prepare the payloads that will be carried in its cargo bay. A 4500m runway was also built, so that the Space Shuttle could return directly to Cape Canaveral after each mission.

Vandenberg Air Force Base

Vandenberg Air Force Base lies on the coast of California, about half way between Los Angeles and San Francisco. The base was established in the late 1950s as a site from which ballistic missiles could be launched under simulated operational conditions, and the first such launch was a Thor Intermediate Range Ballistic Missile test fired on 16 December 1958. Since then the base has grown considerably and a wide variety of launch pads and underground silos have been built. The silos are used for training military personnel in the procedures used to launch nuclear missiles and are also used to test fire missiles from time to time. The conventional launch pads have been used for a variety of launch vehicles including versions of the Atlas, Delta, Titan and Scout rockets.

One important advantage of Vandenberg is that it is an ideal site from which to launch satellites into polar orbit. If a rocket from Vandenberg is launched towards the south then the early part of its flight is over water and there is no risk of discarded rocket stages falling on to populated areas. The first orbital launch from Vandenberg was Discoverer 1, a military satellite, on 2 February 1959. Discoverer 1 was the first object ever placed into a polar orbit and was the first of many such launches from Vandenberg.

Since many polar orbiting satellites are intended for military reconnaissance, Vandenberg is a much more secret establishment than Cape Canaveral. It is not, however, used exclusively for military purposes. The civilian missions launched from Vandenberg include the Landsat and Seasat Earth observation missions and astronomical satellites such as IRAS and COBE. All of these satellites were placed into polar orbits which would have been difficult to reach from Cape Canaveral.

It was intended that Vandenberg would serve as a launch centre for Space Shuttle missions destined for polar orbits. Construction of a launch facility for the Space Shuttle was begun, but work was suspended when the Space Shuttle Challenger exploded in 1986 and has never been restarted. It is now almost certain that the Space Shuttle will never be launched from Vandenberg, since the US Department of Defense have placed orders for a number of powerful Titan rockets which it believes are more suitable for high-priority military payloads than the Shuttle.

Wallops Island

Wallops Island is a small NASA launch centre on the coast of Virginia. The site was founded in 1945 by the National Advisory Committee for Aeronautics and was later transferred to NASA. Wallops Island is mostly used for launches of sounding rockets on sub-orbital trajectories. These flights are used to study the upper atmosphere or to perform experiments which only require a few minutes of exposure to the space environment. Wallops Island is also able to place small satellites into orbit using the solid-fuelled Scout rocket, the smallest of NASA's launch vehicles. About two dozen satellites have been launched from Wallops, but over 10 000 sounding rockets have been launched since 1945.

White Sands Missile Range

Located in New Mexico, this was the site from which the USA fired most of the V2 rockets captured from the Germans at the end of World War II. It was never used to launch satellites, but the third orbital flight of the Space Shuttle landed here when the planned landing site at the Edwards Air Force Base was unavailable due to flooding of the dry lakebed.

So many early missiles crashed that one wit referred to the 'Snark infested waters off Cape Canaveral'.

A Saturn 5 rocket at launch complex 39, Kennedy Space Center, Florida

US launch vehicles

Jupiter C 1958

Trying to close the missile gap

The USA developed a host of different rockets in an attempt to match early Soviet space achievements.

In 1957, shocked by the propaganda impact of the Sputnik satellites, the USA found itself behind the USSR in this area of the space race and when attempts to orbit a satellite with the Vanguard rocket failed, converted missiles were hastily pressed into service to restore US prestige. Over the next two decades a variety of launchers were developed and although some soon became obsolete, others remained in service for many years. Most of these launch vehicles used a modified missile as a first stage with specially developed upper stages added to provide the extra thrust needed to place a satellite into orbit. Some upper stages (most notably the Agena and Centaur) were used in combination with several different missiles and some were developed in a series of different versions. These many combinations make a complete description of US launch vehicles difficult and only the most important are described here. Improved versions of launchers originally developed in the 1960s and which remain in service today are described on page $238.

Jupiter C/Juno 1

The Jupiter C, the rocket which launched the USA's first satellite, was originally intended for testing the protective nose cones of nuclear missiles. It was based on the single stage Redstone missile developed by the US Army. The Jupiter C (C for Composite) comprised a Redstone with enlarged propellant tanks to which upper stages were added. The second stage was a ring of 11 Sergeant solid-fuelled rockets enclosed in a metal cylinder and the third stage comprised three more Sergeants fitted inside this outer ring. The upper stages were spun to provide stability and to average out the thrust from the various solid motors. A four-stage version, using a single Sergeant as a final stage, sent a test payload on a sub-orbital flight of over 2 000km in 1957.

The satellite launching version, which was about 22m high, was also known as Juno 1. It was prepared by adding a small scientific payload to the final Sergeant rocket and altering the flightpath so that the final stage went into orbit. A Jupiter C/Juno 1 launched Explorer 1 on 31 January 1958, Explorer 3 on 26 March 1958 and Explorer 4 on 26 July 1958. Three other attempts (Explorer 2, Explorer 5 and Beacon 1) failed.

Vanguard

Vanguard was intended to launch a scientific satellite as part of the USA's contribution to the International Geophysical Year (1957–8). Rather than using a converted missile, the Vanguard used a modified version of the Viking high altitude research rocket as a first stage. The second stage burned nitric acid and unsymmetrical dimethylhydrazine and the third stage was solid fuelled. The result was a thin (1.1m diameter), but tall (22m-high), rocket which contrasted sharply with its rather less elegant rival the Jupiter C. The first two attempts to launch a satellite with the Vanguard rocket failed and it was not until 17 March 1958 that a Vanguard satellite reached orbit. The programme suffered a further six failures although it did succeed twice more, orbiting Vanguard 2 on 17 February 1959 and Vanguard 3 on 18 September 1959.

Juno II

The Juno II was based on the US Army's Jupiter Intermediate Range Ballistic Missile (IRBM) which itself used technology developed for the Redstone missile. The Juno II was a Jupiter IRBM with enlarged fuel tanks to which were added clusters of Sergeant rockets similar to those used in the Jupiter C. Juno II was used to launch some of the early Pioneer lunar probes (see p150) and a number of Earth satellites. The first Juno II launch was in December 1958 and the last, which was unsuccessful, was on 25 May 1961.

Thor

The Thor rocket was an Intermediate Range Ballistic Missile developed by the US Air Force at the same time as the Jupiter IRBM was being developed by the US Army. It used a Rocketdyne LR79 liquid-fuelled motor and was first launched in January 1957. Its first use as a space launcher was the Thor-Able combination which comprised a Thor missile plus the two upper stages from the earlier Vanguard rocket. Thor-Able was used (unsuccessfully) to boost the US Air Force Pioneer probes towards the Moon in 1958. A later version was the Thor-Able-Star which had a single, liquid-fuelled, upper stage and remained in service until 1965. Thor-Able-Star achieved 14 successes in 19 launches.

Thor was also used in conjunction with the Agena, a sophisticated upper stage originally developed for use with the Atlas rocket. The Agena could be stabilized in space and its motor could be restarted several times once in orbit. The Thor-Agena was used by the US Air Force to launch satellites in its Discoverer series and by NASA to launch a variety of unmanned scientific missions including the Echo 1 balloon satellite. The first Thor-Agena A launch was in 1959, the Agena B version followed in 1960 and the Agena D in 1962. Some versions of the Thor had three small solid-fuelled rockets attached to the first stage to provide additional thrust at lift-off. These versions included Thor-Agena D, the Thrust Augmented Thor and the Long Tank Thrust Augmented Thor, a version with enlarged first-stage fuel tanks.

A Thor missile also formed the first stage of the Delta rocket, a combination which has been developed through a large number of variants and remains in use today (see p238).

Atlas

The Atlas was the first US Intercontinental Ballistic Missile and was first launched in 1957. The Atlas, versions of which remain in service today, has three engines in an unusual 1.5 stage layout. All three engines are connected to the same propellant tanks and all three are used during the first few minutes of flight. The two outermost engines are then jettisoned and the Atlas continues upwards under the power of its single remaining engine. Modified Atlas rockets were used to launch the orbital flights of the one-man Mercury capsule (see p123).

It is possible for an Atlas to reach low Earth orbit with a small payload, as was done with project SCORE, but various other stages are usually added for satellite launches. A commonly used upper stage throughout the 1960s was the Agena, and various Atlas-Agena combinations were used to send Mariner probes to Mars and to Venus, and Ranger and Lunar Orbiter missions to the Moon. Atlas-Agena was also used for a variety of Earth orbiting missions. The Atlas-Burner 2 and Atlas F versions used solid-fuelled upper stages.

The most powerful Atlas variant is the Atlas-Centaur. This uses a liquid hydrogen and liquid oxygen powered second stage which was first flown in 1962. Although the early flights of the Centaur stage suffered a number of failures, the technical problems were eventually resolved and the Atlas-Centaur became a popular launcher. Versions of Atlas-Centaur sent Surveyor probes to the Moon, Mariners to Mars and placed large communications satellites into Earth orbit. A special three-stage version, featuring an extra solid-fuelled stage, was used to send the Pioneer 10 and 11 spacecraft to Jupiter. New commercial versions of the Atlas-Centaur are being developed for use in the 1990s (see p238).

Titan

The Titan family began with the Titan I Intercontinental Ballistic Missile (ICBM), which was first tested in 1959. From this developed the Titan II ICBM series, 54 of which formed part of the US strategic nuclear arsenal. A modified version of Titan II was used to launch the two-man Gemini spacecraft (see p63) in the mid 1960s. Although not used for unmanned satellites when first developed, surplus Titan II missiles are now being converted for use as satellite launchers (see p239).

The Titan IIIA was similar to the two-stage Titan II, but featured an extra stage, called the Transtage, which could be restarted in orbit. All three stages used liquid propellants. The Titan IIIB used the basic Titan II core to which an Agena upper stage was added in place of the the Transtage. The Titan IIIC version was similar to the Titan IIIA, but with the addition of two large solid rocket boosters. The boosters, similar in diameter to the central core and extending about two-thirds of the way up the vehicle, are composed of five segments each. The Titan IIIC was first launched on 18 June 1965 and was mostly used for military satellites. It was, at the time, the USA's most powerful unmanned launch vehicle. The Titan IIID was similar to the IIIC, but without the Transtage.

The Titan IIIE was a special version of Titan developed specifically for NASA deep space missions. It differed from the IIIC by the use of a Centaur upper stage in place of the usual Transtage. The Centaur was wider than the basic Titan core and so the Titan IIIE had an unusual, hammerhead, configuration. The first test flight of the Titan IIIE failed, but the remaining six successfully launched the two Helios solar probes, the two Viking Mars missions and the two Voyager spacecraft. The Titan IIIM was intended to launch the US Manned Orbiting Laboratory (see p129), but this project was cancelled and only one Titan IIIM was ever launched.

Titan 34D was a military version of the Titan with an Inertial Upper Stage (a small solid-fuelled booster designed for use with the Space Shuttle) in place of the Transtage. The first Titan 34D was launched on 30 October 1982. The Titan IV, a still more powerful version, remains in use (see p239).

A Thor-Able rocket is prepared for launch in 1960. The payload is the Explorer 6 satellite

US Launch vehicles today

Pegasus 5 April 1990

Rockets for profit

A new US industry has been born: commercial launches of satellites by private companies.

In the late 1970s NASA expected that its Space Shuttle would eventually replace almost all US expendable launch vehicles, and by the mid 1980s the Space Shuttle was regularly launching satellites and production of conventional rockets was ending. However, the commercial success of the European Ariane rocket, and the explosion of the Space Shuttle Challenger in 1986, forced NASA to reconsider its policy. It was soon decided that the Space Shuttle would no longer carry satellites that could be launched by unmanned rockets and that, instead of restarting its own unmanned launcher programme, NASA would buy launch services from US industry. The result was the reopening of production lines for rockets such as Atlas-Centaur, Delta and Titan. These vehicles, all 1960s designs modified to enable them to compete for business in the 1990s, form the backbone of the present US rocket fleet.

Atlas

At the time of the Challenger disaster production of the Atlas Centaur rocket was about to be stopped but, in 1987 the manufacturer, General Dynamics, decided to risk investing $100 million to restart production in the hope of attracting commercial orders. It was a gamble that paid off because the US Air Force soon placed a contract for the launch of 11 new DSCS communication satellites. This established Atlas as a serious commercial proposition and General Dynamics now offer four different versions of the Atlas to potential customers.

Atlas I is the basic Atlas-Centaur rocket and was first launched in May 1962. Atlas I can place a single satellite weighing up to 2 245kg into geostationary transfer orbit. The Atlas II and Atlas IIA versions use a modified Centaur stage with larger fuel tanks enabling the payload to be increased by about 500kg. The fourth and most powerful version is the Atlas IIAS which has four small solid-fuelled booster rockets to provide additional thrust at launch. This version can lift 3 500kg into a geostationary transfer orbit.

Delta

Over 200 Delta rockets have been launched since the first Thor-Delta failed after only a few minutes of flight on 13 May 1960. Since then the vehicle has gone through a series of developments which make the latest version, the Delta II, almost totally different in appearance from the Delta which exploded over Cape Canaveral three decades ago.

The original Thor-Delta was based on the Thor Intermediate Range Ballistic Missile to which a liquid-fuelled second stage and solid-fuelled third stage were added. The result was a slim, gently tapering rocket which could orbit a satellite weighing about 45kg and whose first successful launch was the Echo 1 balloon satellite on 12 August 1960. The Delta's capability soon began to grow as the basic rocket was modified again and again. The Delta B, launched 1962, had enlarged second-stage fuel tanks, the Delta C had a new second-stage engine, and the Delta D was the first version with solid-fuelled booster rockets added to the base of the Thor first stage. These improvements increased the payload to over 100kg. The Delta E and J versions, launched in 1965 and 1968 respectively, had further improved second stages and the Delta M featured an enlarged first stage. In 1969 the number of solid boosters attached to the first stage had risen to six and this rose again to nine in 1971, by which time the payload had increased to 635kg. New versions of the Delta continued to be developed throughout the 1970s, with larger solid rocket boosters and a new second stage which had by now expanded in diameter until it matched the 2.4m diameter of the first stage.

By 1986, the year in which Challenger was destroyed, only two Delta rockets remained. There were also with three sets of spare parts which were soon assembled into complete vehicles and duly launched. Production of Delta was restarted and a new version, the Delta II, was developed for commercial launches during the 1990s. The Delta II has an enlarged first stage, a more powerful first-stage engine and more powerful solid rocket boosters. It was selected by the US Air Force to launch a series of military navigation satellites and soon began to attract a number of commercial customers. The Delta II can lift 1 450kg into geostationary transfer orbit. The 200th Delta rocket was launched on 30 October 1990.

Pegasus

The Pegasus is a new, winged rocket, able to deliver small satellites into low orbit. The rocket, which is 15m long and 1.3m in diameter has a delta wing spanning about 7m. Pegasus is carried to an altitude of 13 000m by a large jet aircraft and then released. The solid-fuelled motor is then fired and the rocket climbs into space. The maximum payload Pegasus can lift into orbit is 400kg. The first Pegasus launch was on 5 April 1990 when the rocket orbited a test vehicle called Pegsat. A second flight on 17 July 1991 carried seven small satellites although, due to a problem with the booster, the satellites did not go into exactly the planned orbit.

Pegasus offers a flexible launch service since, by varying the point at which it is released from its carrier aircraft, it can fly around bad weather and achieve a wide variety of different orbital inclinations.

Scout

The four-stage solid-fuelled Scout is NASA's smallest launch vehicle. It was developed by the Langley Research Center in 1958 and built by LTV Aerospace in Dallas, Texas. It has been used to launch many small satellites including scientific missions for a number of European countries. The 114th Scout was launched on 29 July 1990. An improved Scout rocket, which will feature two solid rocket boosters to increase takeoff thrust, is expected to be developed in conjunction with Italy.

Titan

Three versions of the Titan rocket remain in service. The smallest is the Titan 23G which is based on the Titan II missile which once served as part of the USA's nuclear deterrent. The Titan II missiles have now been retired, but a number of them are being modified for use as space launchers able to carry about 2 000kg into low Earth orbit. The first flight of a modified Titan II was on 5 September 1988 from the Vandenberg Air Force Base.

The Commercial Titan is a civilian version of the Titan III rocket able to place 2 270kg into geostationary transfer orbit. The first launch was on 1 January 1990 and carried a dual payload: a British Skynet military communication satellite and the Japanese JCSat 2. The second commercial launch on 23 June 1990 was less successful, stranding an INTELSAT 6 communications satellite in a useless low orbit. Launches were then suspended while modifications were made to the launch pad at Cape Canaveral, but are expected to resume in 1992.

The Titan IV is the most powerful US expendable rocket and has a lifting capability similar to that of the Space Shuttle. It has been developed by Martin-Marietta Aerospace for the US Air Force and can place about 4 500kg into the geosynchronous Earth orbit favoured by US military satellites. Titan IV is an enlarged version of the Titan 34D rocket, itself an outgrowth of the Titan III which was first flown in 1965. The primary mission of the Titan IV will be to launch large military satellites. Compared to its predecessor (see p237) the Titan IV has lengthened first and second stages and enlarged solid rocket boosters. Depending on its mission, it can be fitted with one of two upper stages, a liquid-fuelled Centaur or a solid-fuelled Inertial Upper Stage developed originally for use with the Space Shuttle. The first Titan IV launch was on 14 June 1989 from Cape Canaveral.

Thor-Delta was so named because it was the fourth launcher based on the Thor rocket and Delta is the fourth letter of the Greek alphabet.

The first Titan IV rocket launch on 14 June 1989

US space centres

Space technology on Earth

NASA HQ is in Washington, but technical development takes place in various research centres, many of which were transferred to NASA when it was set up on 1 October 1958.

Ames Research Center

The Ames Research Center is located next to the US Naval Air Station at Moffat Field, California. It was set up in 1940 and was originally used for aeronautical research. Since its incorporation into NASA, Ames has been involved in programmes such as the Pioneer deep space probes, life sciences research and the development of the Space Shuttle. Ames is also the base for a fleet of NASA aircraft which carry telescopes to high altitudes so they can make astronomical observations that are impossible from the ground. The centre is named after Joseph Ames, a former president of the National Advisory Committee on Aeronautics (NACA).

Dryden Flight Research Center

Named after a former deputy administrator of NASA, Hugh Dryden, this is the home of some of NASA's most exotic aircraft programmes. The centre was formed by NASA's predecessor organization, NACA, in 1947 during the programme to break the sound barrier with the X-1 research aircraft. The centre is located at Edwards Air Force Base in California which provides both conventional runways and a huge expanse of dry salt lakebed suitable for landing aircraft with unusual handling characteristics. Since 1947 Dryden has been used by a variety of high-speed aircraft programmes including the D-558 Skyrocket, the X-3 Stilleto, the X-15 rocket plane and lifting bodies such as the HL-10, M2-F2 and X-24. The early landings of the Space Shuttle were made here and, although many Shuttle flights are now returning directly to the Kennedy Space Center, some Shuttle missions are still ending at Dryden.

Goddard Space Flight Center

Named after the rocket pioneer Robert Goddard, the Goddard Space Flight Centre (GSFC) is set in a campus-like park about 20km from Washington DC. The centre was established in 1959 as a scientific laboratory devoted entirely to space exploration and its activities are concentrated on the development and operation of unmanned scientific satellites. Goddard scientists have been responsible for a huge number of scientific satellites, including orbiting observatories to study the Earth, the Sun and the stars, astronomical experiments in the Space Shuttle programme and the International Cometary Explorer mission sent to comet Giacobini–Zinner in 1985.

Other activities at the GSFC include the provision of tracking and data acquisition systems to control and receive data from Earth orbiting satellites and acting as the home for the National Space Science Data Centre, a huge archive of information returned by dozens of satellites.

Jet Propulsion Laboratory

The Jet Propulsion Laboratory (JPL) in Pasadena, California is operated for NASA by the California Institute of Technology. It was founded by Professor Theodore von

Karmen who carried out a series of rocket experiments in the Arroyo Seco near Pasadena during the mid 1930s. In 1957 the JPL provided the Explorer 1 satellite that was launched in response to the Soviet satellite Sputnik 1 and went on to develop some of the early Pioneer Moon probes. The laboratory worked in close cooperation with the US Army until the formation of NASA in 1958. Since about 1960 the JPL has concentrated on the exploration of the solar system and was responsible for projects such as the Ranger and Surveyor Moon probes and the Mariner, Viking, and Voyager interplanetary missions. Today the JPL is responsible for deep space missions such as Galileo, Ulysses, Magellan and the Mars Observer as well as the Deep Space Tracking Network used to receive data from these spacecraft.

Closer to home JPL is involved in Earth orbiting missions such as the TOPEX/Poseidon ocean observation satellite and satellites used for infrared astronomy. The laboratory also operates the Table Mountain observatory which is used for ground-based astronomy. In recent years the JPL has been carrying out increasing amounts of work for the US Department of Defense.

Johnson Space Center

The Johnson Space Center (JSC) in Houston is the home of mission control for almost all US manned spaceflights and was named after President Lyndon B Johnson. Originally known as the Manned Spacecraft Center, the facility was opened in 1963 on land about 35km from Houston town centre. The JSC is the main base for the USA's astronauts and is almost entirely devoted to manned spaceflight. In addition to planning and directing space missions, the JSC is responsible for astronaut selection and training and for looking into the future requirements of the USA's manned space programme. The Lunar Receiving Laboratory, where the first lunar samples were quarantined before being distributed to scientists around the world, is located here.

Langley Research Center

The oldest of NASA's centres is the Langley Research Centre in Hampton, Virginia. The centre, which is named after aviation pioneer Samuel Langley, was established in 1917 and has provided a focal point for the USA's aviation research ever since. The centre was involved in the development of military aircraft during World War II and then in the programmes which led to the first supersonic aeroplane, the Bell X-1. Staff from Langley formed the Space Task Force which was responsible for the Mercury programme that put the first American in space, although responsibility for this programme was later transferred to the Manned Spacecraft Center at Houston. Langley was also involved in investigating the best shape for the Gemini and Apollo spacecraft so that they could survive re-entry. The centre has also been involved in unmanned space projects including the Echo balloon satellite, the Pegasus meteoroid detection satellites, the Lunar Orbiter programme and the Viking missions to Mars.

Lewis Research Center

Located in Cleveland, Ohio this facility was founded by NACA in 1941 and is named after former NACA director of Aeronautics Dr George Lewis. Originally responsible for research into aeronautics and jet propulsion the centre later became involved in the development of medium-sized rockets such as the Agena and Centaur. It was also active in research into nuclear and ionic rockets. Today, Lewis is a centre for research and development into new propulsion systems, power systems and satellite communications.

Marshall Space Flight Centre

NASA's Marshall Space Flight Centre, named after General of the Army George C Marshall, is located in the midst of the US Army Redstone Arsenal at Huntsville Alabama. The centre was formed on 1 July 1960 by the transfer of buildings and staff (including rocket designer Wernher von Braun) from what was then the Army Ballistic Missile Agency. In view of its heritage it is not surprising that Marshall has been responsible for the USA's most famous rockets, beginning with the Jupiter C used to launch the first US satellite and the Redstone used to launch Alan Shepard on his sub-orbital mission in 1961. Development of the Saturn 1 and 5 rockets which were used to send men to the Moon was managed from NASA Marshall. Some of the huge test stands built during these programmes were later converted for use during the development of the Space Shuttle.

The centre's activities are not restricted to rocket development; it has been involved in projects such as Skylab, Spacelab and unmanned scientific missions. It is also the location of a huge water tank used for developing equipment needed by space-walking astronauts.

Michoud Assembly Facility

Operated by the Marshall Spaceflight Center, this facility near New Orleans was once used to assemble the first stage of the Saturn 5 moonrocket. Today it is used for engineering, design and manufacturing connected with the external tank for the Space Shuttle.

The huge wind tunnel complex at the NASA Ames Research Center

Vega

Vega 1 15 December 1985

Venus balloons

The Soviet Vega craft flew to Venus and then went on to explore Halley's Comet.

The two Vega probes were the first Soviet spacecraft to explore more than one celestial body; after flying past the planet Venus they used Venus's gravity to send them on to intercept Halley's Comet. During their brief flyby of Venus each probe released a landing capsule and a French atmospheric balloon.

The basic Vega spacecraft was a modified version of the Venera probe which had been in use since 1975. The main spacecraft had a cylindrical central body with a bulbous extension at one end. The spherical landing capsule was carried at the other end of the body and electricity was provided by solar panels which folded out on opposite sides of the central section. A dish-shaped, high-gain antenna was fixed to the central section and various experiments were attached around the bulbous base. The most obvious modification to the basic Venera craft was the addition of an experiment platform carrying cameras and other instruments required for observations of Halley's Comet. Many of the experiments aboard the two craft were provided by countries outside the USSR.

The Venus landing capsules were similar to those used on the Venera 9-14 missions, but with the addition of the atmospheric balloon experiment. Each balloon was released by the landing module during its descent and was slowed by a parachute system. As the balloons entered the middle Venusian atmosphere they were inflated with helium gas. Each balloon was 3m in diameter and carried a set of scientific instruments in a gondola about 1.5m long which was suspended 12m below the balloon. The instruments were designed to determine the temperature and pressure of the atmosphere, the frequency of lightning bolts and the velocities of the Venusian winds. The gondolas were powered by batteries with a lifetime of 60 hours.

Vega 1 was launched by a Proton rocket on 15 December 1985 and Vega 2 followed six days later. Vega 1 released its landing capsule on 9 June 1985. The capsule made a safe touchdown on 11 June 1985 and returned data for two hours. The Vega 1 balloon deployed as planned and for two days floated about 50km above the Venusian surface. In an example of the international cooperation that was the hallmark of the missions to Halley's Comet, precise data on the balloon's flight was provided by tracking it using a network of radio telescopes in the USSR, Europe, Brazil and Australia. This data showed that the planet's winds carried the balloon over 10 000km from its entry point. Vega 2 released its lander and balloon successfully a few days later, with similar results.

The two Vega spacecraft continued on course to intercept Halley's Comet, joining the Japanese and European spacecraft, which had been launched directly towards the comet. Vega 1 was the first member of this international fleet to arrive when it flew within 9 000km of the comet's nucleus on 6 March 1986. During the three hours Vega 1 was closest to the comet, it returned 500 photographs and a stream of other scientific data. Vega 2 encountered the comet on 9 March and returned more photographs and scientific data. Information from the two Vega probes was passed to the European Space Agency (ESA) who used it to make final adjustments to the course of the Giotto probe, which reached the comet on 13 March.

Vega is a contraction of the Russian words for Venus and Halley.

Venera

Venera 1 1961, Venera 16 1983

Soviet Venus probes

Despite setbacks, the Soviet Venera probes provided exciting information about our sister planet.

Although Venus comes closer to Earth than any other planet, very little was known about it until the space age. This is because Venus is permanently shrouded by clouds and astronomers can only ever see the upper layers of its atmosphere. Even such basic information as the planet's surface temperature and its rotation rate were poorly understood until the 1960s when a combination of Earth-based radar observations and spacecraft flybys began to reveal details of the mystery below the clouds. The USA sent a series of Mariner and Pioneer spacecraft to Venus between 1962 and 1970, but by far the most comprehensive assault on the planet has been by the Soviet Venera series, the first of which was launched in 1961.

Venera 1

The first Soviet Venus probe was launched on 4 February 1961, but failed to escape from its Earth parking orbit and was given the cover name of Sputnik 7. Sputnik 8, launched 12 February 1961 was more successful and boosted what was called an Automatic Interplanetary Station towards Venus. The probe, which was also known as Venera 1, weighed 644kg and had a cylindrical body with a domed top, two solar panels and an umbrella-shaped radio antenna. At first all went well, but unfortunately radio contact with the probe was lost on 19 February and was never regained. It was the first of many disappointments in the Venera programme.

Zond 1 and Venera 2,3

The next Soviet mission launched in the direction of Venus was named Zond 1 and was described as a deep space engineering test. After lifting off on 2 April 1964 Zond 1 made several course corrections, but by the time the probe passed Venus radio contact had been lost and no data was returned. Venera 2, launched on 12 November 1965, was intended to fly past Venus on 12 November 1965. All went well until almost the moment it arrived at Venus when, for a still unexplained reason, radio contact was lost and no data was ever received. Venera 3, which was a carrier spacecraft intended to release a landing probe, suffered a similar fate. Launched on 15 November 1965 its course was adjusted to aim it straight at the planet, but radio contact was lost at about the time of entry into the atmosphere.

Success at last

Venera 4 was launched on 12 June 1967 and when it arrived at Venus on 18 October it released a 383kg landing capsule into the Venusian atmosphere. After entry, the probe deployed a parachute to lower itself to the surface and data was received for 94 minutes. Although it was first thought that the probe had reached the surface, analysis of the data showed that this was not the case. As the probe descended, the temperature and pressure increased far more than anticipated and the landing capsule was crushed by the atmospheric pressure before it reached the surface. When Venera 4 stopped transmitting it was still about 25km high, yet the temperature was already 280°C and the atmospheric pressure was 22 times that of the Earth's.

Following the results from Venera 4, the next two probes were strengthened and their parachutes reduced in size so that they would descend more quickly and reach the surface before the atmosphere could heat them up too much. Venera 5 was launched on 5 January 1969 with Venera 6 following five days later. Both probes arrived in the middle of May and both suffered the same fate. Venera 5, which arrived on 16 May, descended to 12km before it was crushed and the next day Venera 6 stopped transmitting about 16km from the surface.

Venera 7 was launched on 17 August 1970. It too arrived at Venus and parachuted into the atmosphere and once again contact was lost before touchdown. However, several weeks afterwards ground controllers re-examined the data which had been recorded and found that, amongst the radio static, there were other faint signals from Venera 7. These indicated that the temperature around the probe had reached 475°C and that the pressure was 90 times that of the Earth's atmosphere. More importantly, the pressure and temperature being recorded were not changing, meaning that the probe must have come to rest on the surface. It was the first ever transmission from the surface of another planet.

Venera 8, launched on 31 March 1972, was similar to its predecessors, but with an improved radio system so that contact could be maintained more easily. Venera 8 arrived at Venus and landed on 22 July 1972. This time there were no problems with the radio and data was received from the surface until the probe succumbed to the intense temperatures after 50 minutes. One of the results from Venera 8 showed that, despite the clouds, the level of light at the surface was similar to that on a rainy day back on Earth. The importance of this would soon become clear.

Pictures from the surface

In June 1975 the next Venera craft, numbers 9 and 10 were sent towards Venus. They were of new design with an improved landing capsule. As each one approached the planet it released a probe and then the carrier spacecraft fired its own rocket motor to go into orbit around Venus. The orbiting sections were designed to act as radio relay stations for the landers. The probes entered the atmosphere and then each deployed a small parachute to slow it down and stabilize it as it descended through the upper atmosphere. Then, as the air became thicker and the temperature began to rise, each probe cast off its parachute and fell to the surface using only airbrakes to moderate its final descent. Both probes landed successfully, Venera 9 on 20 October and Venera 10 five days later. Each probe returned a small number of 180°, black and white, panoramic photographs showing a rock-strewn surface. These were the first ever pictures from Venus's surface.

Venera 11 and 12, launched on 9 and 14 September both reached the planet in December and released probes. No pictures were returned from the surface, but the probes were able to sample the chemical composition of the atmosphere during the descent and they also recorded many lightning flashes.

Venera 13 was launched on 30 October 1981 and was followed by Venera 14 on 4 November. The landing capsules touched down on 1 March and 5 March 1982 respectively. Both returned colour pictures and used a series of drills and other probes to sample the surface rocks, sending back the first details about the chemical composition of the planet's crust.

Mapping by radar

The final Venera craft, numbers 15 and 16, were launched on 2 and 17 June 1983. These 4000kg craft did not carry landing capsules, but were instead equipped with large radio antennas which formed part of a complex Synthetic Aperture Radar system (see p107) intended to bounce radio signals off the surface and use the returned echoes to construct radar images of the surface. In October the two craft both went into elliptical (1000km × 65000km) orbits around Venus and began to return detailed

As well as Sputnik 8, the following Cosmos satellites were probably failed Venus missions: Cosmos 96 (23 November 1965), Cosmos 167 (17 June 1967), Cosmos 359 (22 August 1970), Cosmos 482 (31 March 1972).

radar images of the planet's surface. Each craft operated for about a year and between them they mapped over 120 million sq km of the planet's surface. These radar images were far superior to the data returned by the US Pioneer Venus orbiter (see p152) and were unsurpassed until the US Magellan mission of 1990 (see p105).

Working model of the Venus 15 and 16 automatic stations in the testing building

Viking

Soft landing on Mars 20 July 1976

Searching for life on Mars

The Viking missions were the culmination of the USA's programme to explore Mars with robot spacecraft.

The first US missions to Mars were the Mariner 4, 6 and 7 spacecraft which flew past the planet in 1964 and 1969. Their results suggested that Mars was a cold and lifeless planet but, when Mariner 9 went into orbit around Mars in 1971, it returned photographs which suggested that Mars was much more interesting than first thought. In particular, Mariner 9 data suggested that Mars had a 50 000-year climatic cycle in which cold and dry periods (as at present) alternated with warmer, wetter and more hospitable eras. This raised the possibility that life might have developed on Mars in the distant past and could have evolved to lie dormant during the long dry spells in the present cycle. Since the Mariner images could not show objects smaller than a football field, the only way to find out if there was life on Mars was to land and investigate. This was the main objective of the two Viking spacecraft launched towards the red planet in 1975.

The spacecraft

Each Viking spacecraft comprised two parts, an orbiter and a lander. Unlike the Soviet Mars missions (see p118) which released their landers before they arrived at Mars, each Viking was designed to go into orbit intact so that the cameras on the orbiter could check the nature of the landing site before releasing the lander. This gave the Viking control team a great deal of flexibility in selecting a suitable landing site. After landing, the orbiter was designed to relay data from the lander back to Earth.

The Viking orbiter was based on the Mariner 9 spacecraft, but was enlarged so that it could carry the lander to Mars. The body was an octagonal structure 2.4m across which contained most of the essential electronic systems. On the rear face of the structure were the rocket motor and fuel tanks of the propulsion system required to brake Viking into orbit around Mars. The total height of the orbiter with the lander attached was 4.9m. Solar panels were attached to the body of the orbiter and unfolded into a cross-shaped array after launch. The span across the panels was 9.7m.

The orbiter carried a set of scientific instruments mounted on a movable platform. The most important of these were two television cameras designed to photograph potential landing sites, but other instruments were carried to determine the planet's temperature and to search for evidence of water.

The Viking lander was a six-sided box with sides alternately 109cm and 59cm long. Landing legs, each with a circular footpad, were attached to the short sides. On top of the body were two television cameras, a mast supporting meteorological instruments, two small nuclear power systems, a dish-shaped radio antenna and a robot arm used to sample the surface. Three rockets, used to slow the spacecraft during the final stages of its descent, were attached to the outside of the body. Inside the body were various pieces of electronic equipment and a biology experiment. During the flight to Mars the lander, inside a protective shell, was attached to the orbiter by a truss structure. The shell included an outer biological shield to prevent earthly microbes from making the journey to Mars and a clam-like aeroshell which served to protect the lander during its entry into the Martian atmosphere.

The mission

It had originally been planned to launch the two Vikings in 1973, but the mission ran into a number of problems which caused it to be delayed until 1975. Each Viking was launched by a Titan IIIE (Titan-Centaur) rocket from Cape Canaveral, the first on 20 August 1975 and the second on 9 September 1975.

Viking 1 entered Martian orbit on 19 June 1976 and soon began to return pictures of its planned landing site. After studying the photographs, and examining radar data from Earth-based radio telescopes, the mission controllers decided that the originally planned landing site was too rough and an attempt to land on 4 July 1976, the bicentennial of American independence, was postponed. A new site was soon found and on 20 July 1976, exactly seven years after the first Moon landing, Viking 1 was commanded to land.

The landing sequence, which was completely automatic, began when the lander was separated from the orbiter and fired thrusters on its aeroshell to start the entry. About two hours after separation the aeroshell encountered the upper atmosphere and was slowed by friction with the thin Martian air. At a height of about 6 200m a parachute was deployed to slow the descent still further and the lander was pulled out of the aeroshell. The lander descended under the parachute until it reached a height of about 1 200m at which point the parachute was released and the lander began to fall, braked only by its three landing rocket engines. All went well and Viking 1 landed on Chryse Planitia (The Plains of Gold) at 22.3° North and 48.0° West. Within a few minutes the lander's television cameras were switched on and began to return wonderfully clear images of the surface of Mars. These soon showed that the landing site was quite flat, but was covered in many small angular rocks and a few quite large boulders. The overall colour of the site was red and the sky was pink, probably because of fine red dust suspended in the atmosphere.

The experience of Viking 2 was similar. The spacecraft arrived at Mars on 7 August 1976 and its original landing site was also ruled out for safety reasons. Eventually a new site was found at Utopia Planitia, 47.7° North and 225.8° West and the lander touched down safely on 3 September 1976.

The search for life

Each Viking lander carried three biological experiments. These were carried out on samples of Martian soil which were picked up by a scoop on the robot sampler arm and delivered to the biology package inside the body of the lander. One experiment tested for photosynthesis by exposing the soil sample to radioactive carbon dioxide gas, waiting and then looking to see if any of the radioactive gas had been absorbed by tiny plants in the soil. The second experiment treated a sample of Martian soil with a liquid nutrient which contained molecules 'labelled' with radioactive carbon. It was expected that any organisms in the soil would consume the radioactive nutrient and convert some of it to carbon dioxide gas which would then be released into the chamber. To determine if this happened the chamber was monitored for the appearance of radioactive carbon dioxide. The third experiment also involved moistening a soil sample with a nutrient solution but, in this case, the chamber was monitored to see if there was any evidence for changes in the atmosphere around the sample, for example the appearance of gases such as methane or carbon dioxide that might have been by-products of the metabolism of living creatures in the soil.

Although all three experiments produced interesting and unusual results, they did not prove that there was life on Mars. Even though the Martian soil did react when nutrient solutions were added, it was concluded that this was most probably due to unusual chemical reactions, not to biology. In particular, the inability of the gas chromatography experiment to find any evidence of organic (carbon-bearing) molecules in the soil was a strong argument against the presence of life.

Other results

Apart from the failure of the seismometer (earthquake detector) on Lander 1, all of the scientific instruments worked well. The Martian weather was monitored by both landers, recording temperatures as high as

–14°C and as low as –120°C. A thin layer of frost formed on the ground around Lander 2 during the winter. The wind speeds were found to be lower than expected, although gusts of up to about 100km per hour were recorded from time to time. Both landers returned thousands of photographs (more than 4 500 in all) showing the rocky terrain around each landing site in great detail and revealing that the Martian soil is a type of iron-rich clay.

As well as fulfilling their roles as radio relay stations, the two orbiters carried out their own comprehensive research programmes. Between them, the two orbiters returned over 52 000 photographs, including images in colour and in stereo. They observed interesting new features on the planet and re-observed known areas in greater detail than possible from Mariner 9. The two orbiters also studied the distribution of the planet's water vapour and showed that the residual North polar cap, which does not vanish even during the Martian summer, is comprised of frozen water, not solid carbon dioxide.

End of mission

The planned duration of the Viking mission was for each spacecraft to operate for 90 days after touchdown, but all four spacecraft far exceeded this objective. Orbiter 2 was switched off on 25 July 1978 and Orbiter 1 continued in operation until 7 August 1980. The last data was received from Lander 2 on 11 April 1980 and contact with Lander 1 was lost on 11 November 1982 after it had been accidently sent an incorrect command. Attempts to regain contact with Lander 1 continued unsuccessfully for over six months until the mission was formally ended on 21 May 1983.

The original plan to land instruments on Mars envisaged the use of a large spacecraft launched by a Saturn 5 rocket. The name of this abandoned project was Voyager, a name later used for a mission to the outer planets.

The surface of Mars as seen from the Viking 2 spacecraft

Voskhod

Voskhod 1 1964, Voskhod 2 1965

A dangerous gamble

By squeezing three men into a one-man spacecraft, the USSR pre-empted NASA's two-man Gemini programme.

After the success of the one-man Vostok missions (see p253), the Soviet chief designer Sergei Korolov had planned to use the new Soyuz spacecraft for future space projects. However, delays in the development of Soyuz, together with political pressure from the Soviet Premier Nikita Krushchev, forced Korolov to try to develop a multi-seater spacecraft before the first flight of the US two-man Gemini spacecraft. The only way to do this in time was to adapt the Vostok spacecraft so that it could carry up to three men. It was an ingenious idea, but one so fraught with risk that many of his engineers opposed it. According to designer Konstantin Feoktistov, Korolov solved this problem by making a tempting offer. Korolov said that if the engineers could design a three-seater spacecraft in time to beat the Americans, then one of them could fly in it.

Voskhod

To modify Vostok into Voskhod required a number of drastic steps. The single ejection seat in the re-entry capsule of the Vostok was removed and replaced with three couches. To fit them into the available space they were turned through 90° compared with the original layout and the centre seat was displaced slightly in front of the other two. The use of ordinary couches meant that the crew could not eject from the capsule during the descent (as had been done routinely on Vostok) and so a solid-fuel retro-rocket system was added to cushion the impact at the moment of touchdown. The absence of ejection seats also meant that if a problem developed with the launch vehicle during the ascent to orbit, then the crew aboard Voskhod had no means of escape. Lack of room also meant that the cosmonauts could not wear spacesuits, leaving them no protection if the air pressure in the capsule was lost.

Voskhod was fitted with a reserve retro-rocket in case the main system failed. The one-man Vostok capsule did not need a reserve rocket since it was placed in a low orbit and atmospheric drag would cause it to re-enter naturally before the cosmonaut ran out of air, even if the retro-rocket failed. Voskhod would use a more powerful launch vehicle and would be placed into a higher, more stable orbit. The back-up retro-rocket on Voskhod, which was mounted on the top of the re-entry capsule, was to ensure that its crew would not be marooned beyond the limit of their air supply if the main rocket should fail. An unmanned test flight of the Voskhod took place on 6 October 1964 under the name Cosmos 47.

Voskhod 1

The first Voskhod flight was launched on 12 October 1964 and went into an orbit ranging from 178km to 409km. It was commanded by Vladimir Komarov. The other two men in the cramped cabin were medical expert Boris Yegorov and, in keeping with Korolov's promise, design engineer Konstantin Feoktistov. The flight lasted just one day and although various experiments are said to have been carried out, little has ever been revealed about them. Presumably any such work would have been very limited due to the restricted space available. No clear photographs of Voskhod were released for years, so it was impossible for Western scientists to tell that it was merely a converted Vostok. As a result, the world was lead to believe that Voskhod was an Apollo-type spacecraft and that the USSR was well ahead of the USA in the space race.

Voskhod 2

For the second flight of Voskhod, Korolov planned a mission that would, once again, beat the Americans. The objective this time was the first ever spacewalk. By removing one of its three seats Voskhod could carry two men in spacesuits and still leave enough room for one of them to climb out of the spacecraft and walk in space. To allow this, Korolov and his team designed an accordion-like airlock which fitted over the hatch of the Voskhod and which could be folded flat during launch. In this way the collapsed airlock could be fitted beneath the protective shroud of the launch vehicle with only a few modifications. The system was tested during the unmanned Cosmos 57 mission and, although this mission seems to have been less than fully successful, the problems did not delay the launch of Voskhod 2.

Voskhod 2 was launched on 18 March 1965 into a 175km × 512km orbit. The commander was Pavel Belyayev and the spacewalker was Alexei Leonov. Once in orbit, the cosmonauts immediately began to prepare for the spacewalk by extending the airlock and connecting Leonov to a 5m safety line, which also provided a radio telephone link to Voskhod. About 1.5 hours after launch, Leonov climbed out to become the world's first spacewalker. He was using a spacesuit which was a prototype of the suit Soviet cosmonauts hoped to use on the Moon and which had an air supply in a small backpack.

At first the spacewalk went well but, when Leonov tried to climb back into the airlock, he began to have serious problems. His spacesuit had ballooned outwards and become bloated and stiff, making it impossible to get into the airlock. Leonov reduced the pressure in the suit to make it more flexible and struggled inside, closing the airlock door behind him. The airlock was repressurized and Leonov returned to the main Voskhod cabin with Belyayev. The airlock was then jettisoned.

The two men's adventures were, however, not over. After a flight of 24 hours they prepared to return to Earth, but problems developed during attempts to fire the re-entry rocket. Eventually Belyayev was forced to carry out the retro-fire manually (exactly what happened is unclear), but in the end Voskhod 2 made a safe re-entry, although it landed over 1 000km off course. Instead of landing on the steppes, the capsule came down in the middle of a snowy forest. It was cold and lonely, with only wolves for company and to make things worse, the landing had occurred in the afternoon and night would soon fall. The crew lit a fire and camped until a rescue team arrived the following day. The two men, and their rescuers, left the forest travelling on skis until they were be picked up by helicopter and flown home.

Voskhod 3

It is believed that a third Voskhod mission was planned, which would involve two men staying in space for up to two weeks. The crew would probably have been Boris Volynov and Georgi Shonin.

> *Voskhod* means 'sunrise'.

Cosmonaut Vladimir Komarov, commander of Voskhod 1

Vostok

Vostok 1 1961, Vostok 6 1963

The first manned spacecraft

By putting the first man into space, the Vostok programme was a triumph for the USSR.

The Soviet Vostok spacecraft comprised a spherical cabin about 2.3m in diameter attached to a bi-conical instrument module. The cabin was occupied by a single cosmonaut sitting in an ejection seat which could be used if problems developed during launch and was also used during the landing procedure (the Vostok cabin had a tendency to strike the ground quite hard, so to avoid injury on landing the cosmonaut ejected during the final descent and landed by parachute). The entire sphere was coated with ablative heatshield material so that there was no need to stabilize it in any particular attitude during re-entry. The instrument module, which was attached to the cabin by steel bands, contained a single, liquid-fuelled, retro-rocket and smaller attitude control thrusters. Spherical bottles of nitrogen and oxygen were clustered around the instrument module close to where it joined the cabin.

Unmanned tests of the Vostok were described as Sputnik flights. Sputnik 4 was launched on 19 May 1960 and, although it reached orbit, during its landing attempt its retro-rocket fired in the wrong direction and the craft was stranded in orbit for four years. Sputnik 5, which carried two dogs and other biological specimens, was launched on 19 August and landed normally the next day. Sputnik 6, launched on 1 December 1960 re-entered too steeply and burned up. Sputnik 7 and 8 were Venus probes (see p245), but two final Vostok tests, Sputnik 9 and 10, were launched in March 1961. These both succeeded and cleared the way for the first manned mission.

Vostok 1

Vostok 1 was launched on 12 April 1961. The pilot was 27-year-old Yuri Alexeyevich Gagarin who went into a 181km × 327km orbit. Gagarin soon reported that all was well and he described the views through the windows of his spacecraft. While he was still in orbit Radio Moscow announced that a man had finally been sent into space. After a single orbit of the Earth, Vostok 1's retro-rocket was fired to begin re-entry. The instrument section was released and the capsule arced back towards the Earth, its heatshield glowing red. At a height of 8 000m Gagarin ejected from the capsule and parachuted to Earth after a flight lasting 108 minutes. The USSR had beaten the USA's Mercury astronauts into space and Yuri Gagarin's name went into the history books.

Vostok 2

The objective of Vostok 2 was to spend a whole day in space. This much longer flight was required because as Vostok 2 orbited the Earth turned steadily beneath it and after a few hours the spacecraft no longer passed over the USSR. If the cosmonaut was to land within Soviet territory then the mission must either be quite short (no more than five hours) or must be long enough for the Earth to turn completely and bring the landing site back underneath the orbit of the spacecraft. Since a three-orbit flight was not considered sufficiently impressive, it was decided that the mission must continue for at least 24 hours.

The pilot for Vostok 2 was Gherman Titov, a 36-year-old jet pilot. His launch on 6 August 1961 went smoothly and within minutes he was safely in space, travelling in a 178km × 257km orbit. On the third orbit he ate some food pastes, then a few hours later he took manual control and changed the attitude of his spacecraft. About 10 hours after launch Titov prepared to try and sleep, but he became nauseous, the first of many space travellers to experience space sickness. Fortunately, Titov was able to overcome his sickness and sleep for about 7.5 hours. The next day he tidied up his cabin and prepared to return to Earth. Vostok 2 made a perfect re-entry and Titov ejected and landed by parachute after a flight lasting 25 hours and 18 minutes.

Vostok 3 and 4

The next Vostok mission began on 11 August 1962 when Adrian Nikoleyev was launched into a 183km × 251km orbit. Nikoleyev did not suffer from nausea and was allowed to release his straps and float freely within his tiny cabin. Western observers expected Nikoleyev to settle down for a flight of several days and so were surprised when, on 12 August, Pavel Popovitch was launched aboard Vostok 4. The second spacecraft had been launched from the same pad as its predecessor and the initial orbit of Vostok 4 was so accurate that it passed within 6.5km of Vostok 3. Although it was an impressive achievement to place two craft so close together, the Vostok spacecraft did not have the capability to adjust their orbits and so after a few hours the two men began to drift apart. The joint flight continued, with the two men talking to each other and to ground control by radio, until the morning of 15 August. The two spacecraft then conducted simultaneous re-entries and the cosmonauts landed within a few minutes of each other. Nikoleyev had been in space for four days, Popovitch for three.

It is said that during her three-day flight Tereshkova slept so well that mission control had difficulty waking her and began to wonder if she was still alive.

It was claimed for years that Gagarin landed inside his capsule, since to qualify for international aviation records a pilot must land his aircraft normally, not parachute from it.

Vostok 5 and 6

When Valery Byskovsky was launched in Vostok 5 on 14 June 1963, few people in the West realized that another Soviet propaganda coup was imminent. Two days after Byskovsky's launch another Vostok was sent into space on what would have been a repeat of the Vostok 3,4 joint flight, but for one important difference. The pilot of Vostok 6 was a 26-year-old woman, Valentina Tereshkova, one of a small group of female cosmonauts selected in 1962. Tereshkova was not a jet pilot, but she was an experienced parachutist. Vostok 6 entered orbit only 5km or so from Vostok 5 and the two cosmonauts were soon able to establish radio contact. However, as with the earlier group flight it was not possible to manoeuvre the two spacecraft and they soon drifted apart.

All went well aboard both craft and Tereshkova remained in space until 19 June, returning after a flight of almost three days. Byskovsky returned a few hours later, concluding a five-day flight which set a new space endurance record. Thirty years later Byskovsky's flight remains the longest ever mission in a single-seater spacecraft. A rumoured Vostok 7 flight, expected to last for about one week, never took place.

Yuri Gagarin in Vostok before the spaceflight of 12 April 1961

Voyager

Voyager 1 and 2 1977

Mission to the outer planets

The USA's Voyager spacecraft explored all four of the giant planets and continue to return new data from the edge of the solar system.

The Voyager mission originated in studies during the 1960s which showed how a spacecraft could use the gravitational attraction of one planet to send it further into space without using extra fuel. This led to the concept of a mission called the Grand Tour which was intended to use a rare planetary line-up in the late 1970s to send a single spacecraft to the giant planets Jupiter, Saturn, Uranus and Neptune. Numerous Grand Tour trajectories were calculated, and a study was begun into a new spacecraft, called the Thermoelectric Outer Planets Spacecraft (TOPS), which could undertake such a mission. As these studies proceeded, the complexity of the proposed mission, which now featured a visit to Pluto and the release of probes into the atmospheres of Jupiter and Saturn, grew steadily, increasing the expected cost to almost $1 billion. This was too expensive and on 24 January 1972, the project was cancelled.

Cancelling the Grand Tour did not alter the favourable planetary alignment and so NASA developed a cheaper mission, aimed at only Jupiter and Saturn, based on a Mariner-type spacecraft. This project, known as Mariner Jupiter/Saturn envisaged the launch of two spacecraft towards Jupiter in 1977. If all went well, both craft would arrive at Jupiter in 1979 and then continue on to encounter Saturn in 1981. Although not officially part of the mission plan, at least some senior NASA officials were aware that, with luck, at least one of the spacecraft could continue on to Uranus and perhaps Neptune.

The spacecraft

Although drawing considerably on experience obtained with the earlier Mariner missions, the new spacecraft were considerably different in appearance. One obvious change was the use of nuclear power generators (see p154) in place of solar panels. This change was made because Jupiter is about 10 times further away from Earth than Mars and at such great distances the sunlight is so weak that solar panels cannot generate electricity efficiently. The nuclear generators were mounted on a boom so that their radiation would not affect the sensitive scientific instruments on the spacecraft. The great distances to the outer planets also meant that the spacecraft's radio antenna had to be large enough to concentrate its radio signals into a beam aimed precisely at the Earth, rather than wasting power by broadcasting the signals out over a large volume of space, so the Mariner Jupiter/Saturn had a considerably larger high-gain antenna dish than earlier Mariner craft.

The heart of the Mariner Jupiter/Saturn spacecraft was an octagonal equipment section, containing most of the spacecraft's vital electronic systems, which was mounted on the back of the radio dish. In the centre of the octagon was a large spherical tank of hydrazine propellants for the 16 small rocket motors which were used for attitude control and for course corrections. Most of the scientific instruments were mounted on a boom which projected from the opposite side of the spacecraft to the nuclear generators. Several of the instruments were attached directly to the boom, but others, including the two television cameras and the ultraviolet and infrared spectrometers, were mounted on a movable scan platform so that they could be pointed accurately at specific targets such as a planet or one of its satellites. The instruments on the boom itself, such as those designed to measure charged particles trapped by magnetic fields, could not be pointed without manoeuvring the whole spacecraft. The spacecraft also had two whip antennae used for a planetary radio astronomy experiment and a long boom carrying magnetometers.

Although the launch vehicle for the mission was the powerful Titan IIIE (Titan-Centaur) rocket, even this could not provide enough energy to send the 810kg spacecraft directly on to its planned trajectory. To provide the extra velocity needed, each spacecraft had a solid rocket motor weighing about 1 210kg attached by struts to the main body. This motor, which provided a thrust of about 71kNewtons, was released once it had been fired.

The result of all the modifications was that the final spacecraft differed so much from the original Mariner design that it was given a new name. A competition was held and the winning nomination 'Voyager' was approved on 4 March 1977.

The Voyager mission

Voyager 2 was launched from Cape Canaveral on 20 August 1977 and Voyager 1, which was placed on a faster trajectory and overtook its sister ship en route, followed on 5 September. Voyager 1 had a fairly uneventful journey to Jupiter, but in April 1982, Voyager 2's main radio receiver failed. After a series of problems over the next few days contact with Voyager 2 was finally re-established via its back-up radio receiver. Unfortunately, this receiver also had problems and could only receive over a very narrow range of frequencies, placing further demands on the mission controllers.

Voyager 1 arrived at Jupiter in the spring of 1979. For several weeks before the closest approach, the spacecraft's cameras returned pictures of the planet that showed in more and more detail the planet's turbulent atmosphere. The closest approach occurred on 5 March 1979 and observations continued for several more weeks as Voyager receded from the planet. Many important measurements were made, including the discovery of a thin ring around Jupiter and the first detailed photographs of the planet's major satellites. The satellites proved especially interesting with Io showing evidence of active volcanoes, Europa a smooth icy surface, Ganymede a fascinating mixture of old and relatively new crust and Callisto a surface in which craters stood shoulder to shoulder. Voyager 2, which made its closest approach on 9 July, followed up some of these discoveries by conducting special observations such as a volcano watch of Io and further studies of Jupiter's ring.

Voyager 1 reached Saturn on 12 November 1980 and once again mission controllers were deluged with a fantastic array of photographs and other scientific data. Pictures of Saturn's rings showed that they were much more complicated than expected, with many thousand of ringlets within each major ring. Some of the outer rings were even kinked, apparently defying the laws of celestial mechanics. A host of new small satellites was discovered and it soon became clear that some of these shepherded the material in the rings, helping to keep the ring system together despite the forces that would otherwise have caused it to disperse. Voyager 1 returned good-quality images of several of Saturn's largest moons and then headed for a critical rendezvous with Saturn's largest satellite, Titan.

The flyby of Titan was important for two reasons. Firstly, Titan is one of the few satellites large enough to have an atmosphere and scientists were very anxious to find out as much about it as possible. Secondly, a flyby of Titan would put Voyager on a course that would make it impossible to go on to Uranus. Mission planners had decided that if Voyager 1 failed to obtain useful observations of Titan, then Voyager 2 would have to repeat the attempt, ending all hopes of a mission to Uranus. In fact all went well and, although Titan's atmosphere was too cloudy to allow the surface to be seen, the other experiments were all successful.

Voyager 2 flew past Saturn on 25 August 1981 and, despite the movable scan platform jamming at a crucial moment and causing some photographs to be lost, the flyby was highly successful. Voyager 2 did not attempt to encounter Titan and left Saturn on course for Uranus. NASA officials then formally extended the mission so that money would be available to support the spacecraft during its five-year voyage to its next target.

Despite budget battles that threatened to cut off the funds to operate it, Voyager 2 survived to fly past Uranus on 24 January 1986. During the encounter the spacecraft returned many images of the planet, its faint ring system and its five largest satellites. Unfortunately, much of the scientific impact of the encounter with Uranus was overshadowed by the fatal explosion of the Space Shuttle Challenger on 28 January 1986. Voyager 2 left Uranus on a path that would take the spacecraft on to Neptune 3.5 years later

Voyager 2 arrived at Neptune in August 1989. During its approach the spacecraft searched for the mysterious, and difficult to explain, partial rings detected a few years earlier from Earth and monitored a huge dark spot in the planet's atmosphere. Scientists also discovered a new large satellite of the planet. The partial rings were eventually found to be dense regions of continuous rings and the dark spot was revealed as a huge storm large enough to swallow the Earth. On 25 August Voyager 2 swept over the North polar region of Neptune and then curved southwards towards Neptune's largest moon, Triton. Triton was smaller than expected, but its surface revealed a host of interesting geological features including geysers which squirt material high into the moon's thin atmosphere.

Its work at Neptune completed, Voyager 2 was then switched into its cruising mode so that it could continue to monitor conditions in interplanetary space. Like Voyager 1, Voyager 2 has completed its planetary encounters and has embarked upon the final phase of its mission: cruising out of the solar system returning data on conditions in deep space until its distance becomes so great that it can no longer communicate with Earth.

The launch of Voyager 1

Weather satellites

TIROS 1 1 April 1960

Improving the forecast

Meteorological satellites have been one of the successes of the space age, aiding both weather forecasts and long-term studies of the climate.

Satellites enable meteorologists to study entire weather systems and have removed the need for forecasters to rely on reports from scattered weather stations and from ships at sea. Satellites can return daytime images of clouds that have been taken in ordinary light and then use infrared sensors to continue their observations throughout the night. As well as watching the clouds develop, meteorologists can determine windspeeds from the rate at which the clouds move, and can use infrared data to deduce the temperatures at different points in the atmosphere. This information can be used for such things as weather forecasting and for monitoring long-term climatic change.

Today's weather satellites operate from two types of orbits. Some satellites are placed in circular, polar or near polar orbits, between 500km and 1 500km high. Such satellites take about two hours to circle the Earth and they observe a strip which lies below and to either side of their orbital path. As the Earth turns, the strip covered by the satellite changes, allowing different regions to be observed on every orbit.

The other satellites are those in geostationary orbit, 36 000km above the equator. These satellites complete one orbit per day and so appear to remain fixed above a single point on the Earth's surface. From their lofty vantage points, geostationary satellites can see an entire hemisphere of our planet, although the Earth's curvature means that the poles are invisible and the view of regions towards the edges of the satellite's field of view is badly distorted.

This combination of satellites can observe the whole of the Earth's atmosphere. The geostationary satellites provide a global perspective and the lower, polar orbiting satellites fill in the regions around the poles and provide more detailed views of particular areas.

The US programme

The US meteorological satellite programme started on 1 April 1960 with the launch of TIROS (Television and InfraRed Observation Satellite) 1. This 119kg, spin-stabilized satellite was drum shaped and had its top and sides covered by solar cells. The satellite's television cameras looked out through the bottom of the drum and, since the satellite's spin axis was fixed in space, for part of each orbit the cameras looked down on the Earth, but for much of the time they pointed towards empty space.

Between 1960 and 1965 10 TIROS satellites were launched, and all of them returned thousands of photographs. The first eight satellites were similar, but Tiros 9 introduced a new layout in which the television cameras looked out of the side of the drum and the satellite was orientated so that it cartwheeled around the Earth. This allowed the cameras to take more pictures of the Earth and less of empty space. The success of TIROS lead to the TIROS operational system (TOS) which used TIROS 9-type satellites placed in polar orbits. These were able to contribute to a regular weather monitoring programme operated by the US Environmental Science Services Administration (ESSA). Once safely in orbit, the TOS satellites were designated ESSA 1, ESSA 2 etc. Nine ESSA satellites were launched between 3 February 1966 and 26 February 1969.

After the ESSA satellites came a new generation of TIROS satellites called the Improved Tiros Operational Satellites (ITOS). These were box shaped, with three solar panels and more advanced sensors. The first ITOS was launched on 23 January 1970. The second and subsequent ITOS spacecraft were operated by the National Oceanic and Atmospheric Administration (NOAA) (which had by now absorbed ESSA) and so were known as NOAA 1, NOAA 2 etc. The NOAA programme has continued and presently uses a new type of TIROS satellite (TIROS N) which weighs 1 000kg and is 1.8m across and almost 5m long. NOAA 11, which was of the TIROS N type, was launched into a polar orbit by an Atlas rocket on 24 September 1988 and NOAA 12 followed on 14 May 1991.

The first US weather satellite placed into geostationary orbit was the Synchronous Meteorological Satellite (SMS 1) which was launched on 17 May 1974. There was one other SMS launch before the first Geostationary Operational Environmental Satellite (GOES) was launched on 16 October 1975. Only two operational GOES satellites are required to cover the regions of interest to US meteorologists, but the need to replace failed satellites meant that seven GOES launches took place between 1975 and 1987. A new generation of satellites, called GOES-Next, is under development, but the launch of the first in the series has been delayed a number of times.

Soviet satellites

Since much of its territory lay at high latitudes which are not easy to observe from geostationary orbit, the USSR concentrated on polar orbiting weather satellites launched as part of a continuing series called Meteor. The first Meteor satellites, which comprise a cylindrical body with two solar panels, were launched as part of the Cosmos programme and the existence of an operational programme was not announced until the basic design had been proven. About 30 Meteor satellites were launched between 1969 and 1978 before the first of an improved version, the Meteor II was orbited on 11 July 1975. A further series, Meteor III, made its appearance with a launch on 24 October 1985. This did not mark the end of the Meteor II series, which continued to be launched in parallel to the newer models.

The USSR did indicate that it would launch a weather satellite into geostationary orbit by 1978 in order to take part in a global atmospheric research programme, but the promised satellite, called GOMS, had not been launched by the end of 1991.

Other weather satellites

The European Space Agency (ESA) has never developed a polar orbiting satellite, but launched the first of a series of Meteosats into geostationary orbit on 23 November 1977. The drum-shaped, spin-stabilized Meteosats can provide both visible and infrared data. After the two prototypes, and one pre-operational version launched on 4 June 1988, the first of three operational Meteosats was launched on 6 March 1989. The Meteosat system is expected to provide Europe with a service that should last well into the 1990s.

Japan and India have also invested in geostationary weather satellites. The first Japanese Geostationary Meteorological Satellite (GMS 1) was orbited on 14 July 1977 and has been followed by three others. The first in a series of Indian satellites, INSAT 1A, which is a combined weather and communications satellite, was launched in April 1982.

China has launched two, near-polar orbiting satellites called Fengyun 1 (which failed after 39 days) and Fengyun 2.

> The new technology for many of the US weather satellites was developed using a series of experimental satellites called Nimbus. Nimbus 1 was launched on 28 August 1963 and Nimbus 7 on 24 August 1978.

A Soviet Meteor weather satellite

X-15

First flight 8 June 1959

A spacecraft with wings

The X-15 was one of NASA's most productive research aircraft and was able to reach the fringes of space.

In November 1955 a contract was placed with North American Aviation for the construction of three examples of a research aircraft to be known as the X-15. These aircraft were intended to provide information on aerodynamics, aircraft structures and physiological stress during high-speed, high altitude flight.

The X-15 aircraft was 15.25m long and had stubby, swept-back wings with a span of 6.7m. The vertical tailplane was wedge shaped and had an extension which projected below the fuselage. The outer skin of the X-15 was made from a nickel alloy called Inconel X and below this was a structure of stainless steel, titanium and aluminium. Since the aircraft was designed to operate on the fringes of space, where the air is too thin for normal aerodynamic controls to be effective, the X-15 was equipped with an additional gas jet attitude control system operated by a separate control stick on the left-hand side of the cockpit.

The first X-15 flights were made with an interim installation of two LR-11-RM-5 rocket motors, each with a thrust of 35.5kNewtons. These were later replaced by a single XLR-99 rocket which used anhydrous ammonia and liquid oxygen propellants and could produce a thrust of 266kNewtons. The XLR-99 rocket motor could be throttled between half and full power and could be stopped and restarted in flight if necessary.

Flight profile

To improve the performance of the X-15 it was air launched from a height of about 15 000m by a converted B-52 bomber aircraft operated from Edwards Air Force Base in California. To accommodate the X-15, the B-52 had a special pylon mounted under the starboard wing and a large liquid oxygen tank fitted into the bomb bay. The liquid oxygen was transferred into the X-15 from the B-52 immediately before launch, a special crew member being carried to supervise this operation. Once the fuelling was complete, and the X-15 pilot was ready, the X-15 was released from the B-52 and, after falling for a few seconds, ignited its rocket motor to begin the test flight.

For high-speed missions, the X-15 pilot would attempt to fly more-or-less level, using the rocket to develop the maximum possible speed before it ran out of fuel. For a high altitude flight the X-15 pilot would pull back on the stick shortly after release from the B-52 and climb steeply in order to gain height as quickly as possible. Once the rocket fuel was exhausted, usually between 80 and 120 seconds after release, the X-15 would continue upwards until it was just outside the atmosphere and the pilot would switch to his rocket-powered attitude control system. Then, as the aircraft's speed began to reduce, the pilot would lower the nose and begin to dive back into the atmosphere. Once the X-15 had descended below about 30 000m the aircraft's aerodynamic controls would regain their effectiveness and the pilot would pull out of the dive and prepare for his approach and landing.

With fuel for the rocket motor exhausted, the X-15 became a high-speed glider and returned to Edwards Air Force Base where the dry lake bed provided a large flat area on which to land. Just before touchdown, the X-15 pilot lowered the aircraft's undercarriage, which comprised a twin nosewheel beneath the cockpit and a pair of metal skids at the rear of the fuselage. The X-15 was then landed on the steel skids in a nose-up attitude prior to the front of the aircraft being lowered on to the nosewheels.

The flight programme

The first X-15 flight was a simple glide test on 8 June 1959 which was flown by North American Aviation test pilot Scott Crossfield. The first rocket-powered flight followed on 17 October 1959 and the first flight by a NASA pilot was on 23 March 1960. Twenty-five flights were made using the interim rocket motors before an XLR-99 equipped aircraft was ready to join the test programme.

The first XLR-99 flight was on 15 November 1960, with Scott Crossfield once again at the controls, and on 30 November a new NASA pilot named Neil Armstrong made his first X-15 flight. With several other pilots in the team, progress was rapid as the X-15 explored the boundaries of high-speed flight. On 9 November 1961 the X-15 reached its design speed of six times the speed of sound and on 17 July 1962 Robert White set a world altitude record of 95 935m.

The X-15A-2

X-15 number 2 was damaged in a crash landing on 9 November 1962 and was taken out of the programme and improved so that it could assault speeds of up to eight times the speed of sound. The modified version, renamed the X-15A-2, featured a longer fuselage, a modified windscreen and undercarriage and provision for the attachment of external fuel tanks. To deal with the very high temperatures expected at such speeds, the X-15A-2 used a specially developed, spray-on protective coating. This coating was made from an ablative material which charred and melted when heated, carrying away heat in the process. After each flight, the charred coating was chipped off and fresh material sprayed over the top. To prevent the windscreen being covered by material that had ablated during flight, the left-hand cockpit window had an eyelid-like cover that was closed during most of the flight and only opened during the final descent so that the pilot could see to land the aircraft.

The first flight of the X-15A-2 was on 3 November 1965 and the first flight with the full ablative coating was made by US Air Force pilot William Knight on 21 August 1967. Two months later Knight used the X-15A-2 to set an unofficial world speed record of 7 273km per hour.

Sadly, as the programme neared its end, the X-15 suffered its first fatal crash. Michael J Adams was killed when severe buffeting during re-entry tore off the tailplane and wings of his aircraft and it span out of control and crashed in the Mojave Desert. There were a further eight flights of the surviving X-15 aircraft after the crash, with the 199th and last flight being made on 24 October 1968.

If an X-15 flight exceeded an altitude of 50 miles (80km), then the pilot was awarded astronauts wings.

The X-15A-2 covered with its white ablative coating to protect the rocket plane against the heat generated by friction during hypersonic flight

X-ray astronomy satellites

Uhuru 1970, ROSAT 1990

The universe at high energies

X-ray satellites have provided a new view of the violent Universe.

X-rays are an energetic form of electromagnetic radiation with wavelengths much shorter than those of ordinary light. They can be thought of as individual packets of energy called photons which arrive at short, but discrete, intervals. It is often possible to detect X-ray photons individually which is why many types of X-ray detectors are called 'counters'. Some types of counter can measure the energy of each X-ray photon, but they cannot, however, provide accurate information on the direction from which the X-ray came. X-rays are produced by material with temperatures of over 1 million°C and by the interaction of atoms and electrons in high temperature plasmas. They are of interest to astronomers because they provide information on exploding stars, black holes and other very violent events. Unfortunately, X-rays cannot penetrate the atmosphere and so can only be observed from space.

X-rays from the Sun were first detected by an experiment launched on a captured V2 rocket in 1949, but it was not until 1962 that an X-ray source outside the solar system was found. The first X-ray star was named Sco X-1 because it lay in the constellation Scorpio. Over the next few years rocket-launched experiments discovered many other X-ray sources, and it soon became clear that a special satellite devoted to X-ray astronomy was needed.

Mapping the X-ray sky

The X-ray astronomers' first priority was to survey the sky and find out how many X-ray stars there were. This was the objective of NASA's first Small Astronomy Satellite 1 (SAS-1), which was also known as Uhuru and as Explorer 42. Uhuru was launched on 12 December 1970 and detected several hundred X-ray sources. However, like many of the missions which followed it, Uhuru had X-ray counters with poor angular resolution and so could only indicate the approximate positions of the sources which it detected. Despite this the mission was a great success and led to the development of small X-ray experiments on a number of other satellites including the US OAO-3, the Dutch ANS and the Soviet Cosmos 428. Uhuru continued to operate until 1974 and was followed by another important X-ray mission, the UK/US Ariel 5 satellite. Ariel 5 was launched on 15 October 1974 and detected several hundred X-ray sources. Some of these were observed repeatedly, enabling their variability to be monitored. Ariel 5 was still operating when it re-entered the atmosphere on 14 March 1980.

Ariel 5 was followed by SAS-3, launched on 7 May 1975, and then by the first of a new class of large NASA satellites, High Energy Astronomical Observatory 1 (HEAO 1). HEAO 1, launched on 12 August 1977, was the last major X-ray satellite which relied on simple X-ray counters with low angular resolution. The next logical step was to launch a telescope able to focus X-rays into high-quality images.

HEAO 2 (Einstein)

Building an X-ray telescope requires new technologies since X-rays cannot be focused by lenses or reflected off mirrors. The only way to focus X-rays is to allow them to bounce off a polished surface at what is known as 'grazing incidence', a principle akin to bouncing a stone off the surface of a pond by throwing it almost parallel to the water's surface. Grazing incidence telescopes look like shiny, bottomless buckets, nested one inside each other. X-rays which glance off the inside of one of the buckets continue out through the bottom and come to a focus on a detector placed some distance behind the telescope. The technical problems of building grazing incidence telescopes are considerable, but the ability to make X-ray images, instead of just counting X-rays from an ill-defined region of sky, makes them essential for detailed studies of X-ray sources.

The HEAO 2 mission, which was named 'The Einstein Observatory' soon after its launch on 13 November 1978, was the first X-ray astronomy satellite with a grazing incidence telescope. Its results were spectacular: for the first time astronomers could see X-ray pictures of extended X-ray sources such as the hot gas around supernovae, and detect individual X-ray stars in other, nearby galaxies. Furthermore, because its telescope concentrated the X-rays from weak sources, Einstein was able to detect stars too faint to be seen by earlier satellites. Almost the whole of astronomy benefited from this tremendous mission, which ended in April 1981 when the satellite ran out of fuel for its attitude control system.

EXOSAT

Although there were other small X-ray satellites after Einstein, including the Japanese Astro A and B missions and the British Ariel 6, the next major project was EXOSAT developed by the European Space Agency (ESA). This X-ray observatory had several experiments, including wide-area counter-type detectors and grazing incidence X-ray telescopes. The satellite was launched on 26 May 1983 into a near-polar elliptical orbit which ranged from 356km to 191 581km. This enabled EXOSAT to make long observations of an individual X-ray source without losing sight of the target as the satellite moved behind the Earth. EXOSAT suffered a number of equipment failures early in its life, but mission controllers were able to keep the satellite in operation and over the next three years it made important contributions to X-ray astronomy.

ROSAT

ROSAT was a German X-ray astronomy satellite to which the UK and the USA both made contributions. The name is a contraction of Roentgensatellit and commemorates the discoverer of X-rays, Wilhelm Roentgen. The objective of ROSAT was twofold: to repeat the X-ray surveys of Uhuru and Ariel 5 with much greater sensitivity and then to perform detailed observations of specific X-ray sources. The main ROSAT experiment is a large, grazing incidence X-ray telescope which has both German and US detectors. A smaller, British telescope used for observations in the extreme ultraviolet region of the spectrum, was also carried. The use of X-ray telescopes on ROSAT meant that its survey was much more sensitive and more precise that those done during the 1970s.

Originally intended to be launched by the Space Shuttle, ROSAT was delayed by the explosion of the Space Shuttle Challenger and was eventually launched by a Delta rocket on 1 June 1990. The mission was successful, although some of the scientific instruments were damaged when ROSAT accidently pointed itself at the Sun in February 1991.

The future

Both the USA and Europe hope to launch new large X-ray telescopes towards the end of the 1990s. The NASA spacecraft will be called AXAF, the Advanced X-ray Astrophysics Facility and the European equivalent is the X-ray Multi Mirror (XMM). Other small missions such as the Japanese Astro D (Astro C was launched in 1987) are also expected in the late 1990s.

The discovery of Sco X-1 relied on some subterfuge by the scientists involved. Most physicists believed that X-rays from distant stars would be too faint to detect and so an X-ray detector was launched on the pretext of observing X-rays that might be emitted when particles from the Sun struck the Moon. No lunar X-rays were detected, but the discovery of Sco X-1 more than justified the flight.

Glossary

Although technical terms and acronyms have been avoided wherever possible, a few such expressions are listed here with short explanations. See also the index for references to pages where these and other expressions are defined.

Ablation The process by which a material melts or chars and thus carries away heat. Ablative material is often used in the heatshields of spacecraft to protect them during re-entry.
ACRV Assured Crew Return Vehicle. A space lifeboat to be fitted to the international space station Freedom.
Aerospatiale A large French aerospace company.
ALSEP Apollo Lunar Surface Experiment Package. A set of instruments placed on the Moon by US Apollo astronauts.
Angular resolution The ability to separate two sources of radiation which appear close together. Usually applied to telescopes.
Apogee The furthest point from the Earth in a satellite's orbit. Also aphelion (furthest point from the Sun), apojove (furthest point from Jupiter) etc.
ATS Applications Technology Satellite. Series of US satellites devoted to proving new technology for later use.
Black Brant A Canadian sounding rocket.
Boeing A large US aerospace company.
Cassegrain A type of reflecting telescope in which light is directed from the main mirror to a secondary mirror which then sends the light back through a hole at the centre of the main mirror. Commonly used for large ground-based telescopes and for astronomical satellites.
COBE Cosmic Background Explorer. A NASA satellite devoted to the study of the radiation left over from the Big Bang which formed the Universe. COBE was launched on 18 November 1989.
Comsat A US company representing US interests in INTELSAT. Also used to mean 'communications satellite'.
CRRES Combined Release and Radiation Effects Satellite, a US scientific project launched 25 July 1990.
Cryogenic At very low temperature. Usually used to describe liquified gasses cooled below 100K (–173°C).
ECS European Communication Satellite
Esrange A rocket launching range in Sweden.
Eureca European Retrievable Carrier. A small satellite designed to be left in orbit by the US Space Shuttle and then collected and returned to Earth on a later mission.
Galaxy A series of communication satellites built by the Hughes Aircraft Company, California.
GARP Global Atmospheric Research Programme. An international scientific project to monitor the world's weather.
Geostationary orbit A circular orbit at an altitude of just below 36 000km in which a satellite takes 24 hours to complete one revolution. Since the Earth turns in 24 hours, the satellite appears to hover over a single spot on the equator.
Geosynchronous orbit An orbit with a period of 24 hours in which a satellite follows the same apparent path across the sky every day.
General Dynamics A large US aerospace company.
Geodesy The study of the shape of the Earth.
Get Away Special A small, self-contained, experiment carried in the payload bay of the US Space Shuttle.
Gravity assist The process of flying close to a planet in such a way as to gain energy and deflect a spacecraft on to a different course. No energy is created during a gravity assist; when the spaceprobe is accelerated the planet is slowed down by a tiny, immeasurable amount.
Glavcosmos Soviet space agency created in

1965 to coordinate Soviet space activities.

GLONASS A system of Soviet satellites used for navigation.

Hipparcos A European Space Agency (ESA) satellite devoted to the precise measurement of star positions. Launched on 8 August 1989 it was stranded in the wrong orbit when its boost motor failed to fire, but it has since been able to meet most of its scientific objectives.

Hughes Aircraft Company A large US aerospace company.

IKI A Soviet space research organization which was part of the Soviet Academy of Sciences.

IRBM Intermediate Range Ballistic Missile.

IUS Inertial Upper Stage. Originally known as the Interim Upper Stage the IUS is a two-stage, solid-fuelled rocket booster used to send satellites out of low Earth orbit and further into space. It can be used in conjunction with both the USA's Space Shuttle and Titan rocket.

Kaliningrad Soviet mission control centre.

LAGEOS A satellite, launched by NASA on 4 May 1976, which was equipped with laser reflectors and used for geodesy.

Launch window A period during which a rocket must be launched in order to complete its mission. A launch window may be set by the need to launch an interplanetary probe when the target planet is in the correct position, or to place a satellite in a specific orbit.

LDEF Long Duration Exposure Facility. A large NASA satellite which was intended to provide an opportunity to expose materials and experiments to space for a period of about one year. LDEF was deployed from the Space Shuttle in 1984, but its recovery was delayed until 1989 by changes in the Space Shuttle programme following the Challenger accident in 1986.

LLTV Lunar Landing Training Vehicle. A jet-propelled wingless aircraft used to train Apollo astronauts.

Lockheed A large US aerospace company.

LSAT European large communications satellite project renamed Olympus.

Magnetometer A scientific instrument used to measure magnetic fields.

McDonnell-Douglas A large US aerospace company.

MARECS A European maritime communications satellite based on the European Communication Satellite (ECS).

MAROTS A proposed precursor of the MARECS satellites, MAROTS was to have been based on the European Orbital Test Satellite (OTS).

Martin-Marietta A large US aerospace company.

Matra A large French aerospace company.

MBB-ERNO A large German aerospace company.

Mirak A German rocket of the 1930s.

MMS Multimission Modular Spacecraft. An attempt by NASA to reduce costs by developing a standard spacecraft which, with minor modifications, could be used for a variety of missions.

NERVA Nuclear Engine for Rocket Vehicle Applications. US nuclear rocket programme abandoned in the early 1970s.

Newton A unit of force, that is, the force required to accelerate 1kg at 1m per second^2.

NORAD North American Aerospace Defence. US missile and satellite monitoring organization.

Nova A star which suddenly increases in brightness and then fades away again. Novae usually occur in systems where two stars are orbiting each other and material is transferred from one to another, causing a sudden outburst of energy.

Olympus A large European communications satellite which was launched on 12 July 1989.

OMV Orbital Manoeuvring Vehicle. A proposed reusable, unmanned spacecraft intended to move satellites from one orbit to another. At present, plans for its development have been suspended.

OSS US Office of Space Science

OSSA US Office of Space Science Applications

OSTA US Office of Space and Terrestrial Applications

OTS Orbital Test Satellite. An experimental European communications satellite launched on 11 May 1980 after the first example was destroyed in a launch failure in September 1977.

PAGEOS Passive Geoditic Satellite. A balloon satellite used for geodesy and launched on 23 June 1966.

PAM Payload Assist Module. A solid-fuelled

rocket stage which can be used in conjunction with the US Space Shuttle.
Perigee Nearest point to the Earth in a satellite's orbit. Also perihelion (nearest point to the Sun), perijove (nearest point to Jupiter) etc.
Prospector A proposed US unmanned lunar explorer able to move around the Moon's surface after landing. It was never built.
RCS Reaction Control System. Small rocket motors used to control the attitude of a spacecraft.
RMS Remote Manipulator System. The US Space Shuttle's robot arm.
Rockwell International A large US Aerospace company which now incorporates North American Aviation.
San Marco A small rocket launch facility operated by Italy and based on platforms in the sea off the coast of Kenya. Also used as the name of Italian satellites launched from this facility.
SARSAT Search and rescue satellite system based on US, Soviet and French satellites. The Soviet part of the system was known as COSPAS.
SBS Satellite Business Systems. Commercial communication satellite system.
SIR Shuttle Imaging Radar. US Space Shuttle payload used for Earth observations.
Skylark A popular British sounding rocket.
Solar flare A sudden outburst which occurs over a small region of the Sun and lasts for between a few minutes and a few hours. During the flare considerable quantities of energy are released as ultraviolet and X-ray wavelengths together with streams of charged particles such as electrons, protons and heavy ions. This radiation can affect the Earth's magnetic fields, disrupt the ionosphere and can be harmful to astronauts.
Solar wind A stream of charged particles, electrons and ions, which escape from the Sun and flow outwards across the solar system.
Sounding rocket A rocket launched on a sub-orbital trajectory to sample the upper atmosphere or to carry experiments into space for a few minutes.
Spacehab A small laboratory module, similar to the European Spacelab in concept, which is designed to be carried in the payload bay of the US Space Shuttle to provide extra working space in which astronauts can carry out experiments.
SPAR Aerospace A Canadian aerospace company.
SPARTAN Small scientific satellite deployed from and recovered by the US Space Shuttle after a flight of a few days.
STDN Space Tracking and Data Network. US space tracking system.
Supernova The sudden explosion and complete destruction of star.
TDRS(S) Tracking and Data Relay Satellite (System). Geostationary satellites developed by USA used to receive data from low-flying satellites and then re-broadcast the data to the ground.
Telemetry The process of transmitting data from a satellite to a ground station by radio.
VAB Vehicle Assembly Building. Located at Launch Complex 39 at the Kennedy Space Center in the USA this is where the Saturn 5 rockets were assembled and is now used for the Space Shuttle.
Van Allen belts Two zones which surround the Earth and which contain high concentrations of protons and electrons trapped by the Earth's magnetic field. The precise position of the belts vary but the inner belt is about 1000km to 6000km above the Earth and the outer belt can range from 15000km to 25000km. Prolonged exposure to the radiation trapped in the belts can be harmful for humans and may damage electronic equipment.
Vernier A small rocket engine used to make adjustments to the trajectory of a large rocket. Most commonly fitted to early ballistic missiles.
VfR Verein für Raumschiffahrt. The German space travel society of the 1930s.
Woomera A rocket launch site in Australia used as a missile test centre and for the launch of the British Prospero and Australian WRESAT satellites.
WRESAT A small satellite orbited by the Australian Weapons Research Establishment (WRE) on 29 November 1967 using a modified US Redstone missile.

Index

Note: Entries in **bold type** refer to article headings on the pages referred to in the book.